D0946870

RECENT ADVANCES IN ECONOMETRICS

General Editors

C. W. J. GRANGER G. MIZON

CONTINUOUS TIME ECONOMETRIC MODELLING

A. R. Bergstrom

OXFORD UNIVERSITY PRESS
1990

Oxford University Press, Walton Street, Oxford OX2 6DP
Oxford New York Toronto
Delhi Bombay Calcutta Madras Karachi
Petaling Jaya Singapore Hong Kong Tokyo
Nairobi Dar es Salaam Cape Town
Melbourne Auckland
and associated companies in
Berlin Ibadan

Oxford is a trade mark of Oxford University Press

Published in the United States by
Oxford University Press (USA)

British Library Cataloguing in Publication Data
Bergstrom, A. R.
Continuous time econometric modelling.
1. Continuous time econometric modelling
I. Title II. Series
330.0724
ISBN 0–19–828340–7
ISBN 0–19–828367–9

Library of Congress Cataloging in Publication Data
Bergstrom, A. R. (Abram R.)
Continuous time econometric modelling/A. R. Bergstrom.
(Recent advances in econometrics)
Includes bibliographical references and index.
1. Econometric models. I. Title. II. Series.
HB141.B465 1990 330'.01'5195–dc20 90–7550
ISBN 0–19–828340–7
ISBN 0–19–828367–9

Typeset by KEYTEC, Bridport, Dorset

Printed in Great Britain by Biddles Ltd,
Guildford and King's Lynn

Preface

THIS volume contains a collection of my papers written during the last thirty years, on various aspects of continuous time econometric modelling. It includes papers on theoretical models, econometric methodology, and applications. Although the papers have been written as independent contributions and all except two of them have been previously published, there is extensive cross-referencing between them, and I believe that they provide a connected development of the subject. The previously unpublished papers include a new paper (Chapter 7) on hypothesis testing, which has been written especially for this volume, and a very recent empirical study (Chapter 12, with M. J. Chambers), which provides the first demonstration of the application of the new exact Gaussian methods of estimation that I have developed during the last few years. I have written a long introduction linking the various papers in the volume and discussing their relation to the work of other economists and econometricians.

In editing the previously published papers for this volume I have corrected a few minor errors. I have also made a minor change of notation (indicated by footnotes) in Chapters 5, 6, and 8.

Two of the early papers (Chapters 1 and 3) were written during 1962–4 at the London School of Economics where A. W. (Bill) Phillips was an important influence. Another early paper (Chapter 2) was written shortly afterwards, at the University of Auckland. All of the other papers have been written during the last twenty years, at the University of Essex. Throughout this period I have enjoyed the association of a series of colleagues and Ph.D. students who have been actively engaged in research on the theory and application of continuous time econometric modelling. The first of these was Peter Phillips who wrote several important articles on continuous time econometric methodology, while he was a member of the Department of Economics at Essex, during the early 1970s. During that period I was working with Clifford Wymer (who was at the LSE) on the development of the first continuous time macroeconometric model (described in Chapter 10).

Finally, I would like to thank Penelope Martin and Phyllis Pattenden who skilfully typed most of the papers in this volume.

December 1989 A.R.B.

Acknowledgements

CHAPTER 1, 'A Model of Technical Progress, the Production Function and Cyclical Growth', first appeared in *Economica*, 29 (1962), published by the London School of Economics and Political Science. Chapter 2, 'Monetary Phenomena and Economic Growth: A Synthesis of Neoclassical and Keynesian Theories', first appeared in *The Economic Studies Quarterly*, 17 (1966), published by the Japan Association of Economics and Econometrics. Chapter 3, 'Nonrecursive Models as Discrete Approximations to Systems of Stochastic Differential Equations', first appeared in *Econometrica*, 34 (1966) published by the Econometric Society. Chapter 4, 'Gaussian Estimation of the Structural Parameters in Higher Order Continuous Time Dynamic Models', first appeared in *Econometrica*, 51 (1983), published by the Econometric Society. Chapter 5, 'The Estimation of Parameters in Nonstationary Higher-Order Continuous-Time Dynamic Models', first appeared in *Econometric Theory*, 1 (1985), published by Cambridge University Press. Chapter 6, 'The Estimation of Open Higher Order Continuous Time Dynamic Models with Mixed Stock and Flow Data', first appeared in *Econometric Theory*, 2 (1986), published by Cambridge University Press. Chapter 8, 'Optimal Forecasting of Discrete Stock and Flow Data Generated by a Higher Order Continuous Time System', first appeared in *Computers and Mathematics with Applications*, 17 (1989), published by Pergamon Press. Chapter 9, 'Optimal Control in Wide-Sense Stationary Continuous Time Stochastic Models', first appeared in *The Journal of Economic Dynamics and Control*, published by Elsevier Science Publishers (North-Holland). Chapter 10, 'A Model of Disequilibrium Neoclassical growth and its Application to the United Kingdom', first appeared in A. R. Bergstrom (ed.), *Statistical Inference in Continuous Time Economic Models*, published by Elsevier Science Publishers (North-Holland). Chapter 11, 'Monetary, Fiscal and Exchange Rate Policy in a Continuous Time Econometric Model of the United Kingdom', first appeared in P. Malgrange and P. Muet (eds.), *Contemporary Macroeconomic Modelling*, published by Basil Blackwell Publishers Ltd. I am grateful for permission to reproduce these articles in this volume.

Contents

Introduction

A MAJOR source of difficulty in the construction of dynamic econometric models is the fact that almost all of the variables in macroeconomic models and models of competitive industries are measured at much longer intervals of time than the intervals between the microeconomic decisions that they reflect. Although such variables as consumption, exports, imports, and the gross domestic product are measured only at quarterly intervals (in earlier years only at annual intervals), they are the outcome of millions of decisions taken by different individuals at different points of time. For most such variables there will be thousands of small changes at random intervals of time on a single day, and the changes can occur at any time during that day. A realistic aggregate model which, accurately, takes account of the microeconomic decision processes must, therefore, be formulated in continuous time.

Even if the sole purpose of an econometric model is to provide a basis for forecasting the future discrete observations, there are important gains from formulating it in continuous time. For it is only through a continuous time model that we can, accurately, take account of the restrictions on the probability distribution of the discrete data implied by economic theory and other a priori information about the causal relations between variables. The incorporation of restrictions of this sort can result in a great reduction in the standard errors of the parameter estimates and a corresponding increase in the accuracy of the forecasts.

There are a number of more specific advantages of continuous time econometric modelling. One of the most important advantages relates to the treatment of stock and flow data. Most econometric models contain two types of variable. The first type, which we call stock variables, are measured at points of time and include such variables as the money supply, the stock of fixed capital, and the level of inventories. The second type, which we call flow variables, are measured as integrals over the unit observation period and include such variables as output, consumption, exports, and imports, as well as certain price indices which are intended to measure average prices over the unit observation period. Standard procedures for the estimation of econometric models formulated in discrete time make no distinction between the treatment of these two types of variable, and, consequently, the estimates can be seriously biased because of the specification error resulting from the aggregation over time implicit in the definition of the flow variables. Procedures for estimating continuous timemodels (such as the methods developed in Chapter 6 of this volume) make a clear distinction between the two types of variable and yield consistent estimates under suitable regularity conditions.

A second advantage of continuous time econometric modelling is that the form of a continuous time model does not depend on the unit of time or the frequency of the observations. Discrete time models are much less flexible. Indeed, the form of even quite simple discrete time models is dependent on the unit observation period. For example, if the monthly observations of a certain variable satisfy a second order autoregressive model, the quarterly observations of the same variables satisfy an autoregressive moving average model. Since the econometrician can seldom choose the observation period, but must work with the available data, the dependence of the form of a model on the unit observation period is a serious disadvantage. In order to estimate the parameters of a continuous time model we may, of course, derive a discrete model satisfied by the observations. But the form of this model, which for linear systems is a restricted vector autoregressive moving average model, does not depend on the unit observation period.

A third advantage is that a continuous time model formulated as a differential equation system can be interpreted as a causal chain model, in which each of the variables responds directly to the stimulus provided by a subset only of the other variables in the model, and yet there is indirect interaction between all of the variables in the model during the unit observation period. Most of the coefficients in the differential equation system are restricted a priori to zero and the causal ordering of the variables is represented by the pattern of these zero restrictions. The assumption that the variables can be arranged in a causal chain of this type need not depend on any elaborate or controversial economic theory, but only on our knowledge of the information available to various economic agents at particular points of time. It is obvious, for example, that aggregate consumers' expenditure on a particular day can be influenced by the levels (on that day) of only those variables that are known to consumers, particularly personal income, personal assets, and prices, and not by the levels (on that day) of such variables as exports, imports, and investment. This is very strong information which can be used to reduce the variances of the parameter estimates. But it can be used, efficiently, only if the model is formulated in continuous time. For, because of the interaction of all variables during the unit observation period (say a quarter), the conditional expectation of the value of each variable in quarter t, conditional on information up to the end of quarter $t - 1$, will be a function of the values in period $t - 1$ (and, generally, periods $t - 2, t - 3, \ldots$) of all the variables in the model, not just a subset of them.

A final advantage is that a continuous time model can be used to generate forecasts of the continuous time paths of the variables. Such forecasts can be of considerable value, in spite of the fact that the variables are observable only at discrete intervals of time. For example, a forecast of the continuous time path of the gross domestic product could be used by businessmen for the purpose of sales forecasting and by the government and central bank for dynamic policy formulation.

In view of these advantages the reader may ask why continuous time econometric modelling has not been more extensively used. The answer to this question, undoubtedly, lies in the technical difficulty of developing methods for the efficient estimation of complex continuous time models from discrete data, the intellectual effort required for understanding these methods, and the computing power required for their implementation. The enormous developments in computing technology during the last few years have, however, resulted in a strong resurgence of activity in both the theory and application of continuous time econometric methods.

The problem of estimating the parameters of a continuous time stochastic model from discrete data was first considered (in the literature) by the eminent statistician M. S. Bartlett over forty years ago (see Bartlett 1946). Moreover, by this time the fundamental mathematical theory of continuous time stochastic models was already well developed, major contributions having been made by some of the leading mathematicians of the twentieth century, including Einstein, Wiener, and Kolmogorov. (See my historical survey Bergstrom (1988) for a brief account of this work.) The paper by Bartlett (1946) was presented at a meeting of the Royal Statistical Society and stimulated a lively discussion by M. G. Kendall and other leading statisticians, particularly concerning the potential importance of continuous time stochastic models for econometrics. But it seems to have been overlooked by econometricians. A possible reason for this is that, three years earlier, Haavelmo (1943) had introduced his simultaneous equations methodology which became the dominant econometric methodology for the next thirty years.

The first econometrician to recognize the potential importance of continuous time models in econometrics was Koopmans (1950) who had, himself, played an important role in the development of the theory of statistical inference in simultaneous equations models (see Koopman, Rubin, and Leipnik 1950). But it was Phillips (1959) who developed the first detailed algorithm for estimating a continuous time model of sufficient generality to be of use in applied econometric work. Although I had read the articles by Bartlett (1946) and Koopmans (1950) during the 1950s and was, already, convinced of the potential importance of continuous time models in econometrics, it was only after reading Phillips's article that I was convinced of the practicability of using continuous time models in applied econometric work, with the computing power that was then available.

In 1961–2 I commenced working on the problem of estimating continuous time econometric models using a completely different approach from that of Phillips (1959). At the same time I commenced working on theoretical models of cyclical growth formulated as systems of nonlinear differential equations. This work also was stimulated by that of Phillips, particularly his article Phillips (1961). I have continued to work on various aspects of continuous time econometric modelling up to the present time, and this volume includes my main contributions in that field.

My principal aim in this introduction is to provide a link between the various papers in the volume and discuss their relation to the work of other economists and econometricians. The papers have been divided into three parts. Those in Part I are concerned with theoretical models of cyclical growth which provide the basis for my empirical work on continuous time macroeconomic modelling. Part II is concerned with econometric methodology and includes papers on estimation, testing, forecasting, and optimal control. Part III is concerned with empirical applications.

Part I Theoretical Models of Cyclical Growth

Chapter 1 (Bergstrom 1962) was stimulated by the article of Phillips (1961). At the time when Phillips's article was written theoretical models of cyclical growth were rather mechanical, with cycles being generated by the multiplier–accelerator mechanism and growth introduced through exogenous trends in demand (see, for example, Hicks (1950). There was clearly a need for a more integrated model which incorporated the price mechanism. Several leading economists, including J. E. Meade and D. H. Robertson, had for some time been trying to persuade Phillips to introduce the price mechanism into his dynamic models; and his 1961 article was strongly influenced by the ideas of Meade who had himself been working on this problem. I, too, had been thinking about the problem and regarded Phillips's paper as an important advance. It provided the first mathematical analysis of a model in which real and monetary phenomena, and cycles and growth are synthesized in an integrated system and the price mechanism plays an essential role.

My own paper was an attempt to introduce the influence of the price mechanism in a less *ad hoc* way than in Phillips (1961) by making more use of microeconomic theory. At the same time, I retained his rather extreme technological assumption of a fixed capital–output ratio (no ex ante substitution between labour and capital), with different vintages of capital differing only in the respect of the amount of labour required per unit of capital. From this assumption I developed a pseudo-production function (which works through variations in the age distribution of the stock of capital) and incorporated this into a cyclical growth model. The model is similar to that of Phillips except that the influence of the price mechanism is derived from economic theory using the pseudo-production function. A full discussion of the two models was provided later by Allen (1967, ch. 20).

A natural development of the model described in Chapter 1 would be to relax the assumption of a single technology and allow for ex ante substitution between labour and capital, that is to incorporate a putty–clay technology into a cyclical growth model of the type that Phillips and I had been working on. But in the early 1960s I had already commenced work on the problems of estimating continuous time models, and I was interested in developing theoretical cyclical growth models of the type that could serve as prototype econometric models, as well as providing insights into the working of the economy. It seemed unlikely that reliable estimates of the parameters of a cyclical growth model incorporating a putty–clay technology could be obtained from aggregate time series data. Moreover, the fact that it was possible, through variations in the age distribution of the stock of capital, to get the degree of substitutability between labour and capital represented by my pseudo-production function suggested that (allowing for further substitution which could be obtained through variations in the choice of techniques and design of new equipment in response to factor price changes) the neoclassical production function might not be such a bad approximation, for econometric purposes, as some of its critics had suggested.

My next model, which is described in Chapter 2 (Bergstrom 1966*a*), incorporated a Cobb-Douglas production function and is a development of the pure neoclassical growth model introduced by Solow (1956) and Swan (1956). Indeed, it is obtained from the pure neoclassical model by discarding the assumption of full employment (or, more precisely, the assumption that employment follows an exogenous exponential trend) and introducing a price feedback mechanism which, under suitable conditions on the parameters, causes output, capital, and employment to converge (with oscillations) to their neoclassical growth paths, if undisturbed. This model was also discussed and analysed extensively in my book Bergstrom (1967), and it later served as the prototype for the first continuous time macroeconometric model, the disequilibrium neoclassical growth model of the United Kingdom described in Chapter 10.

The price feedback mechanism incorporated in this model (and also in my earlier model described in Chapter 1 and that of Phillips 1961) is essentially that suggested by Keynes (1936). It relies on the effect of wage and price adjustments on the real liquidity of the economy, and, hence, on interests rates, investment, and aggregate demand. The same mechanism is implicit in the more formal comparative static analysis presented by Pigou (1945) in his restatement of the classical theory. But nothing can be deduced from comparative static analysis about the stability of a complex dynamic system, and Keynes, himself, was very sceptical concerning the ability of the price mechanism to keep the economy operating at near full employment.

The first rigorous analysis of the stability of the neoclassical growth model was that of Solow (1956). The analysis of Solow is greatly simplified by assuming that employment follows an exogenous exponential path (which can be regarded as the full employment path), thus excluding the difficult problem associated with fluctuations in unemployment. He, essentially, proved (under certain assumptions concerning the production function) that, if all prices adjust instantaneously and continuously to their full employment equilibrium levels, then the price mechanism ensures that output and capital converge to steady state exponential growth paths.

In the model described in Chapter 2 prices adjust gradually rather than instantaneously, and the level of employment is treated as an endogenous variable. The analysis in that paper (and also in Bergstrom 1967, ch. 5) was the first rigorous mathematical investigation of the conditions under which a price feedback mechanism of the type suggested by Keynes would keep the economy oscillating about its neoclassical growth path and converging to the steady state neoclassical growth path if undisturbed. The model used for this purpose could be, aptly, described as a neoclassical–Keynesian cyclical growth model.

The methodology introduced in Chapter 2 (and in Bergstrom 1967) for the steady state and stability analysis of the model was later used in the analysis of the continuous time econometric model of the United Kingdom described in Chapter 10, and it has since been used in the analysis of nearly all the continuous time models of other countries for which the United Kingdom model served as a prototype. The methodology involves designing the model in such a way that it is possible to obtain an explicit differential equation system in the logarithms of the ratios of the variables to their steady state exponential growth paths and so that the nonlinear part of this system satisfies the Poincaré–Liapounou–Perron conditions (see Bellman 1953, p. 93, and Coddington and Levinson 1955, p. 134). A clear and detailed exposition of this methodology has recently been given by Gandolfo (1981, ch. 2).

Part II Estimation, Forecasting, and Control

Chapter 3 (Bergstrom 1966*b*) introduced the type of simultaneous equations approximate discrete model that has been the basis of the estimation procedures for most of the continuous time econometric models developed during the last fifteen years. It was also the first study of the sampling properties of estimates of the parameters of a multivariate continuous time model obtained from discrete data.

As I have mentioned earlier, my decision to commence working on the problems of estimating the parameters of continuous time models was a result of reading the article of Phillips (1959). But my method of approach was completely different from his and was closer to that of Bartlett (1946). Phillips used Fourier transforms in order to obtain the autocorrelation properties of the discrete data from the spectral density of the continuous time model, a route which was also followed by Durbin (1961). My own approach was to use the exact discrete model in the time domain in order to obtain the sampling properties of estimates derived by applying standard simultaneous equations estimation procedures to an approximate discrete model. The latter model was obtained by integrating the assumed stochastic differential equation system over the unit observation period and replacing integrals by trapezoidal approximations.

The algorithm developed by Phillips (1959) was, in fact, never used. I believe that the main reason for this was that it was not designed to take account of a priori restrictions on the coefficient matrices of the continuous time model; for example, the restriction that certain elements of these matrices are zero. Given the limited sample sizes with which econometricians must work, such restrictions can be very important in reducing the standard errors of the parameter estimates and increasing the accuracy of the post sample forecasts. Moreover, if our ultimate aim is to obtain the best forecasts of the future discrete observations there is nothing to be gained from the estimation of an unrestricted continuous time model of the type used by Phillips. For the discrete data generated by such a model satisfy a vector autoregressive moving average model with unrestricted coefficient matrices, and this can be estimated directly and used for predictive purposes.

The a priori restrictions played a central role in my own study. Indeed, the main idea put forward in Chapter 3 and further developed in Bergstrom 1967, ch. 9, was that the restrictions on the reduced form of an approximate simultaneous equations model of the form that I used could be regarded as convenient approximations to the restrictions on the exact discrete model derived from the continuous time model. In particular, if the restrictions on the matrices of structural parameters of the approximate simultaneous equations model (which are identical with the restrictions on the matrices of structural parameters of the continuous time model) are linear, then the restrictions on the matrices of coefficients of the reduced form of the approximate discrete model can be represented in terms of rational functions, whereas the restrictions on the matrices of coefficients of the exact discrete model can be represented only in terms of complicated transcendental functions. In Chapter 3 I show how to compute the exact asymptotic bias of parameter estimates obtained by using the approximate discrete model, and in a numerical example, show that this bias is very small compared with the reduction in the standard errors obtained by taking account of the a priori restrictions.

The publication of Chapter 3, in 1966, initiated a series of contributions by other econometricians on the problem of estimating the parameters of continuous time models from discrete data. The first of these was Sargan (1974, 1976) (a mimeographed version of the latter article being available in 1970). He extended my approximate discrete model to include exogenous variables and investigated the behaviour of the asymptotic bias of the estimates of the parameters of the continuous time model as the unit observation period tends to zero. He showed that, under suitable regularity conditions, the asymptotic bias of estimates obtained by applying standard simultaneous equations estimation procedures to the approximate discrete model is of order δ^2, where δ is the unit observation period. Sargan's results were extended to higher order systems by Wymer (1972). By this time Wymer had also developed a computer program for the full information maximum likelihood (FIML) estimation of the approximate discrete model.

A comparison of estimates obtained by using the approximate and exact discrete models was made in an important Monte Carlo study by P. C. B. Phillips (1972). His study showed that there is a spectacular gain in efficiency from the use of the exact discrete model (rather than the approximate discrete model) and that this results mainly from the large reduction in the standard deviations of the estimates achieved by taking account of the exact restrictions on the distribution of the discrete data implied by the continuous time model. Further contributions by Phillips were concerned with the problem of identification (see Phillips 1973), the extension of the exact discrete model to take account of exogenous variables (see Phillips 1974, 1976), and the estimation of models with flow data (see Phillips 1978).

During the same period Robinson (1976*a, b, c*) developed powerful Fourier methods of estimating a very general open continuous time model which includes, as a special case, a system of higher order stochastic differential equations. He showed that his method yields asymptotically efficient estimates when the exogenous variables are generated by a stationary random process and satisfy certain aliasing conditions. (See Bergstrom 1984*a*, p. 1206, for a brief exposition of Robinson's method.)

My own contributions to continuous time econometrics during the 1970s were concerned, mainly, with applications, particularly the development of the first continuous time macroeconometric model described in Chapter 10. I returned to the theoretical problems of estimation in the early 1980s. My work with empirical applications had convinced me of the need to formulate continuous time macroeconometric models as systems of higher order (at least second order) differential equations in order to obtain a sufficiently realistic dynamic specification of the adjustment processes in the economy to make the best use of the data. Moreover, the enormous advances in computing technology that had taken place during the preceding decade were such that it now seemed worth while developing procedures for obtaining exact Gaussian estimates (estimates which would be exact maximum likelihood estimates if the innovations were Gaussian) of the parameters of a higher order continuous time dynamic model from discrete data. Such estimates would take account of the exact restrictions on the probability distribution of the discrete data implied by the continuous time model and could be expected to have much smaller errors and result in much more accurate forecasts than estimates obtained by using approximate discrete models.

Chapter 4 (Bergstrom [1983]) was the first of a series of papers on this subject which I have written during the last few years. It contains a number of fundamental contributions which provide the basis for much of my later work, including Chapters 5,6, 7, and 8.

The most important of these contributions is the proof of a very general existence and uniqueness theorem for the solution of a system of higher order linear stochastic differential equations under very weak assumptions about the white noise innovations, and without assuming that the system is stationary or even stable. In most of the earlier literature on this subject, starting with the pioneer work of Ito (1946, 1951), it is assumed that the integrated white noise innovation process is Brownian motion (a Wiener process), which implies that its sample paths are almost all continuous. This is an unnecessarily restrictive assumption (at least when we are concerned with linear systems) and will often be inappropriate in economic models. As M. G. Kendall pointed out in his discussion of the paper by Bartlett (1946), it will often be more appropriate, when modelling economic phenomena, to assume that the innovations come as discrete jumps at random intervals of time. An important contribution to the literature on the solution of stochastic differential equations was made by Edwards and Moyal (1955) who used an approach suggested by D. G. Kendall. But they dealt with only a single second order differential equation, and an extension of their argument to a system of several differential equations of any order would not be easy. The proof of my existence and uniqueness theorem (Ch. 4, Theorem 1) for the solution of a system of any dimension and order uses a completely different mathematical argument from that of Edwards and Moyal (1955) and is based on the same weak assumptions about the innovations as their theorem.

The second main contribution of Chapter 4 is to show how to obtain (on the basis of Theorem 1) an exact discrete model for stock and flow data generated by a higher order continuous time system. This exact discrete model is in the form of a system of difference equations with autocorrelated disturbances, and both the coefficient matrices of the difference equations and the covariance and autocovariance matrices of the disturbances are obtained as explicit functions of the parameters of the continuous time model.

It is then shown how the exact discrete model can be used in order to obtain both the exact Gaussian estimates of the parameters of the continuous time model, and estimates obtained by maximizing a frequency domain approximation to the Gaussian likelihood, of the type introduced by Whittle (1951, 1953). The application of the latter estimation procedure to continuous time models was proposed by Robinson (1977). But it is greatly facilitated by obtaining an exact discrete model in the form given in Chapter 4, since, from such a model, we can immediately obtain the spectral density of the discrete observations, explicitly in terms of the parameters of the continuous time model.

Finally, Chapter 4 provides a brief discussion of the asymptotic properties of both the exact Gaussian estimators and those obtained by using the frequency domain approximation. This discussion is based on the papers by Dunsmuir and Hannan 1976, Deistler, Dunsmuir, and Hannan 1978, and Dunsmuir 1979 which have been synthesized in Hannan and Deistler 1988.

The publication of Chapter 4 in 1983 initiated a further series of contributions by other authors on the problems of estimating continuous time models from discrete data. The first of these was the article of Harvey and Stock (1985) who developed a computationally efficient algorithm for obtaining exact Gaussian estimators of the parameters of a continuous time model, from discrete stock and flow data, using the Kalman filter. Their article provided the stimulus for my next two papers (Chs. 5 and 6) which are concerned, primarily, with the development of efficient computational procedures.

Chapter 5 (Bergstrom 1985) is concerned with the development of a computationally efficient algorithm for the Gaussian estimation of a closed higher order continuous time model under assumptions of nonstationarity. The initial state vector is treated as non-random and the system is not required to have stable roots. It is argued that the treatment of the initial state vector as non-random is, usually, more realistic in econometric work than the assumption of stationarity. The algorithm is based on an exact discrete model of the type derived in Chapter 4, extended to take account of the fixed initial state vector, the unobservable elements of the latter being treated as parameters to be estimated together with the structural parameters of the model. By using the Cholesky factorization of the covariance matrix of the complete vector (over the whole sample period) of the disturbances in the exact discrete model, the Gaussian likelihood is obtained in a very simple form.

The computations required both for the algorithm developed in Chapter 5 and the Kalman filter algorithm of Harvey and Stock (1985) are of the order of the number of observations in the sample (the length of the discrete time series). From a computational point of view the two algorithms are, therefore, of the same order of efficiency, although each has advantages in particular circumstances.

In Chapter 6 (Bergstrom 1986) the algorithm of Chapter 5 is extended to allow for exogenous variables. The extended algorithm is based on a pseudo-exact discrete model which is satisfied exactly over any sub-period of the sample in which the time paths of the exogenous variables are polynomials in t (time) of degree not exceeding two. This type of discrete model was first used by Phillips (1974, 1976) following a suggestion of J. D. Sargan. But Phillips applied it only to the simplest case, a first order system in which all variables are measured at equi-spaced points of time, that is, there are no flow variables. The application of the method to higher order systems with mixed stock and flow data is much more difficult, because of the complexity of the convolution integrals through which the continuous time paths of the exogenous variables enter the exact discrete model, and the difficulty of explicitly evaluating these integrals after replacing the unobservable continuous time paths of the exogenous variables by their quadratic approximations. In Chapter 6 I use a mathematical argument which avoids the explicit evaluation of these convolution integrals and, by making use of the results obtained for a closed model, leads to a great simplification of the algebra and the formulae for the coefficient matrices of the pseudo-exact discrete model.

An important property of the pseudo-exact discrete model is that the coefficient matrices of both the lagged dependent variables and the exogenous variables (as well as the covariance and autocovariance matrices of the disturbances) depend only on the parameters of the continuous time model, and not on the parameters of the quadratic approximations to the time paths of the exogenous variables. For this reason it can be expected to provide a good approximation even when the behaviour of the exogenous variables differs greatly as between different parts of the sample period.

The method of obtaining Gaussian estimates of the parameters of the open continuous time dynamic model is the same as that developed in Chapter 5. It yields exact maximum likelihood estimates when the innovations are Gaussian and the exogenous variables are polynomials in t of degree not exceeding two and can be expected to yield very good estimates under much more general circumstances. Precise formulae for the implementation of the method are obtained, in Chapter 6, for a second order system in which both the endogenous and exogenous variables are a mixture of stock and flow variables. Formulae for the implementation of the method in a closed mixed first and second order system and a first order system with mixed stock and flow data have been derived by Agbeyegbe (1984, 1987, 1988).

An important advantage of the methods developed in Chapters 4 to 6 is that they provide the basis for the exact asymptotic tests of the specification of a continuous time model and of hypotheses relating to its parameters. Practical procedures for the implementation of these tests are developed in Chapter 7, which has been written especially for this volume and makes extensive use of the results obtained in the preceding chapters, especially Chapter 6. The proposed tests are carried out within the framework of the vector autoregressive moving average representation of the exact discrete model, and the chapter contains a much fuller discussion of this representation than has, hitherto, appeared in the literature, including a new convergence theorem relating to the moving average coefficient matrices. It also includes formulae for the asymptotic covariance matrix of the parameter estimates (which are required for a Wald test of restrictions on the parameters) and an efficient computational procedure for obtaining this matrix.

Another advantage of the exact Gaussian estimation procedures developed in Chapters 4 to 6 is that they provide a very convenient basis for obtaining optimal forecasts of the post sample discrete observations. The procedure for obtaining these forecasts is described in Chapter 8 (Bergstrom 1989) which also provides a rigorous proof of the optimality of the forecasts. They are optimal in the sense that, when the innovations are Gaussian, they are exact maximum likelihood estimates of the conditional expectations of the post sample discrete data, conditional on all the information in the sample, and taking account of the exact restrictions on the probability distribution of the data implied by the continuous time model.

Because of the restrictions on the exact discrete model implied by the continuous time model, the forecasts obtained from this model can be expected to have much smaller errors than those obtained from an unrestricted vector autoregressive moving average model. There are important gains from continuous time econometric modelling, therefore, even if our sole aim is to obtain the best predictions of the future discrete observations.

Another important potential use of continuous time econometric models is in the design of policy feedback rules using optimal control methods. Indeed, there have already been some applications of optimal control methods to policy analysis in continuous time econometric models. (See, for example, Stefansson 1981 and Gandolfo and Petit 1986). The aim of Chapter 9 (Bergstrom 1987) is to provide a firmer theoretical basis for such applications through a rigorous treatment of the problem of optimal control in linear continuous time stochastic models, with infinite horizon quadratic cost functions, under weaker assumptions concerning the white noise innovations that have been made in the literature on this subject.

In recent years, the standard approach to continuous time stochastic control problems has been to use stochastic dynamic programming and the Ito calculus. This approach requires the white noise innovations to be generated by Brownian motion which, as I have already mentioned, is often an inappropriate assumption in econometric work. In Chapter 9 I use a completely different type of mathematical argument, based on the general existence and uniqueness theorem proved in Chapter 4. The optimal feedback is derived under the same weak assumptions concerning the white noise innovations as are used in that theorem. It is shown that, even under these weak assumptions, the optimal feedback is identical with that for the corresponding non-stochastic control problem. The basic theorem of Chapter 9 (Theorem 2) and its extension to a higher order system (Theorem 3) can be regarded, therefore, as a dynamic continuous time version of the certainty equivalence theorem which holds under very weak assumptions concerning the white noise innovations.

Part Three Applications

Chapter 10 (Bergstrom and Wymer 1976) describes the first continuous time macroeconometric model, the disequilibrium neoclassical growth model of the United Kingdom which Wymer and I developed in the early 1970s. This is based on the neoclassical-Keynesian cyclical growth model described in Chapter 2 and is obtained from that model by introducing a foreign sector and replacing the Cobb-Douglas production function by a dynamic version of the more general constant elasticity of substitution production function.

In addition to being formulated in continuous time, the model has three major innovative features which distinguish it from earlier macroeconomic models. First, the complete model is formulated as a system of adjustment equations (continuous time error correction equations) in each of which the proportional rate of change in one of the variables depends on the proportional deviation of that variable from its partial equlibrium level (which is a function of other variables in the model). Secondly, the model makes intensive use of economic theory to obtain cross equation restrictions and a parsimonious parametric representation. Thirdly, it is designed in such a way as to permit a rigorous mathematical analysis of its steady state and stability properties, using the methodology introduced in Chapter 2 and Bergstrom 1967. These innovative features of the model have been adopted in nearly all of the continuous time macroeconometric models subsequently developed for other countries, including Italy (see Gandolfo and Padoan 1984; 1987, and Tullio 1981), Australia (see Jonson, Moses, and Wymer 1977 and Jonson, McKibbin, and Trevor 1982), and Germany (see Kirkpatrick 1987). A larger version of the United Kingdom model, with a more detailed financial sector, was developed by Knight and Wymer (1978).

The problem of obtaining exact Gaussian estimates of the parameters of, even a first order, continuous time model is much more complicated when it includes flow variables, such as consumption and income, than when all variables are measured at equi-spaced points of time. This is because the exact discrete model is an autoregressive moving average model with cross restrictions between the moving average and autoregressive coefficient matrices. In estimating the model of Chapter 10 we made no attempt to take account of these restrictions but, instead, used an approximate procedure which is described in Section 5 of Chapter 10 and has since been widely used in the estimation of continuous time models with flow data. We did, however, take account of the exact restrictions on the autoregressive coefficient matrix implied by the continuous time model. This was, itself, a very complex computational task at the time when the model was estimated.

Chapter 11 (Bergstrom 1984*b*), which extends the results of an earlier paper Bergstrom 1978, introduced simple monetary and fiscal policy feedbacks into the model described in Chapter 10. As in the pioneer work of Phillips (1954), the aim of this study was to show how the application of simple policy rules affect the stability of the system. It is shown, for example, that simple rules involving the acceleration of the money supply when employment or output are below their steady state paths tend to be destabilizing. A more sophisticated approach to policy analysis is, of course, to design optimal control feedbacks of the type derived in Chapter 9.

Chapter 12 (with M.J. Chambers) is a new paper which describes the first application of the general methods developed in Chapters 4 to 8. The application is to the demand for consumer durable goods in the United Kingdom, and the paper makes some new methodological developments to deal with the problems arising from the fact that consumers' stocks are unobservable. Dynamic quarterly forecasts up to two years beyond the sample period are shown to be better for two out of three classes of consumer durable goods than one-quarter ahead forecasts from a naïve model.

Recent Developments

During the last two or three years there has been an acceleration of research activity in both the theory and application of continuous time econometric methods. Some of this work is concerned with the implementation of the methods developed in Chapters 4 to 8. Nowman (1990*b*) has developed a general computer program for the exact Gaussian estimation of an open second order continuous time dynamic model with mixed stock and flow data and tested this in a Monte Carlo study described in Nowman 1990*a*. Chambers (1989), following on from the work described in Chapter 12, has estimated a complete system of dynamic demand equations which includes the demand for both durable and non-durable goods in the United Kingdom.

Work on the Kalman filter estimation of continuous time dynamic models from mixed stock and flow data, commenced by Harvey and Stock (1985), is also continuing. A more general algorithm, which is applicable to a continuous time ARMAX model has been developed by Zadrozny (1988) and further methodological developments and applications have been made by Harvey and Stock (1988*a*; 1988*b*).

There have also been important recent methodological developments resulting in the application of continuous time econometric models, particularly asset pricing models, in the field of finance. Whereas my own work, and the other recent work mentioned above, is concerned with higher order linear systems with constant coefficient matrices, the asset pricing models are, generally, formulated as first order systems whose coefficient matrices are nonlinear functions of the variables. A rigorous exposition and development of procedures for the estimation of such models has been provided by Lo (1988).

Finally, Phillips (1989) has commenced work on the development of asymptotic theory for estimates of continuous time models with unstable roots. In this paper he also proposes a method of estimating long-run relations between the variables when the innovations are assumed to be stationary processes, but not of any particular parametric form.

In this brief concluding section I have mentioned only the major recent developments in continuous time econometrics and those most closely related to my own work. There is also an extensive amount of applied work in continuous time econometric modelling now being done in many different countries. Most of this work is still unpublished, and I am sure that much of it is still unknown to me.

References

AGBEYEGBE, T. D. (1984), 'The exact discrete analog to a closed linear mixed-order system', *Journal of Economic Dynamics and Control,* **7**, 363–75.

—— (1987), 'The exact discrete analog to a closed linear first-order system with mixed sample', *Econometric Theory*, **3**, 142–9.

—— (1988), 'An exact discrete analog of an open order linear non-stationary first-order continuous-time system with mixed sample', *Journal of Econometrics*, **39**, 237–50.

ALLEN, R. G. D. (1967), *Macro-Economic Theory* (London, Macmillan).

BARTLETT, M. S. (1946), 'On the theoretical specification and sampling properties of autocorrelated time-series', *Journal of the Royal Statistical Society Supplement*, **8**, 27–41.

BELLMAN, R. (1953), *Stability Theory of Differential Equations* (New York, McGraw-Hill).

BERGSTROM, A. R. (1962), 'A model of technical progress, the production function and cyclical growth', *Economica*, **29**, 357–70.

—— (1966*a*), 'Monetary phenomena and economic growth: A synthesis of neoclassical and keynesian theories', *Economic Studies Quarterly*, **17**, 1–8.

—— (1966*b*), 'Non-recursive models as discrete approximations to systems of stochastic differential equations', *Econometrica*, **34**, 173–82.

—— (1967), *The Construction and Use of Economic Models* (London, English Universities Press).

—— (1978), 'Monetary policy in a model of the United Kingdom', in A. R. Bergstrom, A. J. L. Catt, M. H. Peston and B. D. J. Silverstone (eds.), *Stability and Inflation* (New York, Wiley).

—— (1983), 'Gaussian estimation of structural parameters in higher order continuous time dynamic models', *Econometrica*, **51**, 117–52.

—— (1984*a*), 'Continuous time stochastic models and issues of aggregation over time', in Z. Griliches and M. D. Intriligator (eds.), *Handbook of Econometrics*, vol. 2 (Amsterdam, North-Holland).

—— (1984*b*), 'Monetary fiscal and exchange rate policy in a continuous time model of the United Kingdom', in P. Malgrange and P. Muet (eds.), *Contemporary Macroeconomic Modelling* (Oxford, Blackwell).

—— (1985), 'The estimation of parameters in nonstationary higher-order continuous time dynamic models', *Econometric Theory*, **1**, 369–85.

—— (1986), 'The estimation of open higher-order continuous time dynamic models with mixed stock and flow data', *Econometric Theory*, **2**, 350–73.

—— (1987), 'Optimal control in wide-sense stationary continuous time stochastic models', *Journal of Economic Dynamics and Control*, **11**, 425–43.

—— (1988), 'The history of continuous-time econometric models', *Econometric Theory*, **4**, 365–83.

—— (1989), 'Optimal forecasting of discrete stock and flow data generated by a higher order continuous time system', *Computers and Mathematics with Applications*, **17**, 1203–14.

——, and C. R. Wymer, (1976), 'A model of disequilibrium neoclassical growth and its application to the United Kingdom', in A. R. Bergstrom (ed.), *Statistical Inference in Continuous Time Economic Models* (Amsterdam, North-Holland).

Chambers, M. J. (1989), 'Estimation of a Continuous Time Dynamic Demand System', Discussion Paper No. 350, Department of Economics, University of Essex

Coddington, E. A., and N. Levinson (1955), *Ordinary Differential Equations* (New York, McGraw-Hill).

Deistler, M., W. Dunsmuir and E. J. Hannan (1978), 'Vector linear time series models: corrections and extensions', *Advances in Applied Probability*, **10**, 360–72.

Dunsmuir, W. (1979), 'A central limit theorem for parameter estimation in stationary vector time series and its application to models for a signal observed with noise', *Annals of Statistics*, **7**, 490–506.

——, and E. J. Hannan (1976), 'Vector linear time series models', *Advances in Applied Probability*, **8**, 339–64.

Durbin, J. (1961), 'Efficient fitting of linear models for continuous stationary time series from discrete data', *Bulletin of the International Statistical Institute*, **38**, 273–82.

Edwards, D. A., and J. E. Moyal (1955), 'Stochastic differential equations', *Proceedings of the Cambridge Philosophical Society*, **51**, 663–76.

GANDOLFO, G. (1981), *Quantitative Analysis and Econometric Estimation of Continuous Time Dynamic Models* (Amsterdam, North-Holland).
——, and P. C. PADOAN (1984), *A Disequilibrium Model of Real and Financial Accumulation in an Open Economy* (Berlin, Springer).
——, —— (1987), *The Mark V Version of the Italian Continuous Time Model* (Siena, Instituto di Economia della Facolta di Scienze Economiche e Bancarie).
——, and M. L. PETIT (1986), 'Optimal control in a continuous time macroeconometric model of the Italian economy', C.N.R. Progetto Finalizzato Struttura ed Evoluzione dell'Economia Italiana, Working Paper, November, No. 6.
HAAVELMO, T. (1943), 'The statistical implications of a system of simultaneous equations', *Econometrica*, **11**, 1–12.
HANNAN, E. J., and M. DEISTLER (1988), *The Statistical Theory of Linear Systems* (New York, Wiley).
HARVEY, A. C., and J. H. STOCK (1985), 'The estimation of higher-order continuous time autoregressive models', *Econometric Theory*, **1**, 97–117.
——, and J. H. STOCK (1988*a*), 'Continuous time autoregressive models with common stochastic trends', *Journal of Economic Dynamics and Control*, **12**, 365–84.
——, —— (1988*b*), 'Estimating integrated higher-order continuous time autoregressions with an application to money-income causality', *Journal of Econometrics*, **42**, 319–36.
HICKS, J. R. (1950), *A Contribution to the Theory of the Grade Cycle* (Oxford University Press).
ITO, K. (1946), 'On a stochastic integral equation', *Proceedings of the Japanese Academy*, **1**, 32–5.
—— (1951), 'On stochastic differential equations', *Memoir of the American Mathematical Society*, **4**, 51.
JONSON, P. D., W. J. MCKIBBIN and R. G. TREVOR (1982), 'Exchange rates and capital flows', *Canadian Journal of Economics*, **15**, 669–92.
——, E. R. MOSES, and C. R. WYMER (1977), 'The RBA 76 model of the Australian economy', in *Conference in Applied Economic Research* (Reserve Bank of Australia).
KEYNES, J. M. (1936), *The General Theory of Employment, Interest and Money*, (London, Macmillan).
KIRKPATRICK, G. (1987), *Employment Growth and Economic Policy: An Econometric Model of Germany* (Tübingen, Mohr).
KNIGHT, M. D., and C. R. WYMER (1978), 'A macroeconomic model of the United Kingdom', *IMF Staff Papers*, **25**, 742–78.
KOOPMANS, T. C. (1950), 'Models involving a continuous time variable', in T. C. Koopmans (ed.), *Statistical Inference in Dynamic Economic Models* (New York, Wiley).
——H. RUBIN, and R. B. LEIPNIK (1950), 'Measuring the equation systems of dynamic economics', in T. C. Koopmans (ed.), *Statistical Inference in Dynamic Economic Models*, (New York, Wiley).
LO, A. W. (1988), 'Maximum likelihood estimation of generalized Ito processes with discretely sampled data', *Econometric Theory*, **4**, 231–47.
NOWMAN, K. B. (1990*a*), 'Finite sample properties of the Gaussian estimation of an open higher order continuous time dynamic model with mixed stock and flow data', unpublished paper, University of Essex.

—— (1990b), 'Computer program manual for computing the Gaussian estimates of an open second order continuous time dynamic model with mixed stock and flow data', unpublished paper, University of Essex.

Phillips, A. W. (1954), 'Stabilization policy in a closed economy', *Economic Journal*, **64**, 290–323.

Phillips, A. W. (1959), 'The estimation of parameters in systems of stochastic differential equations', *Biometrika*, **46**, 67–76.

—— (1961), 'A simple model of employment, money and prices in a growing economy', *Economica*, **28**, 360–70.

Phillips, P. C. B. (1972), 'The structural estimation of a stochastic differential equation system', *Econometrica*, **40**, 1021–41.

—— (1973), 'The problem of identification in finite parameter continuous time models', *Journal of Econometrics*, **1**, 351–62.

—— (1974), 'The estimation of some continuous time models', *Econometrica*, **42**, 803–24.

—— (1976), 'The estimation of linear stochastic differential equations with exogenous variables', in A. R. Bergstrom (ed.), *Statistical Inference in Continuous Time Economic Models* (Amsterdam, North-Holland).

—— (1978), 'The treatment of flow data in the estimation of continuous time systems', in A. R. Bergstrom, A. J. L. Catt, M. H. Peston, and B. D. J. Silverstone (eds.), *Stability and Inflation* (New York, Wiley).

—— (1989), 'Error correction and long-run equilibrium in continuous time', Cowels Foundation Discussion Paper No. 882R.

Pigou, A. C. (1945), *Lapses from Full Employment* (London, Macmillan).

Robinson, P. M. (1976*a*), 'Fourier estimation of continuous time models', in A. R. Bergstrom (ed.), *Statistical Inference in Continuous Time Economic Models*, (Amsterdam, North-Holland).

—— (1976*b*), 'The estimation of liner differential equations with constant coefficients', *Econometrica*, **44**, 751–64.

—— (1976*c*), 'Instrumental variables estimation of differential equations', *Econometrica*, **44**, 765–76.

—— (1977), 'The construction and estimation of continuous time models and discrete approximations in econometrics', *Journal of Econometrics*, **6**, 173–98.

Sargan, J. D. (1974), 'Some discrete approximations to continuous time stochastic models', *Journal of the Royal Statistical Society*, Series B, **36**, 74–90.

—— (1976), 'Some discrete approximations to continuous time stochastic models', in A. R. Bergstrom (ed.), *Statistical Inference in Continuous Time Economic Models* (Amsterdam, North-Holland).

Solow, R. M. (1956), 'A contribution to the theory of economic growth', *Quarterly Journal of Economics*, **70**, 65–94.

Stefansson, S. B. (1981), 'Inflation and economic policy in a small open economy: Iceland in the post-war period', Ph.D. Thesis, University of Essex, Colchester.

Swan, T. W. (1956), 'Economic growth and capital accumulation', *Economic Record*, **32**, 334–61.

Tullio, G. (1981), 'Demand management and exchange rate policy: The Italian experience', *IMF Staff Papers*, **28**, 80–117.

Whittle, P. (1951), *Hypothesis Testing in time Series Analysis* (Almqvist and

Wicksell, Stockholm).
—— (1953), 'The analysis of multiple stationary time series', *Journal of the Royal Statistical Society*, Series B, **15**, 125–39.
WYMER, C. R. (1972), 'Econometric estimation of stochastic differential equation systems', *Econometrica*, **40**, 565–77.
ZADROZNY, P. A. (1988), 'Gaussian likelihood of continuous time ARMAX models when data are stock and flow at different frequencies', *Econometric Theory*, **4**, 108–24.

PART I

Theoretical Models of Cyclical Growth

1

A Model of Technical Progress, the Production Function and Cyclical Growth[1]

1. Introduction

ONE of the major problems confronting the builder of dynamic economic models is that of finding a form of technical relation between output and factor inputs that is realistic and yet can be fitted into a manageable dynamic system. It would be surprising if the forms of technical relation that are appropriate for comparative static analysis were equally appropriate for dynamic analysis. For is it realistic to assume that, in a position of stationary equilibrium, capital takes the form that is most suited to the equilibrium ratio of labour employed to the stock of capital and that there is, therefore, no excess capacity. But it would be unrealistic to assume that this condition holds at all points of time in a dynamic system. A technical relation that is appropriate for dynamic analysis must be consistent with the following facts:

(1) In the short run the form of capital cannot be changed without limit as factor proportions vary (i.e. capital equipment is not 'perfectly malleable').[2]

(2) The stock of capital of an economy is seldom, if ever, used to full capacity.

The first aim of this article is to derive a form of production function from technological assumptions that are consistent with these facts. The function derived is not exact, but holds approximately, in the neighbourhood of the steady growth path of a model. It can be used, therefore, for the purpose of analysing the behaviour of a model in the neighbourhood of its steady growth path.

The second aim is to fit the new form of production function into a model that determines the paths of output, employment, consumption, capital formation, the rate of interest, the wage rate and the price level. This is not merely a matter of adding to the production function an independent set of behaviour relations. The parameters of the production function must enter (e.g. via the marginal product of capital) into certain

[1] I am indebted to Professor A. W. Phillips for his comments on an earlier version of this article.

[2] See Meade 1962.

behaviour relations such as the investment function. Moreover, the forms of the various relations must be chosen in such a way that the complete model does not defy analysis. This is part of the art of model-building.

This model has been considerably influenced by Professor Phillips' recent model,[3] and can be regarded as an extension of it. Like his model, it provides a synthesis of real and monetary quantities and of cycles and growth in a single system. A significant difference between the two models is that it is implicitly assumed in his model that the proportion of the labour force employed is uniquely related to the ratio of output to the stock of capital, whereas this model determines the separate paths of these variables, and there is no unique relation between them. This difference between the two models leads to certain differences in the conclusions with regard to the determinants of steady growth rates. But the conclusions with regard to stability are very similar.

2. The Production Function

The technological assumptions on which the remainder of the paper is based are as follows:[4]

Assumption 1

New technical knowledge is continually being incorporated in new types of capital equipment, and this is the only form of technical progress. Thus capital equipment produced at time t is physically different from capital equipment produced at any other time, and will be referred to as capital of type t.

Assumption 2

Capital goods and consumers' goods are perfect substitutes in production. This assumption enables us to define a unit of capital of type t as the quantity of capital of type t that can be produced at an opportunity cost of one unit of consumption.

Assumption 3

There is a fixed maximum net output v, which can be produced with a unit of capital of any type. This means that technical progress is neutral in the sense that the full-capacity output-capital ratio is constant.

[3] See Phillips 1961.

[4] Assumptions 1 and 5 have also been made by Johansen 1959. An important difference between the present model and that of Johansen, is that, unlike Johansen, we do not assume that the stock of capital is always used to full capacity.

Assumption 4

The number of units of labour required in order to produce v units of output, with a unit of the most modern type of capital, decreases at a constant proportional rate, ρ per unit of time.

Without loss of generality we can choose units of measurement so that $e^{-\rho t}$ units of labour are required in order to produce v units of output with a unit of capital of type t.

Assumption 5

The stock of capital of a given type decreases at a constant proportional rate, δ, per unit of time.

This assumption can be interpreted in either of two ways. We can assume that each piece of capital retains its original efficiency throughout its life, but that the number of pieces of a given type that are still in existence decreases at a constant proportional rate. Or we can assume that the life of each piece of capital is infinite, but that its efficiency decreases at a constant proportional rate.

The assumption implies that

$$K = \int_{\infty}^{t} I(x)e^{-\delta(t-x)}\,dx \tag{2.1}$$

where K denotes the total stock of capital of all types, I denotes gross investment, and t denotes time.

Assumption 6

Output is always produced in the most efficient way possible with the given stock of capital. Thus capital of a given age is not used unless all capital of a lower age is being used to its full capacity.

From assumptions 3, 5 and 6 we obtain

$$Y = (v + \delta)\int_{t-q}^{t} I(x)e^{-\delta(t-x)}dx - \delta K \tag{2.2}$$

where Y denotes net output and q denotes the age of the oldest type of capital in use. And, from assumptions 4, 5 and 6 we obtain

$$L = \int_{t-q}^{t} I(x)e^{-\delta(t-x)-\rho x}dx \tag{2.3}$$

where L denotes the quantity of labour employed.

Equations (2.1) to (2.3) summarise the technological assumptions. We now seek a single equation (i.e. a production function 0, which is, more amenable to mathematical manipulation than equations (2.1) to (2.3) and which can be substituted for these equations, for the purpose of analysing the behaviour of a model in the neighbourhood of its steady growth path. We shall confine our attention to models that satisfy the following conditions:

(i) Equations (2.1) to (2.3) are satisfied.
(ii) When income is growing at a constant proportional rate, the saving-income ratio is constant.
(iii) The model has a steady state solution in which output and the stock of capital each grow at a constant proportional rate, and in which the quantity of labour employed is constant.

The steady state growth rate for any model satisfying the above conditions is necessarily equal to ρ. For

$$K_s = K_s(O)e^{k_s t} \tag{2.4}$$

where k denotes DK/K, D denotes the differential operator, d/dt, and x_s (for any x) denotes the value of x in the steady state.

Therefore

$$I_s = DK_s + \delta K_s = (k_s + \delta)\, K_s(O)e^{k_s t} \tag{2.5}$$

Now from (2.2), (2.4) and (2.5) we obtain

$$Y_s/K_s = (v + \delta)\{1 - e^{-q_s(k_s+\delta)}\} - \delta \tag{2.6}$$

But

$$Y_s/K_s = k_s/s \tag{2.7}$$

where s denotes the saving-income ration in the steady rate.

From (2.6) and (2.7) we obtain

$$q_s = \frac{1}{k_s + \delta} \log\left\{1 - \frac{k_s + \delta s}{s(v + \delta)}\right\} \tag{2.8}$$

which shows that q_s is independent of time. And, from (2.3), (2.4) and (2.5) we obtain

$$L_s = \frac{(k_s + \delta)K_s(O)e^{(k_s-\rho)t}}{k_s + \delta - \rho}\{1 - e - {}^{(k_s+\delta-\rho)q_s}\} \tag{2.9}$$

Since by hypothesis (condition iii) L_s is independent of time, it follows from (2.9) that

$$k_s = \rho \tag{2.10}$$

By substituting ρ for k_s in (2.8) it can be seen that, for any model satisfying conditions (i) to (iii), we must have $sv \geq \rho$. If the model satisfies conditions (i) and (ii) but has $sv < \rho$, then it cannot have a steady state solution of the type specified in condition (iii). The explanation of this result is that, if $sv < \rho$, then, even when the stock of capital is being used to full capacity, saving will not be great enough to enable the stock of capital to grow at the proportional rate, ρ.

We turn now to the derivation of the production function. It will be convenient to start by considering the simple, though unrealistic case of a model whose investment function is

$$k = \rho \tag{2.11}$$

We shall show that, in this case, there is an exact relation between output, the quantity of labour employed and the total stock of capital, regardless of the manner in which output varies.[5]

From (2.11) we obtain

$$K = K(O)e^{\rho t} \tag{2.12}$$

so that

$$I = (\rho + \delta)K(O)e^{\rho t} \tag{2.13}$$

Now from (2.2), (2.12) and (2.13) we obtain

$$Y/K = (v + \delta)\{1 - e^{-(\rho+\delta)q}\} - \delta \tag{2.14}$$

and from (2.3), (2.12) and (2.13) we obtain

$$L/K = \frac{p + \delta}{\delta} e^{-\rho t} \{1 - e^{-\delta q}\} \tag{2.15}$$

Eliminating q from (2.14) and (2.15) we obtain

$$Y/K = v - (v + \delta)\left\{1 - \frac{\delta e^{\rho t} L}{(\rho + \delta)K}\right\}^{(\rho+\delta)/\delta} \tag{2.16}$$

so that

$$Y = vK - (v + \delta)K\left\{1 - \frac{\delta e^{\rho t} L}{(\rho + \delta)K}\right\}^{(\rho+\delta)/\delta} \tag{2.17}$$

or

$$Y/L = vK/L - \frac{(v + \delta)K}{L}\left\{1 - \frac{\delta e^{\rho t} L}{(\rho + \delta)K}\right\}^{(\rho+\delta)/\delta} \tag{2.18}$$

Since q must be non-negative, equation (2.15) implies also that

$$L \leqq \frac{\rho + \delta}{\delta} e^{-\rho t} K \tag{2.19}$$

This inequality restricts the area of the (L, K) plane to which the production function (2.17) is applicable.

From (2.17) and (2.19) we obtain (2.20)

$$Y \leqq vK \tag{2.20}$$

which is an obvious consequence of Assumption 3.

Equation (2.17) shows that, in the special case where (2.11) is satisfied, output is a function of the quantity of labour employed, the stock of capital and time. Moreover, this function is homogeneous of the first degree in the

[5] The assumption that the proportional rate of growth of the stock of capital is constant while the proportional rate of growth of output is varying implies, of course, that the saving-income ratio is varying.

quantity of labour employed and the stock of capital, and has negative second partial derivatives with respect to these variables. Thus it has the properties of a production function which shows constant returns to scale and obeys the law of diminishing returns.

It should, perhaps, be emphasized that K is the total stock of capital and not just that part of the stock of capital which is being used. Equation (2.18) expresses the fact that the greater the stock of capital per unit of labour employed, the lower will be the average age of that part of the stock of capital which is being used, the more modern will be the techniques of production used, and the greater, therefore, will be the output per unit of labour employed.

Consider now the more general class of models which satisfy conditions (i) to (iii), but do not have the trivial investment function (2.11). Any deviation of such a system from its steady state will result in variations in the proportional rate of growth of the stock of capital and variations, therefore, in the age-distribution of the stock of capital. But, so long as k does not deviate much from ρ, the age-distribution of the stock of capital will not change much, and it is intuitively obvious that equation (2.17) will still hold approximately.

The argument can be made rigorous by specifying that $|k - \rho| < c$. It can then be shown that

$$Y = vK - (v + \delta)K\left\{1 - \frac{\delta e^{\rho t} L}{(\rho + \delta)K}\right\}^{(\rho+\delta)/\delta} + \phi(t) \qquad (2.21)$$

where $\phi(t) \to O$ as $c \to O$.

We conclude, therefore, that for the purpose of investigating the behaviour of a model of the class being considered in the neighbourhood of its steady state, we can replace equations (2.1) to (2.3) by the approximating function (2.17) and the inequality (2.19).

It is of interest to see how well our production function can be approximated by the Cobb-Douglas function (2.22).

$$Y = Ae^{at} L^{\varepsilon} K^{(1-\varepsilon)} \qquad (2.22)$$

After putting $a = \rho\varepsilon$ and

$$A = v\left\{\frac{\rho + \delta}{\delta}\right\}^{-\varepsilon},$$

equation (2.22) can be written in the form

$$Y/K = v\left\{\frac{\rho + \delta}{\delta}\right\}^{-\varepsilon}\left\{\frac{e^{\rho t} L}{K}\right\}^{\varepsilon} \qquad (2.23)$$

Equations (2.16) and (2.23) each express Y/K as a function of $e^{\rho t} L/K$, and in each case the function has the value v when $e^{\rho t} L/K = (\rho + \delta)/\delta$. For the the purpose of comparing these two functions we shall put ε equal to the value which minimizes the integral of the squared difference between their ordinates. The expression to be minimized is given by

$$\int_0^{\rho+\delta/\delta}\left[v-(v+\delta)\left\{1-\frac{\delta e^{\rho t}L}{(\rho+\delta)K}\right\}^{(\rho+\delta)/\delta} - v\left\{\frac{\rho+\delta}{\delta}\right\}^{-\varepsilon}\left\{\frac{e^{\rho t}L}{K}\right\}^{\varepsilon}\right]^2 d\left\{\frac{e^{\rho t}L}{K}\right\}$$

$$= u + \frac{\rho+\delta}{\delta}\left[\frac{2v(v+\delta)\Gamma(\varepsilon+1)\Gamma\left(\frac{\rho+2\delta}{\delta}\right)}{\Gamma\left(\varepsilon+\frac{\rho+3\delta}{\delta}\right)} - \frac{v^2(3\varepsilon+1)}{(\varepsilon+1)(2\varepsilon+1)}\right] \tag{2.24}$$

where u is a constant and Γ denotes the gamma function.[6]

Assuming, for example, that $v = .3$, $\rho = .03$, and $\delta = .05$, then $\varepsilon = \cdot 8$, and the graphs of equations (2.16) and (2.23) are the curves a and b respectively in Fig. 1. It can be seen that the approximation given by the Cobb-Douglas function is rather poor, and that this is because the latter function cannot allow the marginal product of labour to tend to zero as output approaches the full capacity level, or net output to become negative at very low levels of employment.[7]

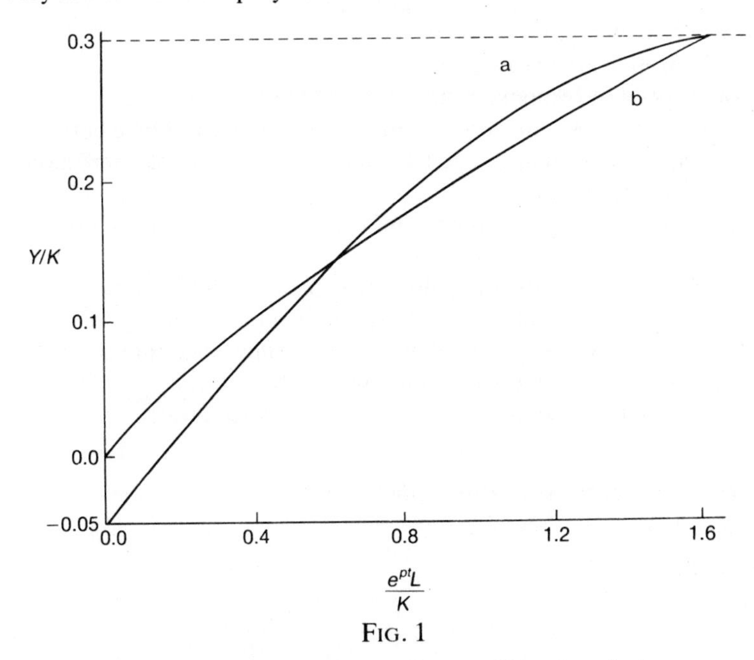

FIG. 1

[6] For a discussion of the gamma function, see Courant 1936, vol. II, ch. 4.

[7] Assumption 5 is unrealistic to the extent that it assumes the rate of capital shrinkage to be independent of the degree of usage. But it would be equally unrealistic to assume that capital which is not being used does not deteriorate. The net output of an actual economy would undoubtedly become negative if employment fell to a sufficiently low level.

3. The Model

The remaining sections will be devoted to an investigation of the behaviour of a complete model belonging to the class considered in section 2. After replacing the technical relations (2.1), (2.2) and (2.3) by (2.17) and (2.19), which have been reproduced as (3.1) and (3.2), the model is as follows:

Endogenous Variables

Y = real net output and income.
L = quantity of labour employed.
K = stock of capital.
C = real consumption.
g = rate of growth expected by entrepreneurs.
r = rate of interest.
W = money wage rate.
P = money price of output.

Exogenous Variables

M = quantity of money.

Constants

v = full capacity output-capital ratio
ρ = rate of technological progress.
δ = rate of capital shrinkage.
s = propensity to save.
λ = speed of response in investment function.
γ = parameter indicating the influence of the difference between the marginal product of capital and the rate of interest on investment.
α = parameter indicating the influence of entrepreneurs' growth expectations on investment.
β = parameter indicating the influence of employment on the proportional rate of change in the money wage rate.
L^* = level of employment at which the money wage rate is constant.
κ, μ = parameters of the demand function for money.
η = speed of response of entrepreneurs' expectations.

Other Symbols

D = differential operator, d/dt.
$x = DX/X$ (for any X).

Technical Relations

$$Y = vK - (v+\delta)K\left\{1 - \frac{\delta e^{\rho t}L}{(\rho+\delta)K}\right\}^{(\rho+\delta)/\delta} \tag{3.1}$$

$$L \leqq \frac{\rho+\delta}{\delta} e^{-\rho t}K \tag{3.2}$$

Behaviour Relations

$$C = (1 - s)Y \tag{3.3}$$

$$k = \left(\frac{N\lambda}{D + N\lambda}\right)^N \left\{\alpha g + \gamma\left(\frac{\partial Y}{\partial K} - r\right)\right\} \tag{3.4}$$

$$W/P = \frac{\partial Y}{\partial L} \tag{3.5}$$

$$w = \beta \log (L/L^*) \tag{3.6}$$

$$r = \kappa + \mu(\log Y + \log P - \log M) \tag{3.7}$$

$$g = \left(\frac{\eta}{D + \eta}\right)y. \tag{3.8}$$

Identity

$$Y = C + DK \tag{3.9}$$

Equation (3.1) has been fully discussed in the preceding section. The inequality (3.2) defines a full capacity boundary.[8] If this boundary is ever reached then the model will be temporarily replaced by a simpler model, obtained by substituting (3.1.1) and (3.2.1) for (3.1) and (3.2):

$$Y = vK \tag{3.1.1}$$

$$L = \frac{\rho + \delta}{\delta} e^{-\rho t} K \tag{3.2.1}$$

Then Y and K will grow at the proportional rate sv, L will grow at the proportional rate $sv - \rho$, w will (because of (3.6)) be increasing and, provided that m is constant, r will be increasing. This mechanism will bring about a reversion to the original model.

Consumption is assumed (equation (3.3)) to be a constant proportion of income. The presence of a time lag in the consumption function would not invalidate the use of (3.1). But, in order to keep the model as simple as possible while preserving its essential features, a time lag of zero has been assumed.

The proportional rate of increase in the stock of capital is assumed (equation (3.4)) to depend, with a distributed time lag, on the rate of growth of output expected by entrepreneurs and the excess of the marginal product of capital over the rate of interest.[9] When $\alpha = 1$, the equation implies that entrepreneurs will plan to increase the stock of capital at a proportional rate which is greater than, equal to, or less than the expected proportional rate of increase in output according as the marginal product

[8] This boundary should not be confused with the 'full employment ceiling'. Because of the flexibility of the labour force, the influence of the latter ceiling is likely to come through the relation (3.6) rather than by imposing a precise limit on output.

[9] The rate of interest, r, is to be interpreted as the money rate of interest minus the expected rate of increase in the price level.

of capital is greater than, equal to, or less than the rate of interest. The marginal product of capital is identified with the partial derivative, $\partial Y/\partial K$, of (3.1). It is evident from the way in which (3.1) has been derived that this derivative is approximately equal to the amount by which output, at time t, would have been greater if the stock of capital had been one unit greater but the age distribution of the stock of capital, and the quantity of labour employed, had been unchanged.

The money price of output is assumed (equation (3.5)) to be adjusted to the money wage rate in such a way as to maintain equality of the real wage rate to the marginal product of labour. This means that the money price of output is just sufficient to cover the labour cost per unit of output in the oldest plant in use. The rate of increase in the money wage rate is assumed (equation (3.6)) to be an increasing function of the level of employment.[10]

It is assumed (equation (3.7)) that the rate of interest is adjusted in such a way as to equate the quantity of money demanded to the quantity of money in existence, and that the quantity of money demanded is a decreasing function of the rate of interest and has unit elasticity of demand with respect to both real income and the price level.[11]

Finally, it is assumed (equation (3.8)) that the rate of growth expected by entrepreneurs depends, with an exponential time lag, on the actual rate of growth.

As a first step in the analysis of the behaviour of the above system, we shall derive from equations (3.1) to (3.9) a differential equation in k. We shall consider first the case in which $N = 1$, so that the lag distribution in the investment function has the simple exponential form.

From (3.3) and (3.9) we obtain

$$Y/K = k/s. \tag{3.10}$$

From (3.1) and (3.10) we then obtain

$$\partial Y/\partial K = v(\rho + \delta)/\delta - \rho k/s\delta - (\rho + \delta)(v + \delta)^{\delta/(\rho+\delta)} (sv - k)^{\rho/(\rho+\delta)}/\delta s^{\rho/(\rho+\delta)}. \tag{3.11}$$

Substituting from (3.11) into (3.4) and putting $N = 1$ we have

$$Dk + \lambda(1+ \gamma\rho/\delta s)k + \lambda\gamma(\rho + \delta)(v + \delta)^{\delta/(\rho+\delta)}(sv - k)^{\rho/(\rho+\delta)}/\delta s^{\rho/(\rho+\delta)} + \lambda\{\gamma r - \alpha g - \gamma v(\rho + \delta)/\delta\} = 0 \tag{3.12}$$

From (3.1), (3.5) and (3.10) we obtain

$$\log P = \log W - \log\{(v + \delta)^{\delta/(\rho+\delta)}/s^{\rho/(\rho+\delta)}\} - \rho \log(sv - k)/(\rho + \delta) - \rho t \tag{3.13}$$

[10] It is implicitly assumed that the population is constant.
[11] Cf. Phillips 1961.

and from (3.10)

$$\log Y = \log K + \log k - \log s. \tag{3.14}$$

Differentiating and combining (3.7), (3.12), (3.13) and (3.14) we obtain

$$D^2k + \lambda[1 + \gamma\rho/\delta s + \gamma\mu\{1/k + \rho/(\rho+\delta)\,(sv-k)\} - \\ -\gamma\rho(v+\delta)^{\delta/(\rho+\delta)}/\delta s^{\rho/(\rho+\delta)}\,(sv-k)^{\delta/(\rho+\delta)}]Dk + \lambda\gamma\mu k \\ + \lambda\gamma\mu(w-\rho-m) - \lambda\alpha Dg = 0 \tag{3.15}$$

From (3.1) and (3.10) we obtain

$$\log L = \log\{(\rho+\delta)/\delta\} - \rho t + \log K \\ + \log\{1 - (sv-k)^{\delta/(\rho+\delta)}/(v+\delta)^{\delta/(\rho+\delta)}s^{\delta/(\rho+\delta)}\}. \tag{3.16}$$

Differentiating and combining (3.6), (3.15) and (3.16) we obtain

$$D^3k + \lambda[1 + \gamma\rho/\delta s + \gamma\mu\{1/k + \rho/(\rho+\delta)(sv-k)\} \\ - \rho\gamma(v+\delta)^{\delta/(\rho+\delta)}/\delta s^{\rho/(\rho+\delta)}(sv-k)^{\delta/(\rho+\delta)}]D^2k + \lambda\gamma\mu[1 + \beta\delta/(\rho+\delta) \\ \times \{(v+\delta)^{\delta/(\rho+\delta)}s^{\delta/(\rho+\delta)}(sv-k)^{\rho/(\rho+\delta)} + k - sv\}]Dk \\ + \lambda\gamma[\mu\rho/(\rho+\delta)(sv-k)^2 - \mu/k^2 \\ - \rho(v+\delta)^{\delta/(\rho+\delta)}/(\rho+\delta)s^{\rho/(\rho+\delta)}(sv-k)^{(\rho+2\delta)/(\rho+\delta)}](Dk)^2 \\ + \lambda\gamma\mu\beta k - \lambda\gamma\mu\beta\rho - \lambda\alpha D^2g = 0 \tag{3.17}$$

Finally, from (3.8), (3.14) and (3.17), we can obtain a fourth order differential equation in k. This and the initial conditions will determine $K(t)$; and from $K(t)$ and the initial conditions we can determine $Y(t)$, C(t), $L(t)$, $W(t)$, $P(t)$ and $r(t)$ by using (3.10), (3.3), (3.16), (3.6), (3.13) and (3.12) respectively.

4. The Steady State

It follows from (3.8), (3.14) and (3.17) that

$$y_s = k_s = g_s = \rho. \tag{4.1}$$

This result is consistent with that obtained for the general class of models discussed in section 2. The conclusion that the steady state growth rate is independent of the propensity to save has been derived from a number of different models by various writers.[12] A different result is yielded by

[12] See, for example, Solow 1956, Kaldor 1957, Johansen 1959 and Meade 1962.

Professor Phillips' model which leads to the conclusion that the steady state growth rate is an increasing function of both the propensity to save and the proportional rate of increase in the quantity of money.[13]

Assuming that $a = 1$, we obtain from (3.12) and (4.1)

$$r_s = v(\rho + \delta)/\delta - \rho^2/\delta s - (\rho + \delta)(v + \delta)^{\delta/(\rho+\delta)}(sv - \rho)^{\rho/(\rho+\delta)}/\delta s^{\rho/(\rho+\delta)} \tag{4.2}$$

It can easily be shown that $\partial r_s/\partial s < 0$ and $\partial r_s/\partial v < 0$, which means that the steady state rate of interest is lower the greater is the propensity to save and the greater is the full capacity output–capital ratio. Moreover, it follows from (3.11), (4.1) and (4.2) that, in the steady state, the marginal product of capital is equal to the rate of interest.

From (3.7), (4.1) and (4.2) we obtain

$$p_s = m - \rho \tag{4.3}$$

and from (3.13), (4.1) and (4.3)

$$w_s = m. \tag{4.4}$$

From (3.6) and (4.4) we then obtain

$$L = L^* e^{m/\beta} \tag{4.5}$$

which shows that, in the steady state, the level of employment will be greater, the greater is the proportional rate of increase in the quantity of money.

We have implicitly assumed that $sv \geq \rho$. If $sv < \rho$ the only sort of steady state that is possible is one in which Y and K grow at the proportional rate sv while L falls at the proportional rate $\rho - sv$.

5. The Stability of the System

We shall assume, for the purpose of this section, that g is constant. This is approximately equivalent to assuming that the speed of response, η, in (3.8) is very small, and is a condition which is likely to be favourable for stability.[14] Introducing the new variable x, defined by (5.1),

$$x = k - \rho \tag{5.1}$$

and assuming that g is constant we obtain from (3.17)

[13] See Phillips 1961.
[14] See, for example, Phillips 1961.

$$D^3x + \lambda[1 + \gamma\rho/\delta s + \gamma\mu\{1/(\rho + x) + \rho/(\rho + \delta)(sv - \rho - x)\}$$
$$- \rho\gamma(v + \delta)^{\delta/(\rho+\delta)}/\delta s^{\rho/(\rho+\delta)}(sv - \rho - x)^{\delta/(\rho+\delta)}]D^2x$$
$$+ \lambda\gamma\mu[1 + \beta\delta/(\rho + \delta)\{(v + \delta)^{\delta/(\rho+\delta)}s^{\delta/(\rho+\delta)}(sv - \rho - x)^{\rho/(\rho+\delta)}$$
$$+ \rho + x - sv\}]Dx + \lambda\gamma[\rho\mu/(\rho + \delta)(sv - \rho - x)^2$$
$$- \mu/(\rho + x)^2 - \rho(v + \delta)^{\delta/(\rho+\delta)}/(\rho + \delta)s^{\rho/(\rho+\delta)}$$
$$\times (sv - \rho - x)^{(\rho+2\delta)/(\rho+\delta)}](Dx)^2 + \lambda\gamma\mu\beta x = 0. \tag{5.2}$$

Let b_1, b_2 and b_3 denote the values, at $t = 0$, of x, Dx and D^2x respectively. Then we shall say that (5.2) is asympototically stable at the origin if, provided that $\Sigma_{i=1}^{3}|b_i|$ is sufficiently small, x, Dx and D^2x tend to zero as $t \to \infty$. By using equations (5.1), (3.14), (3.12), (3.7), (3.13) and (3.6) successively it can be seen that, if x, Dx and D^2x tend to zero as $t \to \infty$, then k, y, r, p, w and L tend to their steady state values as $t \to \infty$. In order to determine the conditions under which (5.2) is asymptotically stable at the origin we shall use the following theorem.[15]

THEOREM

Equation (5.3) is asymptotically stable at the origin if the linear approximation (5.4) is asymptotically stable at the origin and the expression $f(x, Dx, .., D^{n-1}x)/\{|x| + |Dx| + .. + |D^{n-1}x|\}$ tends to zero as $|x| + |Dx| + .. |D^{n-1}x|$ tends to zero.[16]

$$D^n x + a_1 D^{n-1}x + .. + a_{n-1}Dx + a_n x + f(x, Dx, ..., D^{n-1}x) = 0 \tag{5.3}$$

$$D^n x + a_1 D^{n-1}x + .. + a_{n-1}Dx + a_n x = 0. \tag{5.4}$$

Now (5.2) can be written in the form

$$D^3x + \lambda[1 + \gamma\rho/\delta s + \gamma\mu\{1/\rho + \rho/(\rho + \delta)(sv - \rho)\}$$
$$- \rho v(v + \delta)^{\delta/(\rho+\delta)}/\delta s^{\rho/(\rho+\delta)}(sv - \rho)^{\delta/(\rho+\delta)}]D^2x$$
$$+ \lambda\gamma\mu[1 + \beta\delta/(\rho + \delta)\{(v + \delta)^{\delta/(\rho+\delta)}s^{\delta/(\rho+\delta)}(sv - \rho)^{\rho(\rho+\delta)}$$
$$+ \rho - sv\}]Dx + \lambda\gamma\mu\beta x + f(x, D, D^2x) = 0. \tag{5.5}$$

where $f(x, D, D^2x)/\{|x| + |Dx| + |D^2x|\}$ tends to zero as $|x| + |Dx| + |D^2x|$ tends to zero. It follows that (5.2) is asymptotically stable at the origin if all the roots of (5.6) have negative real parts.

[15] This theorem can be proved by transforming (5.3) and (5.4) into systems of n first order differential equations and using theorem 1, Chapter 5 of Struble 1962.

[16] Equations (5.3) and (5.4) are said to be asymptotically stable at the origin if, provided that $\Sigma_{i=1}^{n}|b_i|$ is sufficiently small, x, Dx .., $D^{n-1}x$ tend to zero as $t \to \infty$, b_i being the value of $D^{i-1}x$ at $t = 0$.

$$x^3 + \lambda[1 + \gamma\rho/\delta s + \gamma\mu\{1/\rho + \rho/(\rho + \delta)(sv - \rho)\}$$
$$- \rho v(v + \delta)^{\delta/(\rho+\delta)}/\delta s^{\rho/(\rho+\delta)}(sv - \rho)^{\delta/(\rho+\delta)}]x^2$$
$$+ \lambda\gamma\mu[1 + \beta\delta/(\rho + \delta)\{(v + \delta)^{\delta/(\rho+\delta)}s^{\delta/(\rho+\delta)}$$
$$\times (sv - \rho)^{\rho/(\rho+\delta)} + \rho - sv\}]x + \lambda\gamma\mu\beta = 0. \tag{5.6}$$

Assuming, for example, that $v = 0.3$, $\rho = 0.03$, $\delta = 0.5$, $s = 0.15$, $\delta = 0.25$ and $\lambda = 1$, the stable region is the area to the right of curve *a* in Fig. 2.[17]

So far we have assumed that the lag distribution in the investment function has the simple exponential form obtained by putting $N = 1$ in (3.4). But the forms of lag distribution obtained by putting $N = 2$ or $N = 3$ are perhaps more realistic.[18] If $N = 2$ and v, ρ, δ, s, γ and λ have the same values as in the previous example, then the stable region is the area to the right of curve *b* in Fig. 2. If $N = 3$ the stable region is the area below curve *c*. In each case the path of convergence is oscillatory for almost the entire stable region. These results are very similar to those obtained by Professor Phillips and emphasize the need for empirical estimates, not only of such parameters as μ and β, but also of the forms of the lag distributions in the various relations. It appears, from such evidence as is available, that realistic combinations of values of β and μ may be in either the stable or the unstable regions corresponding to the case where $N = 3$.

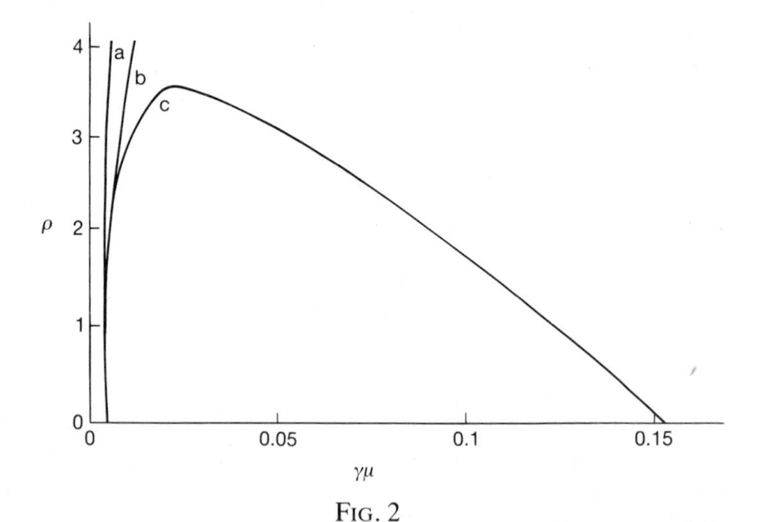

FIG. 2

[17] The parameter β is approximately equal to the increase in the percentage rate of change in the money wage rate caused by a 1 per cent increase in the level of employment, while $-\gamma\mu$ is approximately equal to the increase in the proportional rate of growth of capital caused by a 1 per cent increase in the quantity of money.

[18] See Allen for the forms of these distributions.

References

ALLEN, R. G. D. (1959), *Mathematical Economics*.

COURANT, R. (1936), *Differential and Integral Calculus*.

JOHANSEN, L. (1959), 'Substitution versus fixed production coefficients in the theory of economic growth: A Synthesis ', *Econometrica*, vol. 27.

KALDOR, N. (1957), 'A model of economic growth', *Economic Journal*, vol. LXVII.

MEADE, J. E. (1962), *A Neo-Classical Theory of Economic Growth*.

PHILLIPS, A. W. (1961), 'A simple model of employment money and prices in a growing economy', *Economica*, vol. XXVIII.

SOLOW, R. (1956), 'A contribution to the theory of economic growth', *Quarterly Journal of Economics*, vol. LXX.

STRUBLE, R. A. (1962), *Nonlinear Differential Equations*.

2

Monetary Phenomena and Economic Growth: A Synthesis of Neoclassical and Keynesian Theories

1. Introduction

THE economic growth models discussed in the literature can, with a few exceptions, be classified into two types. The first type includes models in which a continuous state of full employment is assumed and the equilibrium rate of growth of output depends on the rate of increase in the labour supply and the rate of technical progress. As an example we have the simple neoclassical model due to Solow (1956) and Swan (1956). The second type includes models in which the path of output is essentially demand determined and the equilibrium growth rate equals the rate of growth of some sort of autonomous expenditure. As an example we have the model of cyclical growth discussed by Hicks (1950).

In the latter type of model the rate of growth of autonomous expenditure is usually assumed to equal the rate of growth of the full employment ceiling this being the output that could be produced if the labour force were fully employed. Although there is no justification for this assumption, the implication that the equilibrium rate of growth of output equals the rate of growth of the full employment ceiling appears to be fairly realistic. For example, during the last hundred years, the output of the United States economy has grown at approximately the same average rate as the full employment ceiling although output has seldom reached this ceiling. This equality of the two growth rates could be accidental. But a more likely explanation is that the economic system contains a natural feed-back mechanism which tends to keep the proportional level of unemployment within certain bounds.

The purpose of this paper is to develop the simple neoclassical model by discarding the assumption of full employment and introducing such a feed-back mechanism. The mechanism to be introduced involves money, prices and the interest rate, and its discovery is essentially due to Keynes (1950). The first formal model in which this type of mechanism plays an essential role in the synthesis of growth and cycles is due to Phillips (1961). But, whereas Phillips' model is a development of Harrod's model (see Harrod 1948), the following model allows for substitution between labour and capital and is a development of the neoclassical model. It provides a synthesis of Keynesian and neoclassical theories.

In constructing such a model we have two general aims. The first is to formulate a model which is amenable to mathematical and statistical analysis and hence can serve as a basis for econometric work. This is largely a matter of choosing suitable functions to represent the various structural relations. The second aim is to study the properties of the model and ensure that it is capable of providing a good description of reality. This involves an analysis of the time series generated by the model for plausible regions of the parameter space.

2. The Model

The model is:

$$\frac{DC}{C} = \alpha \log \left\{ \frac{(1-s)Y}{C} \right\}, \tag{2.1}$$

$$\frac{DK}{K} = \gamma \log \left\{ \frac{pY - wL}{(1+c)rpk} \right\}, \tag{2.2}$$

$$DY = \lambda(C + DK - Y) + \mu(S^0 - S), \tag{2.3}$$

$$L = Be^{-\rho t} Y^b K^{1-b}, \tag{2.4}$$

$$p = (1+\pi)w \frac{\partial L}{\partial Y} = \frac{b(1+\pi)wL}{Y}, \tag{2.5}$$

$$\frac{Dw}{w} = \beta \log \left\{ \frac{L}{L_s} \right\} + a, \tag{2.6}$$

$$\frac{M_d}{p} = AY^u r^{-v}, \tag{2.7}$$

$$M_d = M_s, \tag{2.8}$$

$$L_s = L_0 e^{lt}, \tag{2.9}$$

$$M_s = M_0 e^{mt}, \tag{2.10}$$

$$S^0 = f(C + DK), \tag{2.11}$$

$$DS = Y - C - DK, \tag{2.12}$$

where:

C = real consumption,
Y = real net income or output,
K = amount of fixed capital,
S = stocks (inventories),
S^0 = desired level of stocks,
L = amount of labour employed,
L_s = supply of labour,
p = price level,
w = wage rate,
r = interest rate,
M_d = demand for money,
M_s = supply of money,
$a, b, c, f, l, m, s, u, v, \alpha, \beta, \gamma, \lambda, \mu, \pi, \rho, A, B, L_0, M_0$ = positive constants ($b > 1, s < 1$).

Equation (2.1) assumes that corresponding to any real income level, Y there is a desired equilibrium consumption level $(1 - s)Y$, and that the proportional rate of increase in consumption is an increasing function of the proportional excess of desired equilibrium consumption over actual consumption. The parameter s can be interpreted as the propensity to save when income is stationary and α as a speed of response parameter.

Equation (2.2) reflects entrepreneurial decisions with respect to investment in fixed capital. It assumes that the proportional rate of increase in the amount of fixed capital is an increasing function of the proportional excess of the rate of profit on fixed capital over the rate of interest. The parameter c is assumed to be positive partly because of the risk associated with investment in fixed capital and partly because profits must cover interest on stocks as well as interest on fixed capital.

Equation (2.3) assumes that the rate of increase in output is greater, the greater is the excess of sales over output and the greater the excess of desired stocks over actual stocks. The introduction of stocks into this equation is not a mere refinement but an essential feature of the model. We shall see later that if μ is too small the model is incapable of providing a good description of reality. Indeed it is clear from (2.3) that if $\mu = 0$ and Y is growing at a constant proportional rate then the excess of sales over output will be positive and increasing and stocks will eventually become negative. In fact this does not happen, because businessmen do take account of stocks when planning production. It is essential that this feature of reality be incorporated in the model.

Equation (2.4) is a Cobb-Douglas production function with a trend term to allow for technical progress. It is convenient in this sort of model to treat L, the amount of labour employed as the dependent variable. We assume that $b > 1$ implying that the marginal product of capital is positive. It should be noticed that fixed capital only has been introduced into the production function, stocks being excluded. This procedure is not only simpler than the alternative of including stocks but may be more realistic also. For a reduction in stocks may affect economic welfare mainly by forcing consumers to buy alternative brands of certain goods when the most preferred brands are out of stock. But this effect will not be reflected in the national income as usually measured.

Equation (2.5) assumes that the price level equals short-run marginal cost plus a constant proportional margin, π which reflects the degree of imperfection of competition. The limiting case in which $\pi = 0$ represents perfect competition.

Equation (2.6) is the labour market adjustment equation and assumes that the proportional rate of increase in the wage rate is an increasing function of the proportional level of employment. It should be noticed that this is essentially a Keynesian relation rather than a classical one since the adjusting variable is the money wage rather than the real wage.

Equation (2.7) is another Keynesian relation. It assumes that the real demand for money is a function of real income and the interest rate, u and $-v$ being the income elasticity and interest elasticity of demand for money respectively. Keynes suggested that the interest elasticity of demand for money is not constant but tends to minus infinity as the interest rate tends to some lower limit. But, so far, there appears to be little empirical support for this hypothesis. Indeed an analysis by Bronfenbrenner and Mayer (1960) of United States data for the period 1919–56 failed to reject the hypothesis that the interest elasticity of demand for money is constant.

Equation (2.8) assumes that the interest rate varies in such a way that the demand for money always equals the supply. Equation (2.9) assumes that the labour supply grows at a constant proportional rate and equation (2.10) that the money supply grows at a constant proportional rate. An interpretation of the latter assumption is that monetary policy is neutral.

Equation (2.11) assumes that the desired level of stocks is proportional to sales and equation (2.12) that the rate of increase in stocks equals the excess of production over sales.

3. The Solution

From (2.4), (2.6) and (2.9) we obtain

$$D \log w = \beta b \log Y - \beta(b-1)\log K - \beta(\rho + l)t + a + \beta \log\left\{\frac{B}{L_0}\right\}, \tag{3.1}$$

from (2.2), (2.4), (2.5), (2.7), (2.8) and (2.10)

$$D \log K = \frac{\gamma}{v}(1 + v - u - b) \log Y - \frac{\gamma}{v}(1 + v - b) \log K$$
$$- \frac{\gamma}{v} \log w + \frac{\gamma}{v}(m + \rho)t \tag{3.2}$$
$$+ \frac{\gamma}{v}\left[v \log \left\{\frac{b(1+\pi) - 1}{b(1+\pi)(1+c)}\right\} + \log \left\{\frac{M_0}{b(1+\pi)AB}\right\}\right],$$

from (2.3) and (2.11)

$$D \log Y = (\lambda + \mu f)\frac{C}{Y} + (\lambda + \mu f)\frac{K}{Y} D \log K - \mu \frac{S}{Y} - \lambda, \tag{3.3}$$

from (2.1)

$$D \log C = \alpha(\log Y - \log C) + \alpha \log (1 - s), \tag{3.4}$$

and from (2.12)

$$D \log S = \frac{Y}{S} - \frac{C}{S} - \frac{K}{S} D \log K. \tag{3.5}$$

The paths of w, K, Y, C and S are determined by their initial values and the system comprising equations (3.1) to (3.5). This system has a particular solution:

$$w = w^* e^{\{(1-u)(\rho+l)+m-l\}t}, \tag{3.6}$$
$$K = K^* e^{(\rho+l)t}, \tag{3.7}$$
$$Y = Y^* e^{(\rho+l)t}, \tag{3.8}$$
$$C = C^* e^{(\rho+l)t}, \tag{3.9}$$
$$S = S^* e^{(\rho+l)t}, \tag{3.10}$$

where:

$$\log w^* = \log \left\{\frac{M_0}{b(1+\pi)AB}\right\} - u \log \left\{\frac{L_0}{B}\right\} + v \log \left\{\frac{b(1+\pi) - 1}{b(1+\pi)(1+c)}\right\}$$
$$- \{v - (1 - u)(b - 1)\} \log q$$
$$+ \frac{u}{\beta}\{a + l - m - (1 - u)(\rho + l)\} - \frac{v}{\gamma}(\rho + l), \tag{3.11}$$

$$\log K^* = \log \left\{\frac{L_0}{B}\right\} + b \log q - \frac{1}{\beta}\{a + l - m - (1 - u)(\rho + l)\}, \tag{3.12}$$

$$\log Y^* = \log \left\{\frac{L_0}{B}\right\} + (b - 1) \log q - \frac{1}{\beta}\{a + l - m - (1 - u)(\rho + l)\}, \tag{3.13}$$

$$\log C^* = \log\left\{\frac{(1-s)L_0}{B}\right\} + (b-1)\log q$$
$$-\frac{1}{\beta}\{a+l-m-(1-u)(\rho+l)\} - \frac{\rho+l}{\alpha}, \tag{3.14}$$

$$\log S^* = \log\left\{\frac{L_0}{B}\right\} + (b-1)\log q - \frac{1}{\beta}\{a+l-m-(1-u)(\rho+l)\}$$
$$+\log\left\{\frac{\mu f-\rho-l}{\mu+(\rho+l)(\lambda+\mu f)}\right\}, \tag{3.15}$$

$$q = \frac{K^*}{Y^*} = \frac{\mu+(\rho+l)(\rho+l+\lambda)}{(\rho+l)\{\mu+(\rho+l)(\lambda+\mu f)\}} - \frac{(1-s)e^{-(\rho+l)/\alpha}}{\rho+l}. \tag{3.16}$$

We assume that:

$$\mu+(\rho+l)(\rho+l+\lambda)-\{\mu+(\rho+l)(\lambda+\mu f)\}(1-s)e^{-(\rho+l)/\alpha}>0, \tag{3.17}$$

$$\mu f-\rho-l>0. \tag{3.18}$$

The inequalities (3.17) and (3.18) are necessary and sufficient conditions for $S^*>0$, while (3.17) is a necessary and sufficient condition for $w^*>0$, $K^*>0$, $Y^*>0$ and $C^*>0$. Both inequalities are satisfied for plausible values of the parameters. The inequality (3.18) expresses the fact that if μ, the speed of response in the adjustment of output to a stock deficiency is too small stocks will eventually disappear, and (3.17) expresses the fact that if s is too small there will be insufficient saving to finance the investment in stocks required for growth at the rate permitted by technical progress and the growth of the labour supply.

Equation (3.8) shows that the equilibrium rate of growth of output equals $\rho+l$. This is also the equilibrium growth rate in the simple neoclassical model comprising equations (2.4), (2.9), $DK=sY$ and $L=L_s$. Although the equilibrium rate of growth of output is independent of s, the level of the equilibrium growth path is higher the greater is s, as can be seen from (3.13) and (3.16). This too is a property of the simple neoclassical model. An interesting new property of the model under discussion is that the level of the equilibrium growth path of output is higher the greater is m, the proportional rate of increase in the money supply. The reason for this is that, in the steady state, the proportion of the labour force employed is an increasing function of m. For, substituting from (3.6) into (2.6) we obtain

$$\log\left\{\frac{L}{L_s}\right\} = -\frac{1}{\beta}\{a+l-m-(1-u)(\rho+l)\}, \tag{3.19}$$

and it follows that the equilibrium growth path of employment is given by

$$L = L^* e^{lt}, \tag{3.20}$$

where

$$\log L^* = \log L_0 - \frac{1}{\beta}\{a + l - m - (1-u)(\rho + l)\}.$$

We implicitly assume that $m < l + a - (1-u)(\rho + l)$. Otherwise the model breaks down, since the amount of labour employed cannot exceed the labour supply.

It follows from (3.6) that the equilibrium rate of increase in the wage rate equals $(1-u)(\rho + l) + m - l$ and from (2.5), (3.6), (3.8) and (3.20) that the equilibrium rate of increase in the price level equals $m - u(\rho + l)$. Hence, if the income elasticity of demand for money is unity the equilibrium rate of increase in the price level equals the rate of increase in the money supply minus the sum of the rate of increase in the labour supply and the rate of increase in the productivity of labour.

The equilibrium interest rate, r^* can be obtained from (2.2), (2.5), (3.7), (3.8) and (3.16) and is given by

$$\log r^* = \log\left\{\frac{b(1+\pi) - 1}{b(1+\pi)(1+c)}\right\} - \frac{\rho + l}{\gamma}$$
$$- \log\left[\frac{\mu + (\rho + l)(\rho + l + \lambda)}{(\rho + l)\{\mu + (\rho + l)(\lambda + \mu f)\}} - \frac{(1-s)e^{-(\rho+l)/\alpha}}{\rho + l}\right]. \tag{3.21}$$

We note that $\partial r^*/\partial b > 0$ and $\partial r^*/\partial s < 0$. The interpretation of these results is that the equilibrium interest rate will be higher the greater are the opportunities for using capital and the lower the propensity to save. Another interesting property of (3.21) is that r^* is independent of the money supply and the liquidity preference parameters A, u and v. The explanation is that in the steady state, the wage rate is so adjusted to the money supply and the liquidity preference parameters as to neutralize their influence on the interest rate and the real variables in the system. Thus, although the model incorporates a Keynesian theory of interest, the equilibrium interest rate can be explained by a classical theory.

We turn now to the properties of the general solution. From equations (3.1) to (3.5) and (3.11) to (3.16) we obtain:

$$Dy_1 = -\beta(b-1)y_2 + \beta b y_3, \tag{3.22}$$

$$Dy_2 = -\frac{\gamma}{v}y_1 - \frac{\gamma}{v}(1 + v - b)y_2 + \frac{\gamma}{v}(1 + v - u - b)y_3, \tag{3.23}$$

$$Dy_3 = q(\lambda + \mu f)e^{y_2 - y_3}Dy_2 + q(\rho + l)(\lambda + \mu f)(e^{y_2 - y_3} - 1)$$
$$+ \{(\lambda + \mu f)(1-s)e^{-(\rho+l)/\alpha}\}(e^{y_4 - y_3} - 1)$$
$$- \frac{\mu(\mu f - \rho - 1)(e^{y_5 - y_3} - 1)}{\mu + (\rho + l)(\lambda + \mu f)}, \tag{3.24}$$

$$Dy_4 = \alpha(y_3 - y_4), \tag{3.25}$$

$$Dy_5 = \frac{\mu + (\rho + l)(\lambda + \mu f)}{\mu f - \rho - l}\{-qe^{y_2 - y_5}Dy_2 - q(\rho + l)(e^{y_2 - y_5} - 1) \tag{3.26}$$

$$+ e^{y_3 - y_5} - 1 - (1 - s)e^{-(\rho + l)/\alpha}(e^{y_4 - y_5} - 1)\},$$

where:

$$y_1 = \{\log\{w/w^* e^{\{(1-u)(\rho+l)+m-l\}t}\},$$

$$y_2 = \{\log K/K^* e^{\{\rho+l\}t}\},$$

$$y_3 = \{\log Y/Y^* e^{\{\rho+l\}t}\},$$

$$y_4 = \{\log C/C^* e^{\{\rho+l\}t}\},$$

$$y_5 = \{\log S/S^* e^{(\rho+l)t}\}.$$

Then from (3.23), (3.24) and (3.26) we obtain:

$$Dy_3 = -\frac{\gamma q}{v}(\lambda + \mu f)y_1 - q(\lambda + \mu f)\left\{\frac{\gamma}{v}((1 + v - b) - \rho - l)\right\}y_2$$

$$+\left[q(\lambda + \mu f)\left\{\frac{\gamma}{v}(1 + v - u - b) - \rho - l\right\}\right.$$

$$- (\lambda + \mu f)(1 - s)e^{-(\rho + l)/\alpha}$$

$$\left. + \frac{\mu(\mu f - \rho - l)}{\mu + (\rho + l)(\lambda + \mu f)}\right]y_3 + \{(\lambda + \mu f)(1 - s)e^{-(\rho + l)/\alpha}\}y_4$$

$$- \left\{\frac{\mu(\mu f - \rho - l)}{\mu + (\rho + l)(\lambda + \mu f)}\right\}y_5 + \phi(y_1, y_2, y_3, y_4, y_5), \tag{3.27}$$

$$Dy_5 = \left\{\frac{\mu + (\rho + l)(\lambda + \mu f)}{\mu f - \rho - l}\right\}\left[\frac{\gamma q}{v}y_1 + \left\{\frac{\gamma q}{v}(1 + v - b) - q(\rho + l)\right\}y_2\right.$$

$$+ \left\{1 - \frac{\gamma q}{v}(1 + v - u - b)\right\}y_3 - \{(1 - s)e^{-(\rho + l)/\alpha}\}y_4$$

$$\left. + \{q(\rho + l) - 1 + (1 - s)e^{-(\rho + l)/\alpha}\}y_5\right] + \psi(y_1, y_2, y_3, y_4, y_5), \tag{3.28}$$

where $\phi(y_1, y_2, y_3, y_4, y_5)/\Sigma_{i=1}^5|y_i|$ and $\psi(y_1, y_2, y_3, y_4, y_5)/\Sigma_{i=1}^5|y_i|$ each tend to zero as the y_1 tend to zero.

The exact paths of the y_i are determined by their initial values and equations (3.22), (3.23), (3.25), (3.27) and (3.28) and the approximate paths by the initial values and the linear system.

$$Dy = Fy \tag{3.29}$$

obtained by omitting $\phi(y_1, \ldots, y_5)$ and $\psi(y_1, \ldots, y_5)$ from those equations. Provided that the characteristic roots of F have negative real parts and the initial values of the y_i are not too large we have $\lim_{t\to\infty} y_i = 0$ $(i = 1, \ldots, 5)$. In this case the proportional deviations of w, K, Y, C and S from their equilibrium growth paths tend to zero as $t \to \infty$.

By using the Routh-Hurwitz conditions we can express the stability conditions (i.e. the conditions for the characteristic roots of F to have negative real parts) in terms of the parameters of the model. But the resulting expressions are rather complicated and will not be discussed. Instead we shall conclude by considering a numerical example in which the parameters assume plausible values.

The assumed values of the parameters are:

$$\alpha = 1.00,\ s = 0.20,\ \gamma = 0.20,\ \lambda = 4.00,\ \mu = 2.00,\ \rho = 0.04,\ b = 1.50,$$

$$\beta = 2.00,\ u = 1.00,\ v = 1.00,\ l = 0.01,\ f = 0.50.$$

The reciprocals of the speed of response parameters, α, λ and μ can be interpreted as the means of exponentially distributed time lags. Hence, if the unit of time is a year, the assumed values of these parameters imply that there is a mean lag of a year in the adjustment of consumption to income, a mean lag of 3 months in the adjustment of production to sales and a mean lag of 6 months in the adjustment of production to a stock deficiency. The parameter γ can be interpreted as the approximate increase in the annual percentage increase in the amount of fixed capital caused by a 1 per cent increase in the rate of profit, assuming that the interest rate is constant. And β can be interpreted as the approximate increase in the annual percentage increase in the wage rate caused by a 1 per cent increase in the level of employment.

For the assumed values of the parameters we have

$$F = \begin{bmatrix} 0.00 & -1.00 & 3.00 & 0.00 & 0.00 \\ -0.20 & -0.10 & -0.10 & 0.00 & 0.00 \\ -4.35 & -1.09 & -6.23 & 3.81 & -0.84 \\ 0.00 & 0.00 & 1.00 & -1.00 & 0.00 \\ 2.06 & 0.52 & 3.39 & -1.80 & -0.05 \end{bmatrix}$$

The characteristic roots of this matrix are: -0.034, -0.45, -3.25, $-1.82 \pm (0.77)i$. Thus the model generates a damped cycle about a trend towards the equilibrium growth path. Moreover, the period of the cycle is approximately 8 years. We conclude therefore that there are plausible values of the parameters for which the model generates paths of output, employment and prices similar to those relating to most industrial economies.

The small negative root, -0.034 is very insensitive to changes in the parameters other than ρ, l and b. It is associated with the slow convergence of $\log K$ to its equilibrium growth path after $\log L$ has approximately reached its equilibrium growth path. It is approximately equal to $(\rho + l)/b$ which is the approximate proportional rate of decrease of $|\log K/K^{**}e^{(\rho+l)t}|$ in the simple neoclassical model, $K^{**}e^{(\rho+l)t}$ being the equilibrium growth path of K in that model.

References

BRONFENBRENNER, M. and T. MAYER, (1960), 'Liquidity functions in the American economy', *Econometrica,* **28**, 810–34.

HARROD, R. F. (1948), *Towards a Dynamic Economics* (London Macmillan).

HICKS, J. R. (1950), *A Contribution to the Theory of the Trade Cycle* (Oxford University Press)

KEYNES, J. M. (1936), *The General Theory of Employment, Interest and Money* (London MacMillan).

PHILLIPS, A. W. (1961), 'A simple model of employment, money and prices in a growing economy', *Economica*, **28**, 360–70.

SOLOW, R. M. (1956), 'A contribution to the theory of economic growth', *Quarterly Journal of Economics,* **70**, 65–94.

SWAN, T. W. (1956), 'Economic growth and capital accumulation', *Economic Record,* **32**, 334–61.

PART II

Estimation, Forecasting, and Control

3

Nonrecursive Models as Discrete Approximations to Systems of Stochastic Differential Equations[1]

1. Introduction

In a recent article Strotz and Wold (1960) argued that the use of a nonrecursive econometric model implies either that the system is in equilibrium whenever observed, or that the model is to be regarded as an approximation to a recursive system. The aim of a separate article by Strotz (1960) was to determine the importance for estimation procedure of the specification error that is implied by the latter assumption. The present paper has a similar aim. But our approach will differ, in one respect, from that of Strotz. We shall assume that the underlying recursive model is a system of stochastic differential equations of the form

$$D^k x_i(t) = \sum_{j=1}^{n} \alpha_{ij}(D)x_j(t) + \zeta_i(t) \qquad (i = 1, \ldots, n) \tag{1}$$

where D denotes the differential operator d/dt, $\alpha_{ij}(D)$ is a polynomial in D of lower order than k, and the $\zeta_i(t)$ are disturbances. It will be assumed that the $\zeta_i(t)$ are white noise elements such that

$$E\left(\int_{t_1}^{t_2} \zeta_i(t)\right) = 0 \qquad (i = 1, \ldots, n),$$

$$E\left(\int_{t_1}^{t_2} \zeta_i(t) \int_{t_1}^{t_2} \zeta_j(t)\right) = \sigma_{ij} \qquad (i, j = 1, \ldots, n),$$

$$E\left(\int_{t_1}^{t_2} \zeta_i(t) \int_{t_3}^{t_4} \zeta_j(t)\right) = 0 \qquad (i, j = 1, \ldots, n),$$

for $t_1 < t_2 < t_3 < t_4$.

Such a system can be given a direct causal interpretation using a notion of causality similar to that used by Strotz and Wold. Alternatively it can be regarded as a transformation of the causal system

[1] This work was supported in part by a grant from the Ford Foundation.

$$x_i(t) = \sum_{j \neq i} \int_0^{\infty} w_{ij}(h) x_j(t-h) dh + \int_0^{\infty} r_i(h) \zeta_i(t-h) dh \qquad (i = 1, \ldots, n) \tag{2}$$

where the $w_{ij}(h)$ and $r_i(h)$ are continuous weighting functions.[1] The advantages of models in which time is treated as a continuous variable have been discussed by Koopmans (1950) who proposed a system similar to (2).

It is known[2] that a sequence of equispaced observations generated by (1) satisfy the system

$$x(t) = B_1 x(t-1) + \ldots + B_k x(t-k) + \xi_t + C_1 \xi_{t-1} + \ldots + C_{k-1} \xi_{t-k+1} \tag{3}$$

where the $x(t-h)$ are vectors of observations, the $\xi(t-h)$ are vectors of disturbances, and the B_h and C_h are $n \times n$ matrices of parameters. The disturbances have zero means and are such that $E(\xi_{t-h} \xi'_{t-h}) = \Omega$ and $E(\xi_{t-h} \xi'_{t-j}) = 0$ for $h \neq j$.

In view of (3) the use of a nonrecursive approximation to (1) may appear unnecessary. The reason for using a nonrecursive model as an approximation to the type of recursive model considered by Strotz and Wold is that the interval between successive observations may be large relative to the time lags in the recursive system. But, if the observations are generated by the system (1), they will satisfy a system of the form (3) regardless of the length of the interval between successive observations. It should be possible, therefore, to obtain better predictions by applying efficient estimation procedures to (3) than by using a nonrecursive approximation to (1).

The use of a nonrecursive approximation to (1) is necessary, not because of the length of the interval between successive observations, but because of the difficulty in obtaining estimators of the parameters of (3) that satisfy the a priori restrictions suggested by economic theory and are asymptotically efficient. The most common form of a priori restriction is that certain variables have no direct causal influence on certain other variables or, more precisely, that certain coefficients in the system (1) are zero. This in turn implies certain restrictions on the parameters of (3). But the form of these restrictions is very complex, and, in any practicable estimation procedure, we must either ignore them of use a disrete approximation to

[1] In order for the converse statement to be true it is necessary to restrict the system (2) by specifying that the Laplace transforms of the weighting functions are rational functions and to widen the system (1) by premultiplying ζ_i by a polynomial in D. The emphasis in this paper is on the treatment of (1) rather than (2) as the underlying causal system. For this reason it has been assumed that the disturbances in (1) are white noise elements. The relation between the weighting functions and the parameters of the stochastic differential equation system, for the general case in which the Laplace transforms of the weighting functions are rational functions, is given by Phillips (1959).

[2] See Bartlett and Rajalakshman 1953, and Phillips 1959.

(1). Whether or not the latter procedure is preferable depends on the importance of the specification error that is involved.

2. The General First Order System

We shall consider first the general first order system

$$Dx(t) = Ax(t) + \zeta(t) \tag{4}$$

where $x(t)$ and $\zeta(t)$ are column vectors and A is an $n \times n$ matrix of parameters some of which are specified, a priori, to be zero. It will be assumed, moreover, that A has distinct characteristic roots $\lambda_1, \ldots, \lambda_n$ all of which have negative real parts.

Let H be a matrix such that

$$HAH^{-1} = \begin{bmatrix} \lambda_1 & 0 & \cdots & 0 \\ 0 & \lambda_2 & \cdots & 0 \\ \cdot & \cdot & \cdots & \cdot \\ 0 & 0 & \cdots & \lambda_n \end{bmatrix} = \Lambda. \tag{5}$$

Then (4) has the solution

$$x(t) = \int_{-\infty}^{t} H^{-1} e^{\Lambda(t-r)} H\zeta(r)dr \tag{6}$$

where

$$e^{\Lambda t} = \begin{bmatrix} e^{\lambda_1 t} & 0 & \cdots & 0 \\ 0 & e^{\lambda_2 t} & \cdots & 0 \\ \cdot & \cdot & \cdots & \cdot \\ 0 & 0 & \cdots & e^{\lambda_n t} \end{bmatrix}.$$

It follows that

$$x(t) = Bx(t-1) + \xi(t) \tag{7}$$

where

$$B = H^{-1} e^{\Lambda} H \tag{8}$$

and

$$\xi(t) = \int_{t-1}^{t} H^{-1} e^{\Lambda(t-r)} H\zeta(r)dr. \tag{9}$$

From (9) we obtain

$$\Omega = E\{\xi(t)\xi'(t)\} = \int_{0}^{1} H^{-1} e^{\Lambda r} H \Sigma H' e^{\Lambda r} H'^{-1} dr \tag{10}$$

where $\Sigma = [\sigma_{ij}]$. It is evident from (10) that, even if the integrals of the $\zeta_i(t)$ are uncorrelated, the ξ_{it} will generally be correlated.

From (7) and (10) we obtain

$$V_0 = BV_0B' + \Omega \tag{11}$$

and

$$V_1 = BV_0 \tag{12}$$

where V_0 and V_1 denote the covariance matrix $E\{x(t)x'(t)\}$ and autocovariance matrix $E\{x(t)x'(t-1)\}$ respectively.

Suppose now that, as an approximation to (4), we use the nonrecursive model

$$x(t) - x(t-1) = A[(0.5)\{x(t) + x(t-1)\}] + u_t \tag{13}$$

together with the assumption that $E\{u(t)u'(t-h)\} = 0$ for $h \neq 0$.[3] Let $\hat{A}$ denote the estimate of A obtained by applying the method of three stage least squares[4] to a sample of T observations which have, in fact, been generated by (4), but are assumed to have been generated by (13). Substituting $\hat{A}$ for A in (13), putting the u_{it} equal to zero and solving for $x(t)$, we obtain the estimated reduced form equation

$$x(t) = \hat{B}x(t-1) \tag{14}$$

where

$$\hat{B} = [I - (0.5)\hat{A}]^{-1}[I + (0.5)\hat{A}]. \tag{15}$$

The approximate model (13) will be useful, as a tool of prediction, only if (14) yields better predictions than

$$x(t) = B^*x(t-1) \tag{16}$$

where B^* is the least squares estimator of B.

If there were no a priori restrictions on A, the restrictions represented by the form of (13) would be exactly identifying so that we would have $\hat{B} = B^*$. In this case there would be nothing to gain by using the model (13). We have assumed, however, that certain specified elements of A are zero. This assumption implies certain restrictions on both B and $\hat{B}$. The restrictions on B are implied by the exact model (4) and those on $\hat{B}$ by the use of the approximate model (13). It follows from (5) that the restrictions on B can be obtained by equating to zero certain elements of $H^{-1}\Lambda H$ and regarding the λ_i as functions of the elements of B. It is evident from (8) that the λ_i are, in fact, the logarithms of the characteristic roots of B. The restrictions on $\hat{B}$ are obtained by eliminating from the equation (15) the nonzero elements of $\hat{A}$. Unlike the restrictions on B, they can be expressed

[3] This approximation can be obtained by integrating each side of (4) from $t-1$ to t and replacing $\int_{t-1}^{t} x(r)dr$ by $(0.5)\{x(t) + x(t-1)\}$.

[4] See Zellner and Theil 1962.

in terms of rational functions. They can be regarded as convenient approximations to the restrictions on B.

Since the restrictions that have been imposed on $\hat{B}$ are only approximately satisfied by B, the former cannot be a consistent estimator of the latter. Hence, since B^* is a consistent estimator of B,[5] it must be preferred to $\hat{B}$ if the sample size is sufficiently large. But the elements of $\hat{B}$ will, presumably, have smaller variances than the corresponding elements of B^*, since the latter are unrestricted. For samples of the size available in practice, this difference between variances may be more important than the asymptotic bias in $\hat{B}$. Whether or not this is likely to be so can be determined only by considering numerical examples. One such example is discussed in the following section.

3. A Numerical Example

For the purpose of this example we shall assume that the parameters of (4) have the following numerical values:

$$A = \begin{bmatrix} -1.0 & 0.8 & 0.0 \\ 0.0 & -0.5 & 0.2 \\ 0.1 & 0.0 & -0.2 \end{bmatrix} \tag{17}$$

and

$$\Sigma = \begin{bmatrix} 1.0 & 0.0 & 0.0 \\ 0.0 & 1.0 & 0.0 \\ 0.0 & 0.0 & 1.0 \end{bmatrix}. \tag{18}$$

This means that x_1 is assumed to be causally dependent on x_2, x_2 on x_3, and x_3 on x_1, the corresponding long-run propensities being 0.8, 0.4, and 0.5, respectively. Assuming the time unit to be 3 months, the numerical values of the diagonal elements of A represent a pattern of time lags that might be expected in a typical econometric model. The mean lags in the first, second, and third relations are 3 months, 6 months, and 15 months, respectively. The integrals of the white noise elements in the different relations are assumed to be uncorrelated and to have equal variances.

The characteristic roots of A are -0.600, -0.953, and -0.147. Using (8), (10), (11), and (12), we obtain

$$B = \begin{bmatrix} 0.369 & 0.382 & 0.046 \\ 0.006 & 0.608 & 0.142 \\ 0.056 & 0.023 & 0.820 \end{bmatrix}, \tag{19}$$

[5] See Mann and Wald 1943. Here, and in the subsequent argument, we assume that the $\zeta_i(t)$ are such that all moments of the ξ_{it} are finite.

$$\Omega = \begin{bmatrix} 0.506 & 0.186 & 0.037 \\ 0.186 & 0.641 & 0.076 \\ 0.037 & 0.076 & 0.776 \end{bmatrix}, \tag{20}$$

$$V_0 = \begin{bmatrix} 1.151 & 0.815 & 0.690 \\ 0.815 & 1.356 & 0.893 \\ 0.690 & 0.893 & 2.689 \end{bmatrix}, \tag{21}$$

$$V_1 = \begin{bmatrix} 0.768 & 0.860 & 0.720 \\ 0.600 & 0.956 & 0.927 \\ 0.650 & 0.809 & 2.265 \end{bmatrix}, \tag{22}$$

Now the nonzero elements of $\hat{A}$ can be expressed as rational functions of the sample moments $\Sigma_{t=2}^{T} x_i(t)x_j(t)/(T-1)$, $\Sigma_{t=2}^{T} x_i(t)x_j(t-1)/(T-1)$ and $\Sigma_{t=2}^{T} x_i(t-1)x_j(t-1)/(T-1)$. And these sample moments converge in probability to the corresponding elements of V_0 and V_1.[6] Hence, the probability limits of the elements of $\hat{A}$ can be expressed as functions of the elements of V_0 and V_1. By using these results we obtain

$$\text{plim}\,\hat{A} = \begin{bmatrix} -0.922 & 0.710 & 0.000 \\ 0.000 & -0.488 & 0.193 \\ 0.098 & 0.000 & -0.199 \end{bmatrix} \tag{23}$$

and hence

$$\text{plim}\,\hat{\Sigma} = \begin{bmatrix} 0.969 & 0.053 & 0.004 \\ 0.053 & 0.981 & 0.011 \\ -0.004 & 0.011 & 0.935 \end{bmatrix} \tag{24}$$

where

$$\hat{\Sigma} = \frac{1}{T-1}\sum_{t=2}^{T} \hat{u}(t)\hat{u}'(t)$$

and $\hat{u}(t) = x(t) - x(t-1) - \hat{A}[(0.5)\{x(t) + x(t-1)\}]$.

It is of some interest to compare $\text{plim}\,\hat{A}$ and $\text{plim}\,\hat{\Sigma}$ with A and Σ, respectively. But, for the purpose of assessing the predictive power of the approximate discrete model, the relevant comparison is between $\text{plim}\,\hat{B}$ and B. From (15) and (23) we obtain

$$\text{plim}\,\hat{B} = \begin{bmatrix} 0.370 & 0.391 & 0.034 \\ 0.005 & 0.609 & 0.141 \\ 0.061 & 0.017 & 0.821 \end{bmatrix} \tag{25}$$

and hence

$$\text{plim}\,\hat{\Omega} = \begin{bmatrix} 0.507 & 0.186 & 0.037 \\ 0.186 & 0.641 & 0.076 \\ 0.037 & 0.076 & 0.776 \end{bmatrix} \tag{26}$$

where

$$\hat{\Omega} = \frac{1}{T-1} \sum_{t=2}^{T} \hat{v}(t)\hat{v}'(t)$$

and $\hat{v}(t) = x(t) - \hat{B}x(t-1)$. Comparing (25) with (19) and (26) with (20), we see that the specification error implicit in the use of the approximate nonrecursive model (13) leads, in this example, to a quite small asymptotic bias in the estimates of B and Ω.[7]

Finally, we wish to compare the asymptotic variances of the corresponding elements of $\hat{B}$ and B^*. Because of the specification error, the determination of the asymptotic variances of the elements of $\hat{B}$ presents some difficulty. This can be seen by considering the models (27) and (28):

$$Dx(t) = \begin{bmatrix} -1.0 & 0.8 & 0.0 \\ 0.0 & -0.5 & 0.2 \\ 0.1 & 0.0 & -0.2 \end{bmatrix} x(t) + \zeta(t) \tag{27}$$

where $\zeta(t)$ is a vector of white noise elements whose integrals over a unit interval have zero means and a covariance matrix given by (18);

$$\begin{aligned} x(t) - x(t-1) &= \begin{bmatrix} -0.922 & 0.710 & 0.000 \\ 0.000 & -0.488 & 0.193 \\ 0.098 & 0.000 & -0.199 \end{bmatrix} \\ &\quad \times [(0.5\{x(t) + x(t-1)\}] + u(t) \\ &= F[(0.5)\{x(t) + x(t-1)\}] + u(t) \end{aligned} \tag{28}$$

where $u(t)$ is a vector of serially uncorrelated disturbances with zero means and a covariance matrix equal to the right-hand side of (24).

From (27) we obtain the autoregressive system

$$x(t) = \begin{bmatrix} 0.369 & 0.382 & 0.046 \\ 0.006 & 0.608 & 0.142 \\ 0.056 & 0.023 & 0.820 \end{bmatrix} x(t-1) + \xi(t) = Bx(t-1) + \xi(t) \tag{29}$$

where $\xi(t)$ is a vector of serially uncorrelated disturbances with zero means and covariance matrix given by (20). And from (28) we obtain the autoregressive system

[7] Each element of $\text{plim}\,\hat{\Omega}$ differs from the corresponding element of Ω at the fourth decimal place.

$$x(t) = \begin{bmatrix} 0.370 & 0.391 & 0.034 \\ 0.005 & 0.609 & 0.141 \\ 0.061 & 0.017 & 0.821 \end{bmatrix} x(t-1) + v(t) = Gx(t-1) + v(t) \tag{30}$$

where $v(t)$ is a vector of serially uncorrelated disturbances with zero means and a covariance matrix equal to the right hand side of (26).

Now, let $\hat{F}$ denote the three stage least squares estimator of F obtained from a sample of T observations generated by (28). And let $\hat{G}$ be defined by

$$\hat{G} = [I - (0.5)\hat{F}]^{-1}[I + (0.5)\hat{F}]. \tag{31}$$

Then $\hat{G}$ is a consistent estimator of G, and the asymptotic variances of its elements can easily be computed from the parameters of (28).[8] But $\hat{B}$ is not a consistent estimator of B, and there is no easy way of obtaining the asymptotic variances of its elements. Nevertheless, the corresponding elements of $\hat{B}$ and $\hat{G}$ are identical rational functions of the sample moments generated by the autoregressive systems (29) and (30), respectively. Moreover, in view of the smallness of the differences between the parameters of the systems (29) and (30), the asymptotic covariance matrices of the sample moments generated by these two systems will be approximately equal. We shall assume, therefore, that, for the purpose of this example, the asymptotic variances of the elements of $\hat{G}$ are sufficiently good approximations to the asymptotic variances of the corresponding elements of $\hat{B}$. These have been computed for $T = 100$ and are shown in Table 1 together with the asymptotic variances of the elements of B^*. The difference between the asymptotic variance of $\hat{b}_{ij}$ and b^*_{ij} is compared (for $i, j = 1, 2, 3$) with the squared asymptotic bias of $\hat{b}_{ij}$.

The results indicate that, in this example, the reduction in variance obtained by taking account of the a priori restrictions greatly outweighs the bias resulting from the error of specification implied by the use of the approximate nonrecursive model. As might be expected, the greatest reductions in variance relate to those reduced form coefficients that occupy the positions in the matrix corresponding to zeros in the matrix of structural coefficients of the continuous model.

4. The Problem of Structural Estimation

We shall consider, finally, the question of how to make the best use of the estimates of the parameters of the approximate model (13) in order to obtain estimates of the parameters of the exact model (4). A crude

[8] See Goldberger, Nagar, and Odeh 1961.

TABLE 1. Comparison of asymptotic variances of elements of B^* and $\hat{B}$ for $T = 100$

(1) b_{ij}	(2) Asymptotic Variance of b^*_{ij}	(3) Approximate Asymptotic Variance of $\hat{b}_{ij}$	(4) (2)–(3)	(5) Squared Asymptotic Bias of $\hat{b}_{ij}$
.369	.0077	.0025	.0052	.0000
.382	.0073	.0073	.0000	.0001
.046	.0024	.0002	.0022	.0001
.006	.0099	.0003	.0096	.0000
.608	.0091	.0041	.0050	.0000
.142	.0031	.0028	.0003	.0000
.056	.0119	.0041	.0078	.0000
.023	.0110	.0003	.0107	.0000
.820	.0038	.0035	.0003	.0000

estimate of A is given by $\hat{A}$. But the example considered in the previous section suggests that a better estimate would be given by

$$\hat{\hat{A}} = \hat{\hat{H}}^{-1}\hat{\hat{\Lambda}}\hat{\hat{H}} \tag{32}$$

where

$$\hat{\hat{\Lambda}} = \begin{bmatrix} \hat{\hat{\lambda}}_1 & & \cdots & 0 \\ 0 & \hat{\hat{\lambda}}_2 & \cdots & 0 \\ \cdot & \cdot & \cdots & \\ 0 & 0 & \cdots & \hat{\hat{\lambda}}_n \end{bmatrix}$$

and the $\hat{\hat{\lambda}}_i$ and $\hat{\hat{H}}$ satisfy

$$|\hat{B} - e^{\mathring{\lambda}_i} I| = 0 \qquad (i = 1, \ldots, n) \tag{33}$$

and

$$\hat{\hat{H}}\hat{B}\hat{\hat{H}}^{-1} = \begin{bmatrix} e^{\mathring{\lambda}_1} & 0 & \cdots & 0 \\ 0 & e^{\mathring{\lambda}_2} & \cdots & 0 \\ \cdot & \cdot & \cdots & \cdot \\ 0 & 0 & \cdots & e^{\mathring{\lambda}_n} \end{bmatrix}. \tag{34}$$

The proposed procedure is, therefore, to first obtain estimates of the parameters of an approximate nonrecursive model, next solve for estimates of the reduced form coefficients, and, from these, deduce estimates of the parameters of the continuous model.

Assuming, for example, that the parameters of the exact model are given by (17) and (18) and that from a particular sample we obtain $\hat{B}$ equal to its probability limit, given by (25), then we obtain

$$\hat{\hat{A}} = \begin{bmatrix} -1.019 & 0.867 & -0.040 \\ -0.015 & -0.473 & 0.192 \\ 0.108 & -0.012 & -0.198 \end{bmatrix}. \tag{35}$$

It can be seen that, in this case, $\hat{\hat{A}}$ gives a somewhat better estimate of A than does $\hat{A}$. The elements of $\hat{\hat{A}}$ corresponding to the zero elements of A can, of course, be ignored.

The above method can, in principle, be applied to the general system (1) which can be written in matrix notation as

$$D^k x(t) = \sum_{r=1}^{k} A_r D^{k-r} x(t) + \zeta(t). \tag{36}$$

As an approximation to (36) we can use, for example, the nonrecursive system

$$\Delta^k x(t) = \sum_{r-1}^{k} A_r \left(\frac{1}{r+1} \sum_{q=0}^{r} \Delta^{k-r} x(t-q) \right) + \frac{1}{k} \sum_{q=0}^{k-1} u_{t-q} \tag{37}$$

where $\Delta x(t) = x(t) - x(t-1)$, together with the assumption that $E\{u_{t-q}u'_{t-h}\} = 0$ for $h \neq q$. From estimates of the parameters of (37) we can obtain estimates of the parameters of (3) and, from these, estimates of the parameters of (36).

The main practical difficulty is that of estimating the parameters of (37). The difficulty arises from the fact that the disturbances are moving averages and, therefore, serially correlated. But the serial correlation coefficients can be deduced from the assumption that the coefficients of the moving average are equal. We can then find a $T \times T$ matrix, L, with zeros above the main diagonal, such that $L^{-1}(L')^{-1}$ equals the matrix of serial correlation coefficients. Now let X be the $T \times n$ matrix whose tth row is $x'(t)$, and let $z'(t)$ be the tth row of $Z = LX$. Then $z(t)$ satisfies a system similar to (37) except that the disturbances are serially uncorrelated. Hence the A_r's can be estimated by applying three stage least squares to the latter system.[9]

References

Bartlett, M. S., and D. V. Rajalakshman (1953), 'Goodness of fit tests for simultaneous autoregressive series', *Journal of the Royal Statistical Society*, B., vol XV, pp. 107–24.

[9] If the sample is sufficiently large it may be better to estimate the coefficients of the moving average disturbances instead of assuming that they are equal. A method of obtaining consistent estimates of the parameters of nonrecursive systems in which the disturbances are moving averages of random elements with unknown coefficients has been proposed by Phillips (1978).

Goldberger, A. S., A. L. Nagar, and H. S. Odeh (1961), 'The covariance matrices of reduced form coefficients and of forecasts for a structural econometric model', *Econometrica*, **29**, 556–73.

Koopmans, T. C. (1950), 'Models involving a continuous time variable', *Statistical Inference in Dynamic Economic Models* (New York), pp. 384–92.

Mann, H. B., and A. Wald (1943), 'On the statistical treatment of linear stochastic difference equations', *Econometrica*, **11**, 173–220.

Phillips, A. W. (1959), 'The estimation of parameters in systems of stochastic differential equations', *Biometrika*, **46** 67–76.

—— (1978), 'The estimation of systems of difference equations with moving average disturbances', in A. R. Bergstrom *et al*. eds. *Stability and Inflation* (New York, Wiley)

Strotz, R. H. (1960), 'Interdependence as a specification error', *Econometrica*, **28** 428–42.

—— and H. O. A. Wold (1960), 'Recursive *vs*. nonrecursive systems', *Econometrica*, **28** 417–27.

Zellner, A., and H. Theil (1962), 'Three stage least squares: simultaneous estimation of simultaneous equations', *Econometrica*, **30** 54–78.

4

Gaussian Estimation of Structural Parameters in Higher Order Continuous Time Dynamic Models[1]

1. Introduction

THE theory of statistical inference in continuous time dynamic models is now well established (see Bergstrom 1976 and the articles contained therein) and has already been widely applied in the construction of macroeconomic models of various countries (see, for example, Bergstrom and Wymer 1976; Bergstrom 1978; Jonson, Moses, and Wymer 1977; Knight and Wymer 1978) and the modelling of commodity and financial markets (see, for example, Richard 1978; Wymer 1973). But both the theoretical and empirical work in this field has been concerned mainly with first order systems of stochastic differential equations. Moreover most work on higher order systems has either ignored the a priori restrictions on the coefficients of the system or been based on the use of approximate discrete models which yield asymptotically biased estimates (see Wymer 1972).

The problem of estimating the parameters of a higher order continuous time dynamic model from discrete data was first discussed by A. W. Phillips (1959) who proposed an estimation procedure, although not one that yields asymptotically efficient estimates. A method of obtaining asymptotically efficient estimates for a single equation model of the form considered by Phillips was proposed later by Durbin (1961). But neither Phillips nor Durbin took account of the a priori restrictions on the coefficients of the continuous system. In econometric work many of these coefficients will be specified a priori to be zero (reflecting the fact that certain variables have no direct causal influence on certain other variables) while the remaining coefficients will be specified functions of a smaller number of structural parameters. Because of the smallness of the samples

[1] I am grateful to two referees for helpful comments on an earlier version of this paper.

with which the econometrician has to work, it is very important that these restrictions should be taken into account in the estimation procedure.

The relation between the a priori restrictions on the coefficients of a continuous time model and the implied restrictions on the coefficients of the exact discrete model (i.e. the model satisfied by the discrete data) was discussed briefly in my article (1966) [Ch. 3 of this volume]. It was shown that, even in the simplest case, where the only a priori restrictions are that certain coefficients of the continuous time model are specified as zero, the implied restriction on the coefficients of the exact discrete model are very complicated and can be represented only in terms of transcendental functions involving all of the coefficients in the exact discrete model. Except for the work of Robinson (1976*a, b, c*) on the estimation of open systems by Fourier methods and his brief discussion in (1977) of the efficient estimation of closed systems, the problem of taking account of such restrictions in the efficient estimation of higher order systems has not been dealt with.

The purpose of this paper is to provide a comprehensive treatment of the problem of obtaining asymptotically efficient estimates of the structural parameters of a closed, higher order, continuous time dynamic model from various types of discrete data. We shall consider, in particular, estimates that would be maximum likelihood estimates if the integral of the white noise disturbance vector were a Gaussian process, although we shall not assume that the disturbance has this property. Following Whittle (1962), we shall call such estimates Gaussian estimates. We shall also consider estimates obtained by maximizing frequency domain approximations to the Gaussian likelihood. This sort of approximation is essentially derived from Whittle's pioneer work (1951; 1953) and is similar to that used by Robinson (1977).

2. The Continuous Time Model and the Problem of Estimation

We shall consider the system

$$d[D^{k-1}x(t)] = [A_1(\theta)D^{k-1}x(t) + \ldots + A_{k-1}(\theta)Dx(t) + A_k(\theta)x(t) + b(\theta)]dt + \zeta(dt), \tag{1}$$

where $x(t)$ is a real n-dimensional, continuous time, random process, θ is a p-dimensional vector of parameters $(p < n(1 + kn))$, $A_1, \ldots, A_k$ are $n \times n$ matrices whose elements are known functions of θ, b is an $n \times 1$ vector whose elements are known functions of θ, and D is the mean square differential operator defined by

$$\lim_{h \to 0} E \left| Dx_i(t) - \frac{x_i(t+h) - x_i(t)}{h} \right|^2 = 0 \quad (i = 1, \ldots, n).$$

With respect to the disturbance vector $\zeta(dt)$ we shall make the following assumption.

ASSUMPTION 1: $\zeta = [\zeta_1, \ldots, \zeta_n]'$ is a vector of random measures, defined on all subsets of the line $-\infty < t < \infty$ with finite Lebesgue measure, such that $E[\zeta(dt)] = 0$, $E[\zeta(dt)\zeta'(dt)] = (dt)\Sigma$, where Σ is a positive definite matrix, and $E[\zeta_i(\Delta_1)\zeta_j(\Delta_2)] = 0$ $(i, j = 1, \ldots, n)$ for any disjoint sets Δ_1 and Δ_2 on the line $-\infty < t < \infty$. (See Rozanov 1967, p. 5 for the definition of a random measure.)

We shall also make the following assumption which implies that if $\zeta(dt)$ were a zero vector the system would have a steady state $x(t) = -A_k^{-1}b$, although this would not necessarily be stable.

ASSUMPTION 2: $A_k(\theta)$ is nonsingular.

The system (1) can be loosely described as a closed linear system of kth order stochastic differential equations. But this is not a very accurate description since, under Assumption 1, the kth order mean square derivative $D^k x(t)$ does not exist. Moreover $d[D^{k-1}x(t)]$ cannot be interpreted as a stochastic differential analogous to the differential of an ordinary nonstochastic function since $[D^k x(t)]dt$ does not exist. The question arises, therefore, as to how (1) should be interpreted.

In most of the literature on continuous time, stochastic, linear models it is assumed that the random process generated by the system is stationary in the wider sense. In that case we can interpret (1) as meaning that the continuous time random process $\{x(t) - Ex(t)\}$ has the spectral density

$$\phi(\lambda) = \frac{1}{2\pi}[(i\lambda)^k I - (i\lambda)^{k-1}A_1 - \ldots - (i\lambda)A_{k-1} - A_k]^{-1} \tag{2}$$
$$\times \Sigma[(i\lambda)^k I - (i\lambda)^{k-1}A_1 - \ldots - (i\lambda)A_{k-1} - A_k]^{*-1}$$

where A^* denotes the transpose of the complex conjugate of a matrix A. This interpretation can be obtained heuristically by considering the proper stochastic differential equation system

$$D^k\eta(t) = A_1 D^{k-1}\eta(t) + \ldots + A_{k-1}D\eta(t) + A_k\eta(t) + \xi(t),$$

where $\{\xi(t)\}$ is an n-dimensional, continuous time, wide sense stationary process with spectral density $\phi^{\xi\xi}(\lambda)$. It can be shown (Rozanov 1967, p. 38) that the spectral density of the continuous time process $\{\eta(t)\}$ is given by

$$\phi^{\eta\eta}(\lambda) = [(i\lambda)^k I - (i\lambda)^{k-1}A_1 - \ldots - (i\lambda)A_{k-1} - A_k]^{-1}$$
$$\times \phi^{\xi\xi}(\lambda)[(i\lambda)^k I - (i\lambda)^{k-1}A_1 - \ldots - (i\lambda)A_{k-1} - A_k]^{*-1}.$$

By considering a sequence of such systems in which $\phi^{\xi\xi}(\lambda)$ tends to a constant matrix $(1/2\pi)\Sigma$ we obtain $\phi(\lambda)$ as the limit of $\phi^{\eta\eta}(\lambda)$. Although a proper continuous time, wide sense, stationary process with constant spectral density does not exist, a continuous time, wide sense, stationary process with the spectral density given by (2) does exist (Rozanov 1967, p. 22, Theorem 5.2), since the integral $\int_\Delta \phi(\lambda)d\lambda$ exists for any Borel set Δ and is a positive definite matrix.

The above interpretation of (1) is not very helpful from the point of view of this paper. For our statistical analysis will be based on exact discrete models specified in the time domain. Moreover we do not, at this stage, wish to impose the restriction of stationarity, which is often an inappropriate assumption in econometric work. We shall assume instead that $x(t)$ satisfies the boundary conditions

$$x(0) = y_1, \qquad Dx(0) = y_2, \qquad \ldots, \qquad D^{k-1}x(0) = y_k, \tag{3}$$

where $y_1, \ldots, y_k$ are random vectors with respect to which we make the following assumption.

ASSUMPTION 3: $E[y_i\zeta'(\Delta)] = 0$ $(i = 1, \ldots, k)$ for any set Δ in the half line $0 < t < \infty$ with finite Lebesgue measure.

We shall interpret (1) as meaning that $x(t)$ satisfies the stochastic integral equation

$$D^{k-1}x(t) - D^{k-1}x(0) = \int_0^t [A_1 D^{k-1}x(r) + \ldots + A_{k-1}Dx(r) \tag{4}$$
$$+ A_k x(r) + b]dr + \int_0^t \zeta(dr),$$

for all $t > 0$, where $\int_0^t \zeta(dr) = \zeta([0, t])$, and the first integral on the right hand side of (4) is defined in the wide sense (Rozanov 1967, p. 10). Our basic continuous time model comprises equations (3) and (4) together with Assumptions 1 to 3 and the specification of the functions defining the elements of $A_1, \ldots, A_k$ and b. We shall be concerned with the problem of estimating θ and Σ from a sample of discrete observations generated by (4).

In the theory of ordinary, linear, differential equations (see, for example, Coddington and Levinson, 1955, chap. 3) a system of equations of order greater than one is dealt with by reducing it to a system of first order equations in the original variables and their derivatives. The reason why the estimation of the parameters of the system (4) is a much more complicated problem than that of estimating the parameters of a first order system is that we cannot observe the derivatives $Dx(t), \ldots, D^{k-1}x(t)$. At best we shall be able to observe $x(t)$ at a sequence of equispaced points of time, in which case the sample will be of the form $[x(1), x(2), \ldots, x(T)]$.

But, when $k > 1$, the exact discrete model satisfied by the sequence of random vectors $x(1), x(2), \ldots$ will, as we shall see, be a vector autoregressive moving average system in which the moving average is of order $k - 1$ and all the coefficient matrices, as well as the covariance matrix of the serially independent vectors in the moving average, will depend on θ. A further complication arises from the fact that only certain variables, such as stocks, and certain prices, will be measured at points of time. All flow variables, such as consumption, income, and investment, will be measured as integrals over the observation period. This produces an additional source of autocorrelation in the residual vector of the exact discrete model.

Because of the interaction of the two sources of autocorrelation discussed in the last paragraph, the problem of estimating the parameters of (1) from a mixed sample, in which some variables are observed at points of time and others as integrals over the observation period, is an extremely complicated one. It is very important, therefore, that the foundations upon which the estimation procedures are based should be laid carefully and rigorously. This will be done in Section 3, where we shall prove the existence and uniqueness of the solution to (1) subject to the boundary conditions (3) and give its precise form. This solution will be used in Sections 4 to 6, in which we shall deal with the problem of estimation under successively more complicated and realistic assumptions about the form in which the data occur. In Section 4 we shall assume that all variables are observed at points of time (stock variables), in Section 5 that all variables are observed as integrals (flow variables), and in Section 6 that there is a mixture of the two types of variable (stock and flow variables). In each of these sections we shall first derive the exact discrete model satisfied by a sequence of discrete observations of the particular form assumed, and then discuss the use of this model in the computation of the exact and approximate Gaussian estimates. The asymptotic sampling properties of the estimates will be discussed briefly in Section 7, particularly in relation to the recent literature on the asymptotic properties of estimates in a very general class of vector time-series models.

3. The Solution of the Model

Before proving the main theorem on which the methods of this paper are based we shall review, very briefly, the existing literature on the solution of stochastic differential equations. In the earlier literature on this subject it is invariably assumed that the disturbance process is Brownian motion (i.e. a process with independent Gaussian increments). For example, Doob [1953, p. 277], in a modification of the work of Ito (1946; 1951), proves the existence and uniqueness of the solution of a single first order, nonlinear, stochastic differential equation assuming that the disturbance is a separable Brownian motion process and that the functions occurring in the equation satisfy a Lipschitz condition and certain other conditions. But the assumption that the disturbance is separable Brownian motion is a very special one, since almost all the sample functions of such a process are continuous (Doob 1953, p. 420, Theorem 7.1). We should expect this property to lead to a considerable simplification of the proofs of existence and uniqueness of the solution of a stochastic differential equation. And, indeed, Doob requires the solution to be a process whose sample functions are almost all continuous, thus avoiding the problems associated with the wide sense integration of a random process.

The first proof of the existence and uniqueness of the solution of a stochastic differential equation under an assumption of comparable generality to our Assumption 1 was given in the fundamental article of Edwards and Moyal (1955) who consider a single second order, linear, differential equation. Their treatment is very general in that they assume that the space of values of the function $x(t)$ is a weakly complete Banach space, which can be given a probabilistic interpretation by assuming that the elements of this space are equivalence classes of measurable functions defining random variables. Their results imply the existence and uniqueness of the solution of (1) in the special case where $k = 2$, $n = 1$.

A heuristic treatment of the first order system (i.e. the case $k = 1$, $n > 1$) was given in Bergstrom 1966. Although questions of existence and uniqueness are not discussed in that article the correct solution is obtained (assuming the matrix of the system has distinct eigenvalues) and used in the derivation of the exact asymptotic bias of estimates obtained from an approximate discrete model.

A more general and rigorous treatment of the first order system was given, later, by Sargan (1976), who also introduced exogenous variables. He shows (1976, Theorem 1) that, if a solution exists, it will be of a particular form, which is identical with that obtained in Bergstrom 1976 when there are no exogenous variables and the eigenvalues of the system are distinct. Sargan's theorem is essentially a uniqueness theorem, including the specification of the form of the solution if one exists. The existence of a solution is assumed rather than proved.

The argument of Edwards and Moyal (1955) could, presumably, be extended to a complete system of any order. But, for the purpose of this paper, such an approach would be excessively general and complicated. In Theorem 1 we shall give a direct proof of the existence and uniqueness of a solution to (1), subject to the boundary conditions (3), together with the precise form of this solution. This theorem includes, as a special case, the probabilistic interpretation of the theorem of Edwards and Moyal (1955).

Definition: We shall say that the n-dimensional, continuous time, random process $x(t)$ is a *solution of the system* (1), *on the interval* $[0, T]$, *subject to the boundary conditions* (3) if: (i) $D^{k-1}x(t)$ exists and is mean square continuous on $[0, T]$ (i.e. $E|D^{k-1}x(t) - D^{k-1}x(t-h)|^2 \to 0$, uniformly in t on $[0, T]$, as $h \to 0$); (ii) $x(t)$ satisfies the boundary conditions (3); and (iii) $x(t)$ satisfies the integral equation (4) for all t in $[0, T]$.

THEOREM 1: *Under Assumptions* 1 *and* 2, (a) *a solution to the system* (1), *on the interval* $[0, T]$, *subject to the boundary conditions* (3) *exists and is given by*

$$x(t) = y_1(t), \tag{5}$$

where $y_1(t)$ *is obtained from the partition of the* $nk \times 1$ *vector* $y(t)$ *into* k $n \times 1$ *vectors* $y_1(t), \ldots, y_k(t)$,

$$y(t) = \int_0^t e^{(t-r)\bar{A}}\bar{\zeta}(dr) + e^{t\bar{A}}[y(0) + \bar{A}^{-1}\bar{b}] - \bar{A}^{-1}\bar{b},$$

$$\bar{A} = \begin{bmatrix} 0 & I & 0 & \cdots & 0 & 0 \\ 0 & 0 & I & \cdots & 0 & 0 \\ \vdots & \vdots & \vdots & & \vdots & \vdots \\ 0 & 0 & 0 & & 0 & I \\ A_k & A_{k-1} & A_{k-2} & \cdots & A_2 & A_1 \end{bmatrix}, \tag{6}$$

$$\bar{b} = \begin{bmatrix} 0 \\ 0 \\ \vdots \\ b \end{bmatrix}, \quad y(0) = \begin{bmatrix} y_1 \\ y_2 \\ \vdots \\ y_k \end{bmatrix},$$

$$\bar{\zeta} = \begin{bmatrix} 0 \\ 0 \\ \vdots \\ \zeta \end{bmatrix},$$

and, for any matrix A,

$$e^A = \sum_{r=0}^{\infty} \frac{1}{r!} A^r;$$

(b) *the solution* (5) *is unique, i.e. if* $\hat{x}(t)$ *is any other solution, then equation* (21) *holds*.

PROOF: (a) It is easy to see that the n-dimensional process $x(t)$ defined by (5) is a solution to (1), on $[0, T]$, subject to the boundary conditions (3) if $y(t)$ is mean square continuous on $[0, T]$ and satisfies

$$y(t) - y(0) = \int_0^t [\bar{A}y(r) + \bar{b}]dr + \int_0^t \bar{\zeta}(dr), \tag{7}$$

for all t on $[0, T]$, where $y(0)$, $\bar{A}$, $\bar{b}$, and ζ are as defined above. For the first n equations of the system (7) can be written

$$x(t) - x(0) = \int_0^t y_2(r)dr. \tag{8}$$

And, because $y_2(t)$ is mean square continuous, $\int_0^t y_2(r)dr$ is mean square differentiable, so that from (8), we obtain

$$Dx(t) = y_2(t) \qquad (0 \leq t \leq T). \tag{9}$$

In a similar way we obtain, recursively, $D^2x(t) = y_3(t), \ldots, D^{k-1}x(t) = y_k(t)$. It then follows from the last n equations of the system (7) that $x(t)$ satisfies (4). Finally the mean square continuity of $y_k(t)$ on $[0, T]$ means that $D^{k-1}x(t)$ is mean square continuous on $[0, T]$. To complete the proof of (a) we are required to show, therefore, that $y(t)$ defined by (6) is mean square continuous on $[0, T]$ and satisfies (7) for all t on $[0, T]$.

We must first verify that the integral $\int_0^t e^{(t-r)\bar{A}}\zeta(dr)$ exists (see Rozanov 1967, p. 7 for the definition of the integral of a measurable function with respect to a random measure.) Since, for any matrix A, the series defining e^A is convergent (Halmos 1958, p. 186), the elements of the matrix $e^{tA}\Sigma e^{tA'}$ are convergent power series in t for any square matrices A and Σ of the same dimension, and they are integrable, therefore (Titchmarsh 1939, p. 39), over any interval on the real line. It follows that

$$\int_0^t e^{(t-r)\bar{A}}\bar{\Sigma}e^{(t-r)\bar{A}'}dr < \infty, \tag{10}$$

where $\bar{\Sigma}$ is the $nk \times nk$ matrix defined by

$$\bar{\Sigma} = \begin{bmatrix} 0 & 0 & \ldots & 0 \\ 0 & 0 & \ldots & 0 \\ \vdots & \vdots & & \vdots \\ 0 & 0 & \ldots & \Sigma \end{bmatrix}.$$

By a multidimensional generalization (for a vector random process) of (Rozanov 1967, p. 6, condition (2.12)), which can be generalized, for our purpose, by diagonalizing Σ and making the corresponding linear transformation of ζ, the condition (10) implies the existence of $\int_0^t e^{(t-r)\bar{A}}\zeta(dr)$.

Let $y(t)$ be defined by (6) and let h be any number such that $0 < h \leq t$. Then, using the identity $e^{(A+B)} = e^A e^B$, which holds for any square matrices A and B of the same dimension (Halmos 1958, p. 186), we obtain

$$\begin{aligned} y(t) - y(t-h) & \qquad\qquad\qquad\qquad\qquad\qquad\qquad\qquad\qquad\qquad (11) \\ &= \int_0^t e^{(t-r)\bar{A}}\zeta(dr) - \int_0^t e^{(t-h-r)\bar{A}}\zeta(dr) \\ &\quad + (e^{t\bar{A}} - e^{(t-h)\bar{A}})(y(0) + \bar{A}^{-1}\bar{b}) + \int_{t-h}^t e^{(t-h-r)\bar{A}}\zeta(dr) \end{aligned}$$

$$= (I - e^{-h\bar{A}})y(t) + (I - e^{-h\bar{A}})\bar{A}^{-1}\bar{b}$$
$$+ e^{-h\bar{A}} \int_{t-h}^{t} (e^{(t-r)\bar{A}} - I)\xi(dr) + (e^{-h\bar{A}} - I)$$
$$\times \int_{t-h}^{t} \xi(dr) + \int_{t-h}^{t} \xi(dr)$$
$$= h\bar{A}y(t) + h\bar{b} + \int_{t-h}^{t} \xi(dr) + u(t, h),$$

where

$$u(t, h) = (-\tfrac{1}{2}h^2\bar{A}^2 + \tfrac{1}{6}h^3\bar{A}^3 - \ldots)(y(t) + \bar{A}^{-1}\bar{b})$$
$$+ e^{-h\bar{A}} \int_{t-h}^{t} (e^{(t-r)\bar{A}} - I)\xi(dr)$$
$$+ (-h\bar{A} + \tfrac{1}{2}h^2\bar{A}^2 - \ldots) \int_{t-h}^{t} \xi(dr).$$

But, using a multidimensional generalization of (Rozanov 1967, p. 6, equation (2.11)), and the fact that a convergent power series can be integrated term by term (Titchmarsh 1939, p. 38), we obtain

$$E\left[\int_{t-h}^{t} (e^{(t-r)\bar{A}} - I)\xi(dr)\right]\left[\int_{t-h}^{t} (e^{(t-r)\bar{A}} - I)\xi(dr)\right]'$$
$$= \int_{t-h}^{t} (e^{(t-r)\bar{A}} - I)\bar{\Sigma}(e^{(t-r)\bar{A}'} - I)dr$$
$$= \int_{0}^{h} (e^{r\bar{A}} - I)\bar{\Sigma}(e^{r\bar{A}'} - I)dr$$
$$= \int_{0}^{h} (r\bar{A} + \tfrac{1}{2}r^2\bar{A}^2 + \ldots)\bar{\Sigma}(r\bar{A}' + \tfrac{1}{2}r^2\bar{A}'^2 + \ldots)dr$$
$$= 0(h^3) \quad \text{as} \quad h \to 0.$$

Therefore

$$E[u'(t, h)u(t, h)] = 0(h^3) \qquad \text{as } h \to 0. \tag{12}$$

Now let $h = t/m$ where m is a positive integer. Then, using (11),

$$y(t) - y(0) = [y(t) - y(t-h)] + [y(t-h) - y(t-2h)] \tag{13}$$
$$+ \ldots + [y(t - (m-1)h) - y(0)]$$
$$= \frac{t}{m}\bar{A}\sum_{r=1}^{m} y\left(\frac{rt}{m}\right) + t\bar{b} + \int_{0}^{t} \xi(dr) + \sum_{r=1}^{m} u\left(\frac{rt}{m}, \frac{t}{m}\right).$$

It is clear from (6), (11), and (12) that $y(t)$ is mean square continuous on $[0, T]$. It follows that, for $0 < t \leq T$, $(t/m)\Sigma_{r=1}^{m} y(rt/m)$ is the integral, over $[0, T]$, of the mth member of a sequence of simple random vector processes which converge uniformly (in mean square) to $y(t)$, on $[0, T]$, as $m \to \infty$. By definition (Rosanov 1967, p. 10) of the integral of a random process which is integrable in the wide sense, we then have

$$\underset{m\to\infty}{\text{l.i.m.}}\ \frac{t}{m}\sum_{r=1}^{m} y\left(\frac{rt}{m}\right) = \int_0^t y(r)dr, \tag{14}$$

where l.i.m. denotes the limit in mean square. Moreover, using (12), we obtain

$$\underset{m\to\infty}{\text{l.i.m.}}\ \sum_{r=1}^{m} u\left(\frac{rt}{m}, \frac{t}{m}\right) = 0. \tag{15}$$

It follows from (13), (14), and (15), by letting $m \to \infty$, that $y(t)$ satisfies (7).

(b) Let $x(t)$ be the solution defined in (a) and $\hat{x}(t)$ any other solution of (1), on $[0, T]$, subject to the boundary conditions (3). Define

$$\hat{y}(t) = \begin{bmatrix} \hat{x}(t) \\ D\hat{x}(t) \\ \vdots \\ D^{k-1}\hat{x}(t) \end{bmatrix} \qquad (0 \leq t \leq T).$$

Then, since $\hat{x}(t)$ satisfies (4),

$$\hat{y}(t) - y(0) = \int_0^t (\bar{A}\hat{y}(r) + \bar{b})dr + \int_0^t \zeta(dr) \qquad (0 \leq t \leq T). \tag{16}$$

Moreover, since $D^{k-1}\hat{x}(t)$ is mean square continuous on $[0, T]$, $\hat{y}(t)$ is mean square continuous on $[0, T]$. Now let

$$\xi(t) = \hat{y}(t) - y(t) \qquad (0 \leq t \leq T),$$

where $y(t)$ is defined by (6). Then, from (6) and (16), we obtain

$$\xi(t) = \bar{A}\int_0^t \xi(r)dr \qquad (0 \leq t \leq T). \tag{17}$$

And, since $y(t)$ and $\hat{y}(t)$ are mean square continuous on $[0, T]$, $\xi(t)$ is mean square continuous on $[0, T]$. Therefore $E[\xi_1(t)]^2, \ldots, E[\xi_{kn}(t)]^2$ are continuous functions of t on $[0, T]$, since

$$\begin{aligned} &|E[\xi_i(t)]^2 - E[\xi_i(t-h)]^2| \\ &\quad \leq E|[\xi_i(t)]^2 - [\xi_i(t-h)]^2| \\ &\quad = E[|\xi_i(t) + \xi_i(t-h)||\xi_i(t) - \xi_i(t-h)|] \\ &\quad \leq \{E[\xi_i(t) + \xi_i(t-h)]^2 E[\xi_i(t) - \xi_i(t-h)]^2\}^{1/2} \\ &\quad \to 0 \quad \text{as } h \to 0 \qquad (i = 1, \ldots, kn). \end{aligned}$$

Let m be any positive integer such that $\tau = (T/m) \leq 1/\|\bar{A}\|$, where $\|\bar{A}\| = \{\operatorname{tr} \bar{A}\bar{A}'\}^{1/2}$. Then, since the functions $E[\xi_1(t)]^2, \ldots, E[\xi_{kn}(t)]^2$ are continuous on $[0, T]$, they have maxima $E[\xi_1(\tau_1)]^2, \ldots, E[\xi_{kn}(\tau_{kn})]^2$ respectively on $[0, \tau]$. From (17) we have

$$E[\xi_i(\tau_i)]^2 = E\left[\sum_{j=1}^{kn} \bar{a}_{ij} \int_0^{\tau_i} \xi_j(r)dr\right]^2 \qquad (i = 1, \ldots, kn), \tag{18}$$

where the coefficients $\bar{a}_{ij} (j = 1, \ldots, kn)$ are the elements of the ith row of $\bar{A}$. And, using (Rozzanov 1967, p. 10, inequality (2.25)), we have

$$E\left[\int_0^{\tau_i} \xi_j(r)dr\right]^2 \leq \left\{\int_0^{\tau_i} (E[\xi_j(r)]^2)^{1/2}dr\right\}^2 \qquad (i, j = 1, \ldots, kn). \tag{19}$$

From (18) and (19), using the Cauchy inequality, we obtain

$$\begin{aligned} E[\xi_i(\tau_i)]^2 &\leq \sum_{j=1}^{kn} \bar{a}_{ij}^2 \sum_{j=1}^{kn} \left\{\int_0^{\tau_i} (E[\xi_j(r)]^2)^{1/2}dr\right\}^2 \\ &\leq \sum_{j=1}^{kn} \bar{a}_{ij}^2 \sum_{j=1}^{kn} \tau_i^2 E[\xi_j(\tau_j)]^2 \\ &\leq \tau^2 \sum_{j=1}^{kn} \bar{a}_{ij}^2 \sum_{j=1}^{kn} E[\xi_j(\tau_j)]^2 \qquad (i = 1, \ldots, kn). \end{aligned}$$

Therefore

$$\sum_{i=1}^{kn} E[\xi_i(\tau_i)^2] \leq \tau^2 \|A\|^2 \sum_{i=1}^{kn} E[\xi_i(\tau_i)]^2. \tag{20}$$

Since $\tau^2\|A\|^2 < 1$, the inequality (20) implies that $\Sigma_{i=1}^{kn} E[\xi_i(\tau_i)]^2 = 0$, and hence, since $E[\xi_1(\tau_1)]^2, \ldots, E[\xi_{kn}(\tau_{kn})]^2$ are the maxima of $E[\xi_1(t)]^2, \ldots, E[\xi_{kn}(t)]^2$ over $[0, \tau]$, that

$$P(\xi(t) = 0) = 1 \qquad (0 \leq t \leq \tau).$$

Since a similar relation holds for each of the m intervals of length τ whose union is $[0, T]$, we obtain

$$P\{(\hat{x}(t) - x(t)) = 0\} = 1 \qquad (0 \leq t \leq T). \quad \text{Q.E.D} \tag{21}$$

REMARK: Doob (1953, p. 000), to whose results we have already referred, proves the uniqueness of the solution of a single first order equation in the sense

$$P\{(\hat{x}(t) - x(t)) = 0, 0 \leq t \leq T\} = 1, \tag{22}$$

which is generally stronger than (21). But, under Doob's assumption that the disturbance process is separable Brownian motion and the requirement that the sample functions of the solution are almost all continuous, the conditions (21) and (22) are equivalent. Condition (21) is equivalent to uniqueness in the sense of Edwards and Moyal (1955), since the probabilistic interpretation of their argument does not distinguish between random variables that are equal with probability 1. Moreover, condition (21) is strong enough for our purpose, since our aim is to derive an exact discrete model satisfied by a sequence of equispaced observations, and (21) certainly implies that

$$P\{\hat{x}(t) - x(t) = 0, t = 1, 2, \ldots, T'\} = 1$$

where T' is the largest integer less than or equal to T.

4. Estimation from a Sample of Observations at Discrete Points of Time

We shall be concerned, in this section, with the problem of estimating θ and Σ from a sample $[x(0), \ldots, x(T)]$ of observations, at unit time intervals, of the solution to (1). There is, of course, no loss of generality in identifying the unit of time with the observation period. The econometrician will normally have to take the observation period as given by the available statistical records.

We shall start by providing a detailed treatment of the second order system, which is likely to be of considerable practical importance. This will be followed by a brief discussion of the extension of the methods to a system of order greater than two. We shall deal with both the exact Gaussian estimates and estimates based on a frequency domain approximation to the Gaussian likelihood function. In either case the estimation procedure is based on an exact discrete model formulated in the time domain.

The second order system can be written

$$d[Dx(t)] = A_1(\theta)Dx(t) + A_2(\theta)x(t) + b(\theta) + \zeta(dt), \tag{23}$$

with boundary conditions

$$x(0) = y_1, \qquad Dx(0) = y_2. \tag{24}$$

With regard to this system, we shall make the following additional assumption.

Assumption 4: The matrix $[e^{\bar{A}}]_{12}$ defined by equation (31) is nonsingular.

The purpose of this assumption is to rule out certain special cases such as would occur if, for example, the period of the cycle generated by a single second order equation coincided with the observation period. In that case the exact discrete model satisfied by the observations would collapse to a first order system. There is no difficulty in dealing with this case, but we shall confine the derivation to the general case in which Assumption 4 is satisfied. Then the exact discrete model is given by Theorem 2.

Theorem 2: *Under Assumption* 1 *to* 4 *the solution of the system* (23), *on the interval* $[0, T]$, *subject to the boundary conditions* (24) *satisfies the system*

$$
\begin{aligned}
x(t) = F_1 x(t-1) + F_2 x(t-2) + g + \eta_t \qquad & (t = 2, \ldots, T), \\
E(\eta_t) = 0, \qquad E(\eta_t \eta_t') = \Omega_0 \quad & (t = 2, \ldots, T), \\
E(\eta_t \eta_{t-1}') = \Omega_1 \qquad & (t = 3, \ldots, T), \\
E(\eta_s \eta_t') = 0 \qquad & (|s-t| > 1;\ s, t = 2, \ldots, T), \\
E(\eta_t y_1') = E(\eta_t y_2') = 0 \qquad & (t = 2, \ldots, T)
\end{aligned}
\tag{25}
$$

where F_1, F_2, g, Ω_0, and Ω_1 are given by:

$$F_1 = [e^{\bar{A}}]_{12}[e^{\bar{A}}]_{22}[e^{\bar{A}}]_{12}^{-1} + [e^{\bar{A}}]_{11}, \tag{26}$$

$$F_2 = [e^{\bar{A}}]_{12}[e^{\bar{A}}]_{21} - [e^{\bar{A}}]_{12}[e^{\bar{A}}]_{22}[e^{\bar{A}}]_{12}^{-1}[e^{\bar{A}}]_{11}, \tag{27}$$

$$g = -(I - F_1 - F_2)A_2^{-1}b, \tag{28}$$

$$\Omega_0 = \int_0^1 P(r)\Sigma P'(r)dr + \int_0^1 [e^{r\bar{A}}]_{12}\Sigma[e^{r\bar{A}}]_{12}' dr, \tag{29}$$

$$\Omega_1 = \int_0^1 P(r)\Sigma[e^{r\bar{A}}]_{12}' dr, \tag{30}$$

$$P(r) = [e^{\bar{A}}]_{12}[e^{r\bar{A}}]_{22} - [e^{\bar{A}}]_{12}[e^{\bar{A}}]_{22}[e^{\bar{A}}]_{12}^{-1}[e^{r\bar{A}}]_{12},$$

$$
\begin{bmatrix} [e^{\bar{A}}]_{11} & [e^{\bar{A}}]_{12} \\ [e^{\bar{A}}]_{21} & [e^{\bar{A}}]_{22} \end{bmatrix} = \begin{bmatrix} I & 0 \\ 0 & I \end{bmatrix} + \begin{bmatrix} 0 & I \\ A_2 & A_1 \end{bmatrix} + \frac{1}{2}\begin{bmatrix} 0 & I \\ A_2 & A_1 \end{bmatrix}^2
$$

$$
+ \frac{1}{3!}\begin{bmatrix} 0 & I \\ A_2 & A_1 \end{bmatrix}^3 + \ldots. \tag{31}
$$

PROOF: Let $x(t)$ be the solution of (23), on $[0, T]$ subject to the boundary conditions (24). Then it follows from Theorem 1, by putting $k = 2$ and using (5), (6), and (9), that $x(t)$ satisfies the equations:

$$x(t) = [e^{\bar{A}}]_{11}x(t-1) + [e^{\bar{A}}]_{12}Dx(t-1) + c_1 + \int_{t-1}^{t} [e^{(t-r)\bar{A}}]_{12}\zeta(dr), \tag{32}$$

$$Dx(t) = [e^{\bar{A}}]_{21}x(t-1) + [e^{\bar{A}}]_{22}Dx(t-1) + c_2 + \int_{t-1}^{t} [e^{(t-r)\bar{A}}]_{22}\zeta(dr), \tag{33}$$

where

$$
\begin{aligned}
\begin{bmatrix} c_1 \\ c_2 \end{bmatrix} &= \begin{bmatrix} [e^{\bar{A}}]_{11} & [e^{\bar{A}}]_{12} \\ [e^{\bar{A}}]_{21} & [e^{\bar{A}}]_{22} \end{bmatrix}\begin{bmatrix} 0 & I \\ A_2 & A_1 \end{bmatrix}^{-1}\begin{bmatrix} 0 \\ b \end{bmatrix} \\
&\quad - \begin{bmatrix} 0 & I \\ A_2 & A_1 \end{bmatrix}^{-1}\begin{bmatrix} 0 \\ b \end{bmatrix} \\
&= \begin{bmatrix} [e^{\bar{A}}]_{11}A_2^{-1}b - A_2^{-1}b \\ [e^{\bar{A}}]_{21}A_2^{-1}b \end{bmatrix}.
\end{aligned}
$$

From (32) we obtain

$$Dx(t-1) = [e^{\bar{A}}]_{12}^{-1}x(t) - [e^{\bar{A}}]_{12}^{-1}[e^{\bar{A}}]_{11}x(t-1) - [e^{\bar{A}}]_{12}^{-1}c_1 - [e^{\bar{A}}]_{12}^{-1}\int_{t-1}^{t}[e^{(t-r)\bar{A}}]_{12}\zeta(dr), \tag{34}$$

$$Dx(t-2) = [e^{\bar{A}}]_{12}^{-1}x(t-1) - [e^{\bar{A}}]_{12}^{-1}[e^{\bar{A}}]_{11}x(t-2) - [e^{\bar{A}}]_{12}^{-1}c_1 - [e^{\bar{A}}]_{12}^{-1}\int_{t-2}^{t-1}[e^{(t-1-r)\bar{A}}]_{12}\zeta(dr), \tag{35}$$

and, from (33)

$$Dx(t-1) = [e^{\bar{A}}]_{21}x(t-2) + [e^{\bar{A}}]_{22}Dx(t-2) + c_2 + \int_{t-2}^{t-1}[e^{(t-1-r)\bar{A}}]_{22}\zeta(dr). \tag{36}$$

Substituting from (34) and (35), for $Dx(t-1)$ and $Dx(t-2)$ respectively, into (36) and premultiplying by $[e^{\bar{A}}]_{12}$ we obtain (25) with

$$\eta_1 = \int_{t-2}^{t-1}P(t-1-r)\zeta(dr) + \int_{t-1}^{t}[e^{(t-r)\bar{A}}]_{12}\zeta(dr). \tag{37}$$

Then, by using the multidimensional generalization of (Rosanov 1967, p. 6, equations (2.11) and (2.14)), we obtain

$$E(\eta_t\eta_t') = \int_{t-2}^{t-1}P(t-1-r)\Sigma p'(t-1-r) + \int_{t-1}^{t}[e^{(t-r)\bar{A}}]_{12}\Sigma[e^{(t-r)\bar{A}}]_{12}'dr$$

$$= \int_0^1 P(r)\Sigma P'(r)dr + \int_0^1 [e^{r\bar{A}}]_{12}\Sigma[e^{r\bar{A}}]_{12}'dr,$$

$$E(\eta_t\eta_{t-1}') = \int_{t-2}^{t-1}P(t-1-r)\Sigma[e^{(t-1-r)\bar{A}}]_{12}'dr$$

$$= \int_0^1 P(r)\Sigma[e^{r\bar{A}}]_{12}'dr,$$

$$E(\eta_s\eta_t') = 0 \quad (|s-t| > 1).$$

The conditions $E(\eta_t) = E(\eta_t y_1') = E(\eta_t y_2') = 0$ $(t = 2, \ldots, T)$ obviously follow from (37) and Assumptions 1 and 3. Q.E.D.

We shall refer to the system (25), together with the conditions on the $\eta_t (t = 2, \ldots, T)$, as the exact discrete model for a sequence of observations, at equispaced points of time, generated by the continuous time model. Although the expression on the right-hand side of (37) is not in the form of a discrete moving average it follows from the implied conditions on $E(\eta_s\eta_t')$ that $\{\eta_t (t = 0, \pm 1, \pm 2, \ldots)\}$ is a wide sense stationary, first order, vector moving average process. Hence there exists random vectors $\varepsilon_1, \ldots, \varepsilon_T$ such that

$$\eta_t = \varepsilon_t + G\varepsilon_{t-1} \qquad (t = 2, \ldots, T), \quad (38)$$

$$E(\varepsilon_t) = 0, \qquad E(\varepsilon_t\varepsilon_t') = K, \qquad E(\varepsilon_s\varepsilon_t') = 0 \qquad (s \neq t),$$

where G and K are $n \times n$ matrices, K being positive definite. Moreover, G and K must satisfy the equations

$$\begin{aligned}\Omega_0 &= E(\varepsilon_t + G\varepsilon_{t-1})(\varepsilon_t + G\varepsilon_{t-1})' \\ &= K + GKG', \\ \Omega_1 &= E(\varepsilon_t + G\varepsilon_{t-1})(\varepsilon_{t-1} + G\varepsilon_{t-2})' \\ &= GK,\end{aligned}$$

from which we obtain

$$K'\Omega_1^{-1}K - K'\Omega_1^{-1}\Omega_0 + \Omega_1' = 0, \quad (39)$$

$$K + \Omega_1 G' - \Omega_0 = 0. \quad (40)$$

It is convenient, at this stage, to write Σ as $\Sigma(\mu)$ where μ is a vector of parameters and the elements of $\Sigma(\mu)$ are functions of μ. If, as will usually be the case in econometric work, Σ is unrestricted then μ will have $n(n+1)/2$ elements. But we could, for example, require Σ to be a diagonal matrix, in which case μ would have n elements. It is clear from (29) and (30) that the elements of Ω_0 and Ω_1 are functions of the complete parameter vector (θ, μ). These functions can be approximated to any required degree of accuracy, by expanding the integrands on the right hand side of (29) and (30) as power series in r (which are convergent for all real r) and integrating term by term. It follows from equations (39) and (40) that the elements of G and K will also be functions of the complete parameter vector (θ, μ), so that we can write these matrices as $G(\theta, \mu)$ and $K(\theta, \mu)$ respectively.

By combining equations (25) and (38) we obtain a representation of the exact discrete model as a vector autoregressive moving average model:

$$x(t) - F_1(\theta)x(t-1) - F_2(\theta)x(t-2) - g(\theta) = \varepsilon_t + G(\theta, \mu)\varepsilon_{t-1}, \quad (41)$$

$$E(\varepsilon_t) = 0, \qquad E(\varepsilon_t\varepsilon_t') = K(\theta, \mu), \qquad E(\varepsilon_s\varepsilon_t') = 0 \qquad (s \neq t).$$

It is important to notice that the parameter vector θ occurs not only in the matrices of coefficients of both the autoregressive and moving average parts of the system (41), but also in the covariance matrix of ε_t, while the parameter vector μ occurs not only in the covariance matrix of ε_t, but also in the matrix of coefficients of the moving average. Thus, although there is a clear separation of θ and μ in the continuous time model, with the coefficient matrices depending only on θ and the covariance matrix of the random measure disturbance only on μ, these two parameter vectors are intimately mixed in the exact discrete model.

Because of the difficulty of solving equations (39) and (40) for G and K, the model (41) is not very convenient as a basis for the computation of the Gaussian estimate of (θ, μ). For this purpose it is easier to work directly from the original form of the exact discrete model, as given by Theorem 2. But the model (41) is useful, nevertheless, as a basis for the discussion of the asymptotic sampling properties of the estimator. In particular, it provides a link with the literature on discrete multiple time series models. In this connection all that we require is that $G(\theta, \mu)$ and $K(\theta, \mu)$ satisfy certain regularity conditions which can be deduced from the properties of $\Omega_0(\theta, \mu)$ and $\Omega_1(\theta, \mu)$ without solving for G and K. We shall return to this discussion in Section 7.

Theorem 2 can be used as a basis for the Gaussian estimation of (θ, μ) under either of two alternative assumptions. The first is that y_1 and y_2 (i.e. $x(0)$ and $Dx(0)$) are arbitrary random vectors satisfying Assumption 3. In this case, since we are interested only in the conditional distribution of our estimator, for given $x(0)$ and $Dx(0)$, we can treat these vectors as nonrandom. The vector $x(0)$ is given as part of the data. But, since $Dx(0)$ is unobservable, it must be estimated. This can be done by adjoining equation (32), with $t = 1$, to the system (25) and treating $Dx(0)$ as an extension of the vector of parameters to be estimated. The second alternative assumption is that $\{x(t), -\infty < t < \infty\}$ is an n-dimensional, wide sense, stationary process. This is equivalent to assuming that the eigenvalues of $\bar{A}$ have negative real parts and the vector $[x(0), Dx(0)]$, in addition to satisfying (3) and Assumption 3, has the covariance matrix implied by the assumption that (1) is satisfied for all $t > -\infty$. In the remainder of the paper we shall omit the words 'wide sense' when referring to the latter assumption. If a random process is assumed to be strictly stationary we shall say so explicitly.

Which of the above alternative assumptions is the more appropriate in applied econometric work will depend on the time interval between the last major structural change in the economy (caused, for example, by a major war) and the beginning of the sample period. Since the computations are lighter under the assumption of stationarity, we shall deal with this case in detail. The sort of modifications to the computational procedure that will be necessary under the assumption of nonstationarity will then be fairly obvious.

It follows from (25), since $E(\eta_t) = 0$, that the mean $m(\theta) = E[x(t)]$ of the stationary process $\{x(t)\}$ satisfies

$$m(\theta) = F_1(\theta)m(\theta) + F_2(\theta)m(\theta) + g(\theta), \tag{42}$$

from which we obtain

$$m(\theta) = (I - F_1(\theta) - F_2(\theta))^{-1}g(\theta) \tag{43}$$

$$= -A_2^{-1}(\theta)b(\theta).$$

From (25) and (42) we obtain

$$[x(t) - m(\theta)] = F_1(\theta)[x(t-1) - m(\theta)] + F_2(\theta)[x(t-2) - m(\theta)] + \eta_t \quad (44)$$

whose solution can be written

$$x(t) - m(\theta) = \eta_t + B_1(\theta)\eta_{t-1} + \ldots + B_r(\theta)\eta_{t-r} + \ldots \quad (45)$$

By equating to zero the coefficient of $\eta_t, \eta_{t-1}, \ldots$ in the identity

$$\begin{aligned}(\eta_t + B_1\eta_{t-1} + \ldots + B_r\eta_{t-r} + \ldots) \\ - F_1(\eta_{t-1} + B_1\eta_{t-2} + \ldots + B_{r-1}\eta_{t-r} + \ldots) \\ - F_2(\eta_{t-2} + B_1\eta_{t-3} + \ldots + B_{r-2}\eta_{t-r} + \ldots) = 0,\end{aligned}$$

we obtain:

$$\begin{aligned}B_1 &= F_1, \\ B_2 &= F_1B_1 + F_2, \\ B_r &= F_1B_{r-1} + F_2B_{r-2} \qquad (r = 3, 4, \ldots),\end{aligned} \quad (46)$$

from which the matrices $B_1(\theta)$, $B_2(\theta)$, ... can be computed recursively.

It follows from (45) and the condition $E(\eta_s\eta_t') = 0\ |s-t| > 1$ that the (matrix valued) correlation function $V_r(\theta, \mu) = E[x(t) - m(\theta)][x(t+r) - m(\theta)]'(r = 0, \ldots, T)$ is given by

$$V_r(\theta, \mu) = \sum_{j=0}^{\infty} B_j(\theta)\Omega_0(\theta, \mu)B'_{j+r}(\theta) + \sum_{j=0}^{\infty} B_j(\theta)\Omega_1(\theta, \mu)B'_{j+r+1}(\theta) + \sum_{j=0}^{\infty} B_{j+1}(\theta)\Omega_1'(\theta, \mu)B'_{j+r}(\theta) \qquad (r, 0, \ldots, T). \quad (47)$$

The Gaussian estimator $(\hat{\theta}, \hat{\mu})$ is then the vector (θ, μ) that minimizes $\hat{L}(\theta, \mu)$, which is defined by

$$\hat{L}(\theta, \mu) = \log|V(\theta, \mu)| + [(x(0) - m(\theta))', \ldots, (x(T) - m(\theta))'] \times [V(\theta, \mu)]^{-1} \begin{bmatrix} x(0) - m(\theta) \\ \vdots \\ x(T) - m(\theta) \end{bmatrix} \quad (48)$$

where $V(\theta, \mu)$ is the $n(T+1) \times n(T+1)$ matrix

$$V = \begin{bmatrix} V_0 & V_1 & V_2 & \ldots & V_T \\ V_1' & V_0 & V_1 & \ldots & V_{T-1} \\ V_2' & V_1' & V_0 & \ldots & V_{T-2} \\ \vdots & \vdots & \vdots & & \\ V_T' & V_{T-1}' & V_{T-2}' & \ldots & V_0 \end{bmatrix}.$$

It is not necessary to invert V in order to compute the value of $\hat{L}$ for given values of the parameters. Since V is a positive definite matrix it can be factorized as

$$V = MM', \tag{50}$$

where M is a lower triangular real matrix with positive elements on the diagonal (see, for example (Phillips and Taylor 1973, p. 215, Theorem 8.2)). Moreover, the elements $m_{ij}(i = 1, \ldots, n(T+1); j = 1, \ldots, i)$ of the matrix M can be computed recursively. If we now define $z' = [z_1, \ldots, z_{n(T+1)}]$ by $z = M^{-1}x$ where $x' = (x'(0) - m', \ldots, x'(T) - m')$, the elements of z can be computed, recursively, from the solution of

$$Mz = x. \tag{51}$$

We can then compute the value of $\hat{L}$ from

$$\hat{L} = \sum_{i=1}^{n(T+1)} \{z_i^2 + 2 \log m_{ii}\}. \tag{52}$$

A complete list of steps for the computation of the value of $\hat{L}$ for given values of the parameters, starting with the continuous time model, is as follows.

(i) Compute A_1, A_2, b, and Σ, using the specified form of the functions defining their elements. Most of the elements of the matrices will be equal to zero or one of the structural parameters.
(ii) Compute F_1, F_2, g, Ω_0, and Ω_1 from equations (26) to (30). The series occurring in the formulae for these matrices converge very rapidly (for example the nth term of $\int_0^1 e^{rA} dr$ is $(1/(n+1)!)A^n$) and should be summed to a sufficient number of terms to achieve the desired degree of accuracy.
(iii) Compute $m(\theta)$ from (43).
(iv) Compute $B_1, B_2, \ldots$, recursively, from (46) until the elements of B_r are sufficiently small to be neglected.
(v) Compute $V_0, \ldots, V_T$ from (47).
(vi) Compute the elements of M recursively from (50).
(vii) Compute the elements of z recursively from (51).
(viii) Compute $\hat{L}$ from (52).

The Gaussian estimate $(\hat{\theta}, \hat{\mu})$ can be obtained by various iterative optimization procedures involving only the repeated computation of $\hat{L}$ for a sequence of vectors of the parameter values. The simplest (although not the fastest) procedure is the maximum gradient method (Malinvaud 1970, p. 343) with the direction of the maximum gradient being approximated, at each stage, by a vector of approximate first derivatives with respect to the elements of (θ, μ) using the rule $f'(x) \simeq \{f(x+h) - f(x)\}/h$, where h is small. There is, of course, no way of ensuring that this (or any other) procedure will lead to the minimum of $\hat{L}$ if there are several local minima. For this reason the procedure should be repeated using a different initial vector of parameter values.

The main disadvantage of the above estimation procedure is that the matrix M, which is an $n(T+1) \times n(T+1)$ triangular matrix with distinct elements on and below the diagonal, would require a large amount of computer storage space. The method would be feasible, nevertheless, for a model of ten to fifteen equations with a sample of forty to fifty observations. Moreover, computer costs have now been reduced to a point where the computation of the exact Gaussian estimates of the parameters of a second order model, of this size, is likely to be justified by the gain in predictive power, as compared with less sophisticated models or less efficient methods of estimation. But, for the estimation of the parameters of larger models, we shall need to find approximations to $\hat{L}$ which avoid the factorization of matrices with dimensions of the order nT.

An elegant and convenient approximation, which involves only $n \times n$ matrices, can be obtained by using the spectral density $f(\lambda)$ of the n-dimensional, discrete time, stationary process $\{x(t)(t = 0, \pm 1, \pm 2, \ldots)\}$ generated by the continuous time model. The use of this sort of approximation in the estimation of the parameters of multiple time series models, by classical least squares methods, was introduced by Whittle (1953). In that article he applies to the analysis of multiple time series the general methods developed in his earlier work (starting from Whittle 1951) on statistical inference in univariate time series models. Whereas this earlier work was based on the decomposition of a stationary process obtained by Wold (1938), the analysis in Whittle 1953 is based on Zasukhin's extension (1941) of this decomposition to an n-dimensional stationary process. Zazukhin's extension of the Wold decomposition implies (Rosanov 1967, p. 56) that any linearly regular, n-dimensional, discrete time, stationary process $\{x(t)(t = 0, \pm 1, \pm 2, \ldots)\}$ whose spectral density matrix has rank n, has the representation

$$x(t) = C_0\varepsilon_t + C_1\varepsilon_{t-1} + C_2\varepsilon_{t-2} + \ldots, \tag{53}$$

$$E(\varepsilon_t) = 0, \qquad E(\varepsilon_t\varepsilon_t') = K, \qquad E(\varepsilon_s\varepsilon_t') = 0, \qquad (s \neq t),$$

where $C_0, C_1, C_2, \ldots$ are $n \times n$ matrices and either $C_0 = I$ or $K = I$. For the purpose of proving theorems it is mathematically convenient to assume $K = I$, and Whittle (1953), following Zasukhin, makes this assumption. But in the specification of the model it is usually more appropriate to assume $C_0 = I$, with the elements of K and the C_i $(i = 1, 2, \ldots)$ being functions of a vector α of structural parameters. It is clear from (41) that our model can be represented in this form.

Whittle (1953) assumes that (53) is invertible so that the process has an autoregressive representation.

$$\varepsilon_t = x(t) + H_1 x(t-1) + H_2 x(t-2) + \ldots. \quad (54)$$

Then, assuming a sample $[x(1), \ldots, x(N)]$ and neglecting end effects (i.e. putting $x(t) = 0$ for $t < 1$), he shows (Whittle 1953, Theorem 6) that the estimates obtained by maximizing the resulting approximation to the Gaussian likelihood can be obtained by minimizing

$$\bar{L}(\alpha) = \frac{1}{2\pi} \int_0^{2\pi} \{\log |f(\lambda, \alpha)| + \operatorname{tr}[f^{-1}(\lambda, \alpha)\psi(\lambda)]\} d\lambda, \quad (55)$$

where $f(\lambda, \alpha)$ is the spectral density of the n-dimensional stationary process $\{x(t)(t = 0, \pm 1, \pm 2, \ldots)\}$ and $\psi(\lambda)$ is its empirical equivalent, i.e., $\psi(\lambda)$ is the matrix whose kjth element ψ_{jk} is defined by

$$\psi_{jk} = \frac{1}{2\pi} \sum_{s=1-N}^{N-1} c_{jk}(s) e^{i\lambda s},$$

$$c_{kj}(-s) = c_{jk}(s) = \frac{1}{N} \sum_{t=1}^{N-s} x_j(t+s) x_k(t) \qquad (s = 0, \ldots, N-1).$$

It is easy to see (using Whittle 1953, Theorem 2) that the function $\bar{L}(\alpha)$ defined by (55) can be written in the alternative form

$$\bar{L}(\alpha) = \log |K(\alpha)| + \frac{1}{2\pi} \int_{-\pi}^{\pi} \operatorname{tr}[f^{-1}(\lambda, \alpha) I(\lambda)] d\lambda, \quad (56)$$

where

$$I(\lambda) = w(\lambda) w^*(\lambda), \quad w(\lambda) = \frac{1}{\sqrt{2\pi N}} \sum_{=1}^{N} x(t) e^{i\lambda t}.$$

A further approximation can be obtained by replacing the integral in (56) by a finite sum, which gives

$$\tilde{L}(\alpha) = \log |K(\alpha)| + N^{-1} \sum_n \operatorname{tr}[f^{-1}(\lambda_n, \alpha) I(\lambda_n)], \quad (57)$$

where

$$\lambda_n = \frac{2\pi n}{N}, \quad -\frac{N}{2} < n \leq \frac{N}{2}.$$

Estimates $\bar{\alpha}$ and $\tilde{\alpha}$ can be obtained by minimizing $\bar{L}(\alpha)$ and $\tilde{L}(\alpha)$ respectively. The estimate $\tilde{\alpha}$ has computational advantages over $\bar{\alpha}$. Moreover, when the model is specified in the form (53) (or easily transformed into the form (53), as it will be in most models specified in discrete time), the estimates $\bar{\alpha}$ and $\tilde{\alpha}$ will both be considerably easier to compute than the exact Gaussian estimate $\hat{\alpha}$.

In order to apply the above methods in obtaining estimates $(\bar{\theta}, \bar{\mu})$ and $(\tilde{\theta}, \tilde{\mu})$ of the parameter vector (θ, μ) of our continuous time model, it is necessary to modify the functions $\bar{L}$ and $\tilde{L}$ by replacing $x(t)$ by

$x(t) - m(\theta)$ in the definition of $w(\lambda)$. It is also necessary to solve (39) for $K(\theta, \mu)$. But we could avoid solving for K by replacing $\log|K|$ by $N^{-1}\Sigma \log|f|$ in (57), as suggested by Robinson (1977). It should, of course, be remembered that (55) is only a large sample approximation to the Gaussian likelihood, and the use of this formula (or approximations derived from it) with small samples could yield estimates that differ considerably from the exact Gaussian estimates.

For the evaluation of $\bar{L}$ and $\tilde{L}$ we need the spectral density $f(\lambda; \theta, \mu)$ of the n-dimensional stationary process $\{x(t) - m(\theta)(t = 0, \pm 1, \pm 2, \ldots)\}$ generated by the exact discrete model derived in Theorem 2 (under the assumption of stationarity). From (44) and the spectral density of the stationary process $\{\eta_t(t = 0, \pm 1, \pm 2, \ldots)\}$ we obtain (using Rosanov 1967, p. 36, Theorem 8.1)

$$\begin{aligned} f(\lambda; \theta, \mu) = \frac{1}{2\pi} & [I - e^{-i\lambda} F_1(\theta) - e^{-2i\lambda} F_2(\theta)]^{-1} \\ & \times [e^{-i\lambda}\Omega_1(\theta, \mu) + \Omega_0(\theta, \mu) + e^{i\lambda}\Omega_1'(\theta, \mu)] \\ & \times [I - e^{-i\lambda} F_1(\theta) - e^{-2i\lambda} F_2(\theta)]^{*-1}. \end{aligned} \tag{58}$$

This is, of course, different from the spectral density $\phi(\lambda; \theta, \mu)$ of the continuous time stationary process $\{x(t) - m(\theta), -\infty < t < \infty\}$ generated by the system (23). The latter is (putting $k = 2$ in (2)) given by

$$\begin{aligned} \phi(\lambda; \theta, \mu) = \frac{1}{2\pi} & [(i\lambda)^2 - (i\lambda)A_1(\theta) - A_2(\theta)]^{-1} \\ & \times \Sigma(\mu)[(i\lambda)^2 - (i\lambda)A_1(\theta) - A_2(\theta)]^{*-1}. \end{aligned} \tag{59}$$

It is interesting to note that we have moved from the spectral density of the continuous time process to the spectral density of the corresponding discrete time process by a different route from that followed by Phillips (1959) and Durbin (1961). They start by expressing the spectral density of the continuous time process in the form of partial fractions and then taking a Fourier transform to obtain the correlation function, whereas we have used an exact discrete model formulated in the time domain. From the point of view of this paper, our method has the advantage that we obtain the elements of the spectral density matrix of the discrete time process as explicit functions of the structural parameters of the continuous time model. This, of course, did not concern Phillips and Durbin, since they placed no restrictions on the coefficients of the continuous time model.

An alternative procedure would be to use the formula

$$f(\lambda; \theta, \mu) = \sum_{j=-\infty}^{\infty} \phi(\lambda + 2\pi j; \theta, \mu)$$

as proposed by Robinson (1977). By using the exact discrete model and the formulae given in Theorem 2, we have, essentially, found a compact formula for the infinite sum on the right hand side of the last equation, thus avoiding the need to carry out a separate summation for each of the frequencies $\lambda_n (-N/2 < n \leq N/2)$.

We shall conclude this section with a brief discussion of the application of the above methods to a system of order $k > 2$. We start again with Theorem 1 from which it follows, using (5), (6), and (9), that the solution $x(t)$ of the system (1) and its derivatives satisfy the nk-dimensional system (60).

$$
\begin{aligned}
x(t) &= [e^{\bar{A}}]_{11}x(t-1) + [e^{\bar{A}}]_{12}Dx(t-1) + \dots \\
&\quad + [e^{\bar{A}}]_{1k}D^{k-1}x(t-1) + c_1 + \int_{t-1}^{t} [e^{(t-r)\bar{A}}]_{1k}\zeta(dr), \\
Dx(t) &= [e^{\bar{A}}]_{21}x(t-1) + [e^{\bar{A}}]_{22}Dx(t-1) + \dots \\
&\quad + [e^{\bar{A}}]_{2k}D^{k-1}x(t-1) + c_2 + \int_{t-1}^{t} [e^{(t-r)\bar{A}}]_{2k}\zeta(dr), \qquad (60) \\
&\vdots \\
D^{k-1}x(t) &= [e^{\bar{A}}]_{k1}x(t-1) + [e^{\bar{A}}]_{k2}Dx(t-1) + \dots \\
&\quad + [e^{\bar{A}}]_{kk}D^{k-1}x(t-1) + c_k + \int_{t-1}^{t} [e^{(t-r)\bar{A}}]_{kk}\zeta(dr).
\end{aligned}
$$

If we now substitute $t = t-1, t = t-2, \dots, t = t-k+1$, successively, in (60) and combine all the resulting equations, together with the first n-dimensional subsystem of (60), we obtain a system of dimension $n\{k(k-1)+1\}$ in the k^2+1 n-dimensional vectors $x(t)$, $x(t-1), \dots, x(t-k)$, $Dx(t-1), \dots, Dx(t-k), \dots, D^{k-1}x(t-1), \dots, D^{k-1}x(t-k)$. By eliminating from this system the last $k(k-1)$ vectors, we obtain an exact discrete model:

$$
\begin{aligned}
x(t) &= F_1(\theta)x(t-1) + F_2(\theta)x(t-2) + \dots \\
&\quad + F_k(\theta)x(t-k) + g(\theta) + \eta_t, \\
E(\eta_t) &= 0, \qquad (61) \\
E(\eta_t\eta'_{t-r}) &= \Omega_r(\theta, \mu) \qquad (r = 0, \dots, k-1), \\
E(\eta_t\eta'_{t-r}) &= 0 \qquad (r > k-1),
\end{aligned}
$$

where the matrix functions $F_1(\theta), \dots, F_k(\theta)$, $g(\theta)$, $\Omega_0(\theta, \mu)$, $\Omega_1(\theta, \mu), \dots, \Omega_{k-1}(\theta, \mu)$ can all be obtained explicitly, although their form will become increasingly complicated as k increases.

The model (61) has a representation as an autoregressive moving average model:

$$
\begin{aligned}
x(t) - F_1(\theta)x(t-1) - \dots - F_k(\theta)x(t-k) - g(\theta) \\
= \varepsilon_t + G_1(\theta, \mu)\varepsilon_{t-1} + \dots + G_{k-1}(\theta, \mu)\varepsilon_{t-k+1}
\end{aligned}
$$

$$
E(\varepsilon_t) = 0, \qquad E(\varepsilon_t\varepsilon'_t) = K(\theta, \mu), \qquad E(\varepsilon_s\varepsilon'_t) = 0 \quad (s \neq t), \qquad (62)
$$

which is a generalization of (41). But for estimation purposes it is easier to work directly from the model (61).

The exact Gaussian estimates can be obtained by the estimation procedure set out in detail for the case $k = 2$. The matrices $V_0, \ldots, V_T$ will, of course, be more complicated, since they will now include terms in $\Omega_2, \ldots, \Omega_{k-1}$. We can also obtain estimates by minimizing either of the approximations $\bar{L}$ and $\tilde{L}$ derived from the Gaussian likelihood function. For this purpose we need the spectral density $f(\lambda; \theta, \mu)$ of the discrete time process. This is given by

$$f(\lambda) = \frac{1}{2\pi}[I - e^{-i\lambda}F_1 - \ldots - e^{-i\lambda k}F_k]^{-1} \tag{63}$$
$$\times [e^{-i\lambda(k-1)}\Omega_{k-1} + \ldots e^{-i\lambda}\Omega_1 + \Omega_0 + e^{i\lambda}\Omega_1' + \ldots$$
$$+ e^{i\lambda(k-1)}\Omega_{k-1}'][I - e^{-i\lambda}F_1 - \ldots - e^{-i\lambda k}F_k]^{*-1}.$$

5. Estimation From a Sample of Integral Observations

We turn now to the problem of obtaining estimates of the structural parameters of the continuous time model from a sample $[x_1, \ldots, x_T]$ of observations of the form

$$x_t = \int_{t-1}^{t} x(r)dr \qquad (t = 1, \ldots, T), \tag{64}$$

where $x(t)$ is the solution of the model. Again we shall start with a detailed treatment of the second order system and conclude with a brief discussion of the extension of the methods to a system of order greater than two.

The most obvious way to proceed would be to integrate (25) over the interval $[t-1, t]$ to obtain an exact discrete model in terms of the observations and a disturbance vector $\int_{t-1}^{t}\eta_t d\tau$, where η_t is given by (37). This procedure was used by Bergstrom and Wymer (1976) and P. C. B. Phillips (1978) in their treatments of the problem of obtaining estimates of the structural parameters of a first order continuous time model from flow data. Its main disadvantage is that, even with a first order model, it yields the (matrix) values of the correlation function of the n-dimensional disturbance process in the form of a triple integral (Phillips 1978, p. 259). These are not very convenient for application, particularly in higher order models, where the integrands are more complicated. Moreover their rigorous derivation requires quite a complex mathematical argument, since it involves a change in the order of three different types of integration: (i) the integration of a measurable function with respect to a random measure, (ii) the integration, in the wide sense, of a random process with respect to the time parameter, (iii) integration over the probability space to obtain expected values. This change of order can be rigorously justified (for example, by redefining the various formulae in such a way that the limits of integration are fixed and using a multidimensional generalization of (Rosanov 1967, p. 12, Theorem 2.4), although its validity was taken for granted in Bergstrom and Wymer 1976 and Phillips 1978.

In the proof of Theorem 3 we shall use a different type of argument which avoids these mathematical difficulties and also leads to more convenient formulae for the values of the correlation function of the disturbance process. For a system of any order, these formulae are obtained as single integrals of expressions that involve nothing more complicated than matrix exponential functions.

THEOREM 3: *Let $x(t)$ be the solution of* (23), *on* $[0, T]$, *subject to the boundary conditions* (24). *Then, under Assumptions* 1 *to* 4, *the random vectors* $x_1, \ldots, x_T$, *defined by* (64) *satisfy the system*

$$x_t = F_1 x_{t-1} + F_2 x_{t-2} + g + \eta_t \qquad (t = 3, \ldots, T),$$

$$E(\eta_t) = 0, \quad E(\eta_t \eta_t') = \Omega_0 \qquad (t = 3, \ldots, T),$$

$$E(\eta_t \eta_{t-1}') = \Omega_1 \qquad (t = 4, \ldots, T),$$

$$E(\eta_t \eta_{t-2}') = \Omega_2 \qquad (t = 5, \ldots, T),$$

$$E(\eta_s \eta_t') = 0 \quad (|s - t| > 2; s, t = 3, \ldots, T),$$

$$E(\eta_t y_1') = E(\eta_t y_2') = 0 \qquad (t = 3, \ldots, T),$$

where F_1, F_2 and g are defined as in Theorem 2 and Ω_0, Ω_1, and Ω_2 are given by:

$$\begin{aligned}\Omega_0 = &\int_0^1 \{A_2^{-1}Q(r)\}\Sigma\{A_2^{-1}Q(r)\}'dr \\ &+ \int_0^1 \{A_2^{-1}S(r) + [e^{\bar{A}}]_{12}[e^{r\bar{A}}]_{12} + WQ(r)\} \\ &\times \Sigma\{A_2^{-1}S(r) + [e^{\bar{A}}]_{12}[e^{r\bar{A}}]_{12} + WQ(r)\}'dr \\ &+ \int_0^1 \{WS(r) + [e^{\bar{A}}]_{12}([e^{r\bar{A}}]_{12} - [e^{\bar{A}}]_{12})\} \\ &\times \Sigma\{WS(r) + [e^{\bar{A}}]_{12}([e^{r\bar{A}}]_{12} - [e^{\bar{A}}]_{12})\}'dr,\end{aligned} \tag{66}$$

$$\begin{aligned}\Omega_1 = &-\int_0^1 \{A_2^{-1}S(r) + [e^{\bar{A}}]_{12}[e^{r\bar{A}}]_{12} + WQ(r)\}\Sigma\{A_2^{-1}Q(r)\}'dr \\ &- \int_0^1 \{WS(r) + [e^{\bar{A}}]_{12}([e^{r\bar{A}}]_{12} - [e^{\bar{A}}]_{12})\} \\ &\times \Sigma\{A_2^{-1}S(r) + [e^{\bar{A}}]_{12}[e^{r\bar{A}}]_{12} + WQ(r)\}'dr,\end{aligned} \tag{67}$$

$$\Omega_2 = \int_0^1 \{WS(r) + [e^{\bar{A}}]_{12}([e^{r\bar{A}}]_{12} - [e^{\bar{A}}]_{12})\}\Sigma\{A_2^{-1}Q(r)\}'dr, \tag{68}$$

$$Q(r) = A_1[e^{r\bar{A}}]_{12} - [e^{r\bar{A}}]_{22} + I,$$

$$S(r) = A_1[e^{r\bar{A}}]_{12} - [e^{r\bar{A}}]_{22} - A_1[e^{\bar{A}}]_{12} + [e^{\bar{A}}]_{22},$$
$$W = [e^{\bar{A}}]_{12}[e^{\bar{A}}]_{22}[e^{\bar{A}}]_{12}^{-1}A_2^{-1}.$$

PROOF: Let $x(t)$ be the solution of (23), on $[0, T]$, subject to the boundary conditions (24). Then, from the properties of the solution, we have (using the notation introduced below equation (6) with $k = 2$):

$$y(t) - y(0) = \bar{A}\int_0^t y(r)dr + t\bar{b} + \int_0^t \bar{\zeta}(dr), \tag{69}$$

from which we obtain

$$y(t) - y(t-1) = \bar{A}\int_{t-1}^t y(r)dr + \bar{b} + \int_{t-1}^t \bar{\zeta}(dr), \tag{70}$$

and hence

$$\int_{t-1}^t y(r)dr = \bar{A}^{-1}(y(t) - y(t-1)) - \bar{A}^{-1}\bar{b} - \bar{A}^{-1}\int_{t-1}^t \bar{\zeta}(dr). \tag{71}$$

Now applying the first difference operator to equations (32) and (33) and combining the resulting equations with (71) we obtain

$$\begin{aligned}\int_{t-1}^t y(r)dr &= \bar{A}^{-1}[e^{\bar{A}}][y(t-1) - y(t-2)] \\ &\quad + \bar{A}^{-1}\left[\int_{t-1}^t e^{(t-r)\bar{A}}\bar{\zeta}(dr) - \int_{t-2}^{t-1} e^{(t-1-r)\bar{A}}\bar{\zeta}(dr)\right] \\ &\quad - \bar{A}^{-1}\bar{b} - \bar{A}^{-1}\int_{t-1}^t \bar{\zeta}(dr).\end{aligned} \tag{72}$$

Then, substituting $t = t - 1$ in (70) and combining the resulting equation with (72), we obtain

$$\begin{aligned}\int_{t-1}^t y(r)dr &= \bar{A}^{-1}e^{\bar{A}}\bar{A}\int_{t-2}^{t-1} y(r)dr + \bar{A}^{-1}e^{\bar{A}}\bar{b} + \bar{A}^{-1}e^{\bar{A}}\int_{t-2}^{t-1}\bar{\zeta}(dr) \\ &\quad + \bar{A}^{-1}\left[\int_{t-1}^t e^{(t-r)\bar{A}}\bar{\zeta}(dr) - \int_{t-2}^{t-1} e^{(t-1-r)\bar{A}}\bar{\zeta}(dr)\right] \\ &\quad - \bar{A}^{-1}\bar{b} - \bar{A}^{-1}\int_{t-1}^t \bar{\zeta}(dr),\end{aligned} \tag{73}$$

from which, using $\bar{A}^{-1}e^{\bar{A}}\bar{A} = e^{\bar{A}}$ and $\bar{A}^{-1}e^{\bar{A}} = e^{\bar{A}}\bar{A}^{-1}$, we obtain

$$\begin{aligned}\int_{t-1}^t y(r)dr &= e^{\bar{A}}\int_{t-2}^{t-1} y(r)dr + c + \bar{A}^{-1}e^{\bar{A}}\int_{t-2}^{t-1}\bar{\zeta}(dr) \\ &\quad + \bar{A}^{-1}\left[\int_{t-1}^t e^{(t-r)\bar{A}}\bar{\zeta}(dr) - \int_{t-2}^{t-1} e^{(t-1-r)\bar{A}}\bar{\zeta}(dr)\right] \\ &\quad - \bar{A}^{-1}\int_{t-1}^t \bar{\zeta}(dr),\end{aligned} \tag{74}$$

where $c' = [c_1 c_2]$, c_1 and c_2 being defined as in the proof of Theorem 2.

The system (74) can be written as a pair of simultaneous difference equations in the vectors $\int_{t-1}^{t} x(r)dr$ and $\int_{t-1}^{t} Dx(r)dr$:

$$\int_{t-1}^{t} x(r)dr = [e^{\bar{A}}]_{11}\int_{t-2}^{t-1} x(r)dr + [e^{\bar{A}}]_{12}\int_{t-2}^{t-1} Dx(r)dr + c_1 \tag{75}$$
$$- A_2^{-1}\int_{t-1}^{t}(A_1[e^{(t-r)\bar{A}}]_{12} - [e^{(t-r)\bar{A}}]_{22} + I)\zeta(dr)$$
$$+ A_2^{-1}\int_{t-2}^{t-1}(A_1[e^{(t-1-r)\bar{A}}]_{12} - [e^{(t-1-r)\bar{A}}]_{22}$$
$$- A_1[e^{\bar{A}}]_{12} + [e^{\bar{A}}]_{22})\zeta(dr),$$

$$\int_{t-1}^{t} Dx(r)dr = [e^{\bar{A}}]_{21}\int_{t-2}^{t-1} x(r)dr + [e^{\bar{A}}]_{22}\int_{t-2}^{t-1} Dx(r)dr + c_2 \tag{76}$$
$$+ \int_{t-1}^{t}[e^{(t-r)\bar{A}}]_{12}\zeta(dr)$$
$$- \int_{t-2}^{t-1}([e^{(t-1-r)\bar{A}}]_{12} - [e^{\bar{A}}]_{12})\zeta(dr).$$

Equations (75) and (76) are identical with the equations that would be obtained by integrating (32) and (33) over the interval $[t-1, t]$, except that the disturbance vector in each equation is in the form of a single integral with respect to the random measure ζ. The operations that were applied to (32) and (33) in the proof of Theorem 2 to obtain (25) can now be applied to (75) and (76) to obtain (65) with η_t given by

$$\eta_t = -\int_{t-1}^{t}\{A_2^{-1}Q(t-r)\}\zeta(dr) \tag{77}$$
$$+ \int_{t-2}^{t-1}\{A_2^{-1}S(t-1-r) + [e^{\bar{A}}]_{12}[e^{(t-1-r)\bar{A}}]_{12}$$
$$+ WQ(t-1-r)\}\zeta(dr)$$
$$- \int_{t-3}^{t-2}\{[e^{\bar{A}}]_{12}([e^{(t-2-r)\bar{A}}]_{12} - [e^{\bar{A}}]_{12}) + WS(t-2-r)\}\zeta(dr).$$

The formulae (66), (67), and (68) can be obtained from (77) by using the multidimensional generalizations of (Rosanov 1967, p. 6, equations (2.11) and (2.14)) as in the proof of Theorem 2. The conditions $E(\eta_t) = E(\eta_t y_1') = E(\eta_t y_2') = 0$ $(t = 3, \ldots, T)$ and $E(\eta_s \eta_t') = 0$ $(|s - t| > 2;\ s, t = 3, \ldots, T)$ obviously follow from (77) and Assumptions 1 and 3. Q.E.D.

The exact discrete model obtained in Theorem 3 has a representation as a vector autoregressive moving average model:

$$x_t - F_1(\theta)x_{t-1} - F_2(\theta)x_{t-2} - g(\theta) = \varepsilon_t + G_1(\theta,\mu)\varepsilon_{t-1} + G_2(\theta,\mu)\varepsilon_{t-2}, \tag{78}$$

$$E(\varepsilon_t) = 0, \qquad E(\varepsilon_t\varepsilon_t') = K(\theta,\mu), \qquad E(\varepsilon_s\varepsilon_t') = 0 \qquad (s \neq t).$$

The model (78) differs from the model (41), satisfied by observations at points of time, in that the moving average component is of the second order. Moreover, although the elements of $F_1(\theta)$, $F_2(\theta)$, and $g(\theta)$ are identical functions of θ in the two models, the elements of $K(\theta,\mu)$ are different functions of (θ,μ) and, of course, the elements of $G_1(\theta,\mu)$ in (78) are different from the elements of $G(\theta,\mu)$ in (41). Like (41), the model (78) is not a very convenient basis for obtaining the Gaussian estimate of (θ,μ), because of the difficulty of solving for $G_1(\theta,\mu)$, $G_2(\theta,\mu)$, and $K(\theta,\mu)$. For estimation purposes it is easier to work directly from the original form of the exact discrete model given by Theorem 3.

The model given by Theorem 3 can be used as a basis for obtaining the exact Gaussian estimate of (θ,μ) under the assumption that y_1 and y_2 are arbitrary random vectors satisfying Assumption 3 or, alternatively, under the assumption of stationarity. Since, in this section, we assume that neither y_1 nor y_2 is observable, we must, under the former assumption, treat the realizations of these vectors as parameter vectors to be estimated. For this purpose, we can obtain x_1 and x_2 as functions of y_1, y_2, and disturbance vectors whose properties can be derived by the methods used in the proof of Theorem 3. These functions can be adjoined to (65) to provide a basis for the estimation of $(\theta, \mu, y_1', y_2')$.

Under the assumptions of stationarity the exact Gaussian estimate can be computed by the procedure described in Section 4 with the modification that the formula (47) is replaced by

$$\begin{aligned} V_r(\theta,\mu) = {} & \sum_{j=0}^{\infty} B_j(\theta)\Omega_0(\theta,\mu)B_{j+r}'(\theta) + \sum_{j=0}^{\infty} B_j(\theta)\Omega_1(\theta,\mu)B_{j+r+1}'(\theta) \qquad (79) \\ & + \sum_{j=0}^{\infty} B_{j+1}(\theta)\Omega_1'(\theta,\mu)B_{j+r}'(\theta) \\ & + \sum_{j=0}^{\infty} B_j(\theta)\Omega_2(\theta,\mu)B_{j+r+2}'(\theta) \\ & + \sum_{j=0}^{\infty} B_{j+2}(\theta)\Omega_2'(\theta,\mu)B_{j+r}'(\theta) \qquad (r = 0, \ldots, T-1), \end{aligned}$$

where, now, $V_r(\theta,\mu) = E[x_t - m(\theta)][x_{t+r} - m(\theta)]'$. We can also obtain estimates by minimizing either of the frequency domain approximations $\bar{L}(\theta,\mu)$ and $\tilde{L}(\theta,\mu)$. These are given by (56) and (57) where $I(\lambda)$ is now obtained from a discrete Fourier transformation of the integral observations $x_1, \ldots, x_T$, and f is the spectral density of the n-dimensional stationary process $\{x_t\ (t = 0, \pm 1, \pm 2, \ldots)\}$; so that, instead of (58), we have

$$f(\lambda; \theta, \mu) = \frac{1}{2\pi}[I - e^{-i\lambda}F_1(\theta) - e^{-2i\lambda}F_2(\theta)]^{-1} \times [e^{-2i\lambda}\Omega_2(\theta, \mu) + e^{i\lambda}\Omega_1(\theta, \mu) + \Omega_0(\theta, \mu) + e^{i\lambda}\Omega_1'(\theta, \mu) + e^{2i\lambda}\Omega_2'(\theta, \mu)][I - e^{-i\lambda}F_1(\theta) - e^{-2i\lambda}F_2(\theta)]^{*-1}. \tag{80}$$

The methods developed in this section can, like those of Section 4, be applied to a system of order greater than 2. The exact discrete model satisfied by a sequence of integral observations generated by a continuous time model of order k is of the form

$$x_t = F_1(\theta)x_{t-1} + F_2(\theta)x_{t-2} + \ldots + F_k(\theta)x_{t-k} + g(\theta) + \eta_t, \tag{81}$$

where

$$E(\eta_t) = 0,$$
$$E(\eta_t\eta_{t-r}') = \Omega_r(\theta, \mu) \qquad (r = 0, \ldots, k),$$
$$E(\eta_t\eta_{t-r}') = 0 \qquad (r > k).$$

This model has a representation as a vector autoregressive moving average model:

$$x_t - F_1(\theta)x_{t-1} - \ldots - F_k(\theta)x_{t-k} - g(\theta) = \varepsilon_t + G_1(\theta, \mu)\varepsilon_{t-1} + \ldots + G_k(\theta, \mu)\varepsilon_{t-k}, \tag{82}$$

$$E(\varepsilon_t) = 0, \qquad E(\varepsilon_t\varepsilon_t') = K(\theta, \mu), \qquad E(\varepsilon_s\varepsilon_t') = 0 \qquad (s \neq t).$$

The model (82) differs from the model (62), satisfied by the observations at points of time, in that the moving average is of order k rather than $k - 1$. Moreover, although the corresponding elements of $F_1(\theta), \ldots, F_k(\theta)$, and $g(\theta)$ are identical functions of θ in the two models, the corresponding elements of $G_1(\theta, \mu), \ldots, G_k(\theta, \mu)$, and $K(\theta, \mu)$ are different functions of (θ, μ). Again it is convenient for estimation purposes, to work directly from the exact discrete model in the original form (81).

6. Estimation from a Mixed Sample

We shall deal, finally, with the problem of estimation from a mixed sample, in which some variables are measured at points of time and others as integrals. In applied econometric work we shall often find that all except two or three variables are measured as integrals, being either genuine flow variables like output and consumption or variables, such as price indices, which are reported as averages over the observation period. In a typical macroeconomic model, for example, the only variables that are not measured as integrals are the stock of fixed capital, the level of inventories, and the money supply. The simplest procedure, in this case, is to approximate the integrals of these stock variables, using some interpolation formula, and then use the exact discrete model derived in Section 5. Alternatively, we can avoid the asymptotic bias resulting from such approximations by using a somewhat more complicated exact discrete model satisfied by the mixed data. Since the manipulations involved in obtaining this model are similar to those used in the proofs of Theorems 2 and 3 and the resulting formulae are rather complicated, we shall give only a brief outline of their derivation.

Using the superscript s to denote the vector of variables measured at points of time (stock variables) and the superscript f to denote the vector of variables measured as integrals (flow variables), the system (70) can be written in the partitioned form

$$\begin{bmatrix} x^s(t) - x^s(t-1) \\ x^f(t) - x^f(t-1) \\ Dx^s(t) - Dx^s(t-1) \\ Dx^f(t) - Dx^f(t-1) \end{bmatrix} = \begin{bmatrix} 0 & 0 & I & 0 \\ 0 & 0 & 0 & I \\ A_2^{ss} & A_2^{sf} & A_1^{ss} & A_1^{sf} \\ A_2^{fs} & A_2^{ff} & A_1^{fs} & A_1^{ff} \end{bmatrix} \begin{bmatrix} \int_{t-1}^{t} x^s(r)dr \\ \int_{t-1}^{t} x^f(r)dr \\ \int_{t-1}^{t} Dx^s(r)dr \\ \int_{t-1}^{t} Dx^f(r)dr \end{bmatrix} + \begin{bmatrix} 0 \\ 0 \\ b^s \\ b^f \end{bmatrix} + \begin{bmatrix} 0 \\ 0 \\ \int_{t-1}^{t} \zeta^s(dr) \\ \int_{t-1}^{t} \zeta^f(dr) \end{bmatrix} \tag{83}$$

from which we obtain

$$\begin{bmatrix} \int_{t-1}^{t} x^s(r)dr \\ \int_{t-1}^{t} Dx^s(r)dr \end{bmatrix} = \begin{bmatrix} 0 & I \\ A_2^{ss} & A_1^{ss} \end{bmatrix}^{-1} \begin{bmatrix} x^s(t) - x^s(t-1) \\ Dx^s(t) - Dx^s(t-1) \end{bmatrix}$$

$$- \begin{bmatrix} 0 & I \\ A_2^{ss} & A_1^{ss} \end{bmatrix}^{-1} \begin{bmatrix} 0 & 0 \\ A_2^{sf} & A_1^{sf} \end{bmatrix} \begin{bmatrix} \int_{t-1}^{t} x^f(r)dr \\ \int_{t-1}^{t} Dx^f(r)dr \end{bmatrix} \tag{84}$$

$$- \begin{bmatrix} 0 & I \\ A_2^{ss} & A_1^{ss} \end{bmatrix}^{-1} \begin{bmatrix} 0 \\ b^s \end{bmatrix} - \begin{bmatrix} 0 & I \\ A_2^{ss} & A_1^{ss} \end{bmatrix}^{-1} \begin{bmatrix} 0 \\ \int_{t-1}^{t} \zeta^s(dr) \end{bmatrix}.$$

If we now partition (75) and (76) into vectors relating to stock and flow variables and combine these equations with (84), we obtain a system of first order difference equations in the four vectors $x^s(t) - x^s(t-1)$, $Dx^s(t) - Dx^s(t-1)$, $\int_{t-1}^{t} x^f(r)dr$, and $\int_{t-1}^{t} Dx^f(r)dr$, with a disturbance vector in the form of the integral of a certain matrix function, over the interval $[t-2, t]$, with respect to the n-dimensional random measure ζ. By a similar argument to that used in the proofs of Theorems 2 and 3 we can then derive from this an exact discrete model of the form

$$\begin{bmatrix} x^s(t) - x^s(t-1) \\ \int_{t-1}^{t} x^f(r)dr \end{bmatrix} = \begin{bmatrix} F_1^{ss}(\theta) & F_1^{sf}(\theta) \\ F_1^{fs}(\theta) & F_1^{ff}(\theta) \end{bmatrix} \begin{bmatrix} x^s(t-1) - x^s(t-2) \\ \int_{t-2}^{t-1} x^f(r)dr \end{bmatrix}$$

$$+ \begin{bmatrix} F_2^{ss}(\theta) & F_2^{sf}(\theta) \\ F_2^{fs}(\theta) & F_2^{ff}(\theta) \end{bmatrix} \begin{bmatrix} x^s(t-2) - x^s(t-3) \\ \int_{t-3}^{t-2} x^f(r)dr \end{bmatrix}$$

$$+ \begin{bmatrix} g^s(\theta) \\ g^f(\theta) \end{bmatrix} + \begin{bmatrix} \eta_t^s \\ \eta_t^f \end{bmatrix}, \tag{85}$$

$$E(\eta_t^s) = E(\eta_t^f) = 0,$$

$$E\begin{bmatrix} [\eta_t^s][\eta_{t-r}^s]' & [\eta_t^s][\eta_{t-r}^f]' \\ [\eta_t^f][\eta_{t-r}^s]' & [\eta_t^f][\eta_{t-r}^f]' \end{bmatrix} = \begin{bmatrix} \Omega_r^{ss}(\theta, \mu) & \Omega_r^{sf}(\theta, \mu) \\ \Omega_r^{fs}(\theta, \mu) & \Omega_r^{ff}(\theta, \mu) \end{bmatrix} \quad (r = 0, 1, 2),$$

$$E\begin{bmatrix} [\eta_t^s][\eta_{t-r}^s]' & [\eta_t^s][\eta_{t-r}^f]' \\ [\eta_t^f][\eta_{t-r}^s]' & [\eta_t^f][\eta_{t-r}^f]' \end{bmatrix} = \begin{bmatrix} 0 & 0 \\ 0 & 0 \end{bmatrix} \quad (r = 3, 4, \ldots).$$

This model can be used as a basis for the estimation of (θ, μ) in the same way as the models derived in Sections 4 and 5. Moreover, a similar model can be derived for systems of order greater than two.

7. Asymptotic Properties of the Estimators

Before discussing the asymptotic properties of the estimators we shall deal briefly with the more fundamental problem of identification. For, although the procedures described in Sections 4 to 6 will invariably lead to some set of numbers which could be regarded as estimates, these will be rather meaningless if the parameters are not identifiable. The structural parameters are said to be identifiable if they can be uniquely inferred from the discrete spectrum of the observations. It is important to notice that the existence of a one to one correspondence between the set of p-dimensional vectors θ and the set of matrices that can be written in the form $\bar{A}(\theta)$ does not ensure that the true parameter vector θ^0 is identifiable. For it is possible to have $e^{\bar{A}(\theta)} = e^{\bar{A}(\theta^0)}$ when $\bar{A}(\theta) \neq \bar{A}(\theta^0)$ if the complex eigenvalues of $\bar{A}(\theta)$ and $\bar{A}(\theta^0)$ differ by integer multiples of $2\pi i$. In that case θ^0 will not be identifiable because we are unable to distinguish between structures generating oscillations whose frequencies differ by integer multiples of the observation period. This is known as the aliasing phenomenon.

The problem of identification in a first order continuous time model was discussed by P. C. B. Phillips (1972, 1973) who, in the latter article, derives a sufficient condition for identification through Cowles Commission type constraints. Work on this problem has been advanced considerably further by Hansen and Sargent (1979) who show how identification can be achieved through the more complicated nonlinear cross equation restrictions occurring in models with rational expectations. But Robinson (1980*a* and *b*) finds evidence that, even when the continuous time structure is globally identifiable, the discrete spectrum may be relatively insensitive to large changes in the structural parameters, and in moderate samples a structure very different from the true one might be chosen. These considerations must be borne in mind before embarking on estimation.

A detailed study of the asymptotic properties of the estimators discussed in Sections 4 to 6 is beyond the scope of this paper. But certain conclusions can be reached from the existing literature on the estimation of the structural parameters of the general model (53). For this purpose we shall extend the model (53) by introducing a mean vector m and write it in the form

$$x(t) - m(\alpha) = \varepsilon_t + C_1(\alpha)\varepsilon_{t-1} + C_2(\alpha)\varepsilon_{t-2} + \ldots, \tag{86}$$

$$E(\varepsilon_t) = 0, \qquad E(\varepsilon_t\varepsilon_t') = K(\alpha), \qquad E(\varepsilon_s\varepsilon_t') = 0 \qquad (s \neq t),$$

in order to emphasize the dependence of m, K, and the C_i $(i = 1, 2, \ldots)$ on a single structural parameter vector α.

Each of the exact discrete models derived in Sections 4 to 6 can be expressed in the form (86) provided that the eigenvalues of $\bar{A}$ have negative real parts, in which case the autoregressive transformation from $\{x(t)(t = 0, \pm 1, \pm 2, \ldots)\}$ to $\{\eta_t(t = 0, \pm 1, \pm 2, \ldots)\}$ is invertible. For example, if we define the shift operator U by $Ux(t) = x(t-1)$, then from the model (41) we obtain

$$\begin{aligned} x(t) - m(\theta) &= [I - F_1(\theta)U - F_2(\theta)U^2]^{-1}[I + G(\theta, \mu)U]\varepsilon_t \\ &= \varepsilon_t + C_1(\theta, \mu)\varepsilon_{t-1} + C_2(\theta, \mu)\varepsilon_{t-2} + \ldots, \end{aligned}$$

$$E(\varepsilon_t) = 0, \quad E(\varepsilon_t\varepsilon_t') = K(\theta, \mu), \quad E(\varepsilon_s\varepsilon_t') = 0 \quad (s \neq t), \tag{87}$$

where $G(\theta, \mu)$ and $K(\theta, \mu)$ are obtained from the solution of (39) and (40). Moreover, provided that the elements of $A_1(\theta)$, $A_2(\theta)$, and $b(\theta)$ are continuous functions of θ and the elements of $\Sigma(\mu)$ are continuous functions of μ, the elements of $K(\theta, \mu)$ and the $C_i(\theta, \mu)(i = 1, 2, \ldots)$ are continuous functions of (θ, μ) and the elements of $m(\theta)$ are continuous functions of θ.

The investigation of the sampling properties of the Gaussian estimator of the structural parameters of the model (86) started with the influential article of Mann and Wald 1943 who dealt with the special cases of an autoregressive system of finite order and a closed simultaneous equations system with Cowles Commission type restrictions. Notable contributions, dealing with progressively more general cases, have been made by Anderson 1959, Whittle 1962, Walker 1964, Hannan 1973, Dunsmuir and Hannan 1976, and Dunsmuir 1979. We shall state two theorems which follow from the results obtained in Dunsmuir and Hannan 1976 and Dunsmuir 1979. Although these theorems will be formally stated for the estimates defined in Section 4, similar theorems hold for the corresponding estimators using integrated and mixed data. Since it is assumed in Dunsmuir and Hannan 1976 and Dunsmuir 1979 that $m(\alpha) = 0$ we shall incorporate in Assumption 5 the assumption $b(\theta) = 0$, although this restriction can, undoubtedly, be relaxed under appropriate regularity conditions. Theorem 4 then follows from Dunsmuir and Hannan 1976 Theorem 3 and Corollary 1.

ASSUMPTION 5: (i) The true parameter vector (θ^0, μ^0) is identifiable and belongs to a closed bounded set Θ. (ii) The process $\{x(t)$ $(t = 0, \pm 1, \pm 2, \ldots)\}$ is strictly stationary and ergodic. (iii) The elements of $A_1(\theta), A_2(\theta), \ldots, A_k(\theta)$ are continuous functions of θ, and the elements of $\Sigma(\mu)$ are continuous functions of μ at all points of Θ; and $b(\theta) = 0$.

THEOREM 4: *Under Assumptions* 1 *to* 5 *each of the estimators* $(\hat{\theta}, \hat{\mu})$, $(\bar{\theta}, \bar{\mu})$ *and* $(\tilde{\theta}, \tilde{\mu})$, *defined in Section* 4 *converges almost surely to* (θ^0, μ^0) *as* $T \to \infty$.

For a central limit theorem we introduce the following assumptions.

ASSUMPTION 6: The increments in the n-dimensional process $\{\int_0^t \zeta(dr), -\infty < t < \infty\}$ are independent.

ASSUMPTION 7: (i) The elements of $A_1(\theta), A_2(\theta), \ldots, A_k(\theta)$ are twice continuously differentiable functions of θ, and the elements of $\Sigma(\mu)$ are twice continuously differentiable functions of μ at all points of Θ. (ii) $\det \Omega(\theta, \mu) \neq 0$, where

$$\Omega(\theta, \mu) = \left[\frac{1}{2\pi} \int_{-\pi}^{\pi} \operatorname{tr}\left[f_0^{-1}(\lambda) \frac{\partial f_0(\lambda)}{\partial \alpha_i} f_0^{-1}(\lambda) \frac{\partial f_0(\lambda)}{\partial \alpha_j}\right] d\lambda\right], \tag{88}$$

$f_0(\lambda)$ denotes the matrix $f(\lambda; \theta, \mu)$ evaluated at (θ^0, μ^0), $\partial f_0(\lambda)/\partial \alpha_i$ denotes the derivative of $f(\lambda; \theta, \mu)$ with respect to the ith element of (θ, μ) evaluated at (θ^0, μ^0), and $f(\lambda; \theta, \mu)$ is given by equation (63).

Under Assumption 6 the increments in $\int_0^t \zeta(dr)$ are normally distributed as can be seen by dividing the interval $[0, t]$ into m equal subintervals and applying the Lindberg–Levy central limit theorem (Cramér, 1946, p. 215) to the sum $\Sigma_{i=1}^m \int_{(i-1)t/m}^{it/m} \zeta(dr)$ when $m \to \infty$. Although this means that $\{\int_0^t \zeta(dr), -\infty < t < \infty\}$ is a Brownian motion process, it need not be separable (Doob 1953, p. 51 and p. 57, Theorem 2.4) and Assumption 6 does not necessarily imply, therefore, that the sample functions are almost all continuous. The normality of the increments in $\int_0^t \zeta(dr)$ implies that the ε_t $(t = 0, \pm 1, \pm 2, \ldots)$ are normal and hence, since they are mutually orthogonal, that they are independent. Dunsmuir (1979), who relies essentially on the theorem of Billingsley (1961), proves a central limit theorem under milder conditions on the innovations in (86) than independence.

Theorem 5 follows from Dunsmuir 1979, Theorem 2.1 and Corollary 2.2.

THEOREM 5: *Under Assumptions* 1 *to* 7 $\sqrt{T}(\bar{\theta} - \theta^0, \bar{\mu} - \mu^0)$ *and*

$\sqrt{T}(\tilde{\theta} - \theta^0, \tilde{\mu} - \mu^0)$ *each have asymptotic normal distributions with zero mean vectors and covariance matrix* $2\Omega^{-1}(\theta^0, \mu^0)$.

It is shown in Dunsmuir 1979 that

$$\text{plim}\left[\frac{\partial^2 \bar{L}(\alpha)}{\partial\alpha_i \partial\alpha_j}\right] = \Omega(\alpha),$$

where for our model $\alpha = (\theta, \mu)$ and $\Omega(\alpha)$ is defined by (88). Assuming, therefore, that

$$\text{plim}\left[\frac{\partial^2 \hat{L}(\alpha)}{\partial\alpha_i \partial\alpha_j}\right] = \text{plim}\left[\frac{\partial^2 \bar{L}(\alpha)}{\partial\alpha_i \partial\alpha_j}\right],$$

Theorem 5, together with the fact that under Assumption 6 $\hat{L}(\alpha)$ is the true likelihood function, implies that the estimators $(\bar{\theta}, \bar{\mu})$ and $(\tilde{\theta}, \tilde{\mu})$ are asymptotically efficient in the sense defined by Cramer (1946, p. 489).

8. Conclusions

We have been concerned in this paper with the efficient estimation of structural parameters in a continuous time dynamic model when the data are in discrete form, with some variables (stock variables) being observed at equispaced points of time and others (flow variables) as integrals between these points of time. The model that we have considered can be loosely described as a closed linear system of higher order stochastic differential equations, but is precisely defined by the system (4) and the associated assumptions concerning the random disturbance measure, the matrices of coefficients, and the boundary conditions. We have proved a more general existence and uniqueness theorem for the solution of such a system than has appeared in the literature and used this solution in the rigorous derivation of exact discrete models satisfied by the data. And we have shown how these exact discrete models can be used in the computation of both the exact Gaussian estimates and asymptotically equivalent estimates obtained by maximizing a frequency domain approximation to the Gaussian likelihood.

Although we have discussed the estimation problem entirely within the context of a closed model, this is not such a serious limitation as it might appear to be. For, if the data are in discrete form, we can hardly expect to obtain consistent estimates of the structural parameters of a continuous time model without some knowledge of the stochastic processes generating the continuous time paths of the exogenous variables. And in many cases the most realistic assumption will be that the exogenous variables are themselves generated by a system such as (1) with unrestricted coefficient matrices. In that case consistent estimates of the structural parameters of the open models can be obtained by applying the methods of this paper to an extended closed model obtained by adjoining the equations generating the exogenous variables. The main disadvantage of this procedure is its heavy computational cost. An alternative procedure would be to use the Fourier method developed by Robinson (1976*a, b* and *c*). It should be noted, however, that Robinson's method (which is not applicable to a closed model) will yield consistent estimates only if the spectral density of the process generating the exogenous variables is zero outside the frequency band $(-\pi, \pi)$. This condition would not be satisfied if the exogenous variables were really generated by a system such as (1). But Robinson's method could still be applied in this case, and it may be the best practicable procedure if there are many exogenous variables. Moreover it has the advantage that we need make only vague assumptions about the disturbance vector.

The arguments in favour of formulating econometric models in continuous time have been presented (Bergstrom 1967 chap. 1; 1976, chap. 1) in the context of the sort of model considered in this paper and by Sims (1971) in the context of a more general continuous time model. But it is worth emphasizing once again that, even for the purpose of obtaining the best predictions of the future discrete observations, there is an important gain from formulating the model in continuous time and estimating its structural parameters. For, if each of the variables in a dynamic system is adjusting continuously in response to the stimulus provided by a subset of the other variables in the system, any attempt to take account of the a priori restrictions implicit in this causal chain of dependencies without formulating the model in continuous time will result in asymptotically biased predictions of the future discrete observations.

References

Anderson, T. W. (1959), 'On asymptotic distribution of estimates of parameters of stochastic difference equations', *Annals of Mathematical Statistics,* **30**, 676–87.

Bergstrom, A. R. (1966), 'Non-recursive models as discrete approximations to systems of stochastic differential equations', *Econometrica,* **34**, 173–82.

—— (1967), *The Construction and Use of Economic Models.* London: English Universities Press.

—— (ed) (1976), *Statistical Inference in Continuous Time Economic Models.* Amsterdam: North-Holland.

—— (1978), 'Monetary policy in a model of the United Kingdom', in *Stability and Inflation,* ed. by A. R. Bergstrom, A. J. L. Catt, M. H. Peston, and B. D. J. Silverstone (New York; John Wiley & Sons).

—— and C. R. Wymer (1976), 'A model of disequilibrium neoclassical growth and its application to the United Kingdom', in *Statistical Inference in Continuous Time Economic Models,* ed. by A. R. Bergstrom (Amsterdam; North-Holland).

BILLINGSLEY, P. (1961), 'The Lindberg–Levy theorem for Martingales', *Proceedings of the American Mathematical Association,* **12**, 788–92.

CODDINGTON, E. A., and N. LEVINSON (1955), *Ordinary Differential Equations* (New York, McGraw-Hill).

CRAMER, H. (1946), *Mathematical Methods of Statistics* (Princeton: Princeton University Press).

DOOB, J. L. (1953), *Stochastic Processes* (New York, John Wiley & Sons).

DUNSMUIR, W. (1979), 'A central limit theorem for parameter estimation in stationary vector time series and its application to models for a signal observed with noise', *Annals of Statistics,* **7**, 490–506.

—— and E. J. HANNAN (1976) 'Vector linear time series models', *Advances in Applied Probability*, **8**, 339–64.

DURBIN, J. (1961), 'Efficient fitting of linear models for continuous stationary time series from discrete data', *Bulletin of the International Statistical Institute,* **38**, 273–82.

EDWARDS, D. A., and J. E. MOYAL (1955), 'Stochastic differential equations', *Proceedings of the Cambridge Philosophical Society,* **51**, 663–76.

HALMOS, P. R. (1958), *Finite Dimensional Vector Spaces* (Princeton, Van Nostrand).

HANNAN, E. J. (1973), 'The asymptotic theory of linear time series models', *Journal of Applied Probability,* **10**, 130–45.

HANSEN, L. P., and T. J. SARGENT (1979), 'Rational expectations models and the aliasing phenomenon', Federal Reserve Bank of Minneapolis Working Paper.

ITO, K. (1946), 'On a stochastic integral equation', *Proceedings of the Japanese Academy,* **1**, 32–5.

—— (1951), 'On Stochastic differential equations', *Memoir of the American Mathematical Society,* **4**, 51.

JONSON, P. D., E. R. MOSES, and C. R. WYMER (1977) 'The RBA76 model of the Australian economy', in *Conference in Applied Economic Research,* Reserve Bank of Australia.

KNIGHT, M. D., and C. R. WYMER (1978), 'A macroeconomic model of the United Kingdom', *IMF Staff Papers,* **25**, 742–78.

MALINVAUD, E. (1970), *Statistical Methods of Econometrics* (Amsterdam, North-Holland).

MANN, H. B., and A. WALD (1943), 'On the statistical treatment of linear stochastic difference equations', *Econometrica,* **11**, 173–220.

PHILLIPS, A. W. (1959), 'The estimation of parameters in systems of stochastic differential equations', *Biometrica,* **46**, 67–76.

PHILLIPS, G. M., and P. J. TAYLOR (1973), *Theory and Applications of Numerical Analysis* (London, Academic Press).

PHILLIPS, P. C. B. (1972), 'The structural estimation of a stochastic differential equation system', *Econometrica,* **40**, 1021–41.

—— (1973), 'The problem of identification in finite parameter continuous time models', *Journal of Econometrics,* **1**, 351–62.

—— (1978), 'The treatment of flow data in the estimation of continuous time systems', in *Stability and Inflation,* ed. by A. R. Bergstrom, A. J. L. Catt, M. H.

Peston, and B. D. J. Silverstone (New York, John Wiley & Sons).
RICHARD, D. M. (1978), 'A dynamic model of the world copper industry', *IMF Staff Papers,* **25**, 779–833.
ROBINSON, P. M. (1976*a*), 'Fourier estimation of continuous time models', in *Statistical Inference in Continuous Time Economic Models,* ed. by A. R. Bergstrom (Amsterdam, North-Holland).
—— (1976*b*) 'The estimation of linear differential equations with constant coefficients', *Econometrica,* **44**, 751–64.
—— (1976*c*) 'Instrumental variable estimation of differential equations', *Econometrica,* **44**, 765–76.
—— (1977) 'The construction and estimation of continuous time models and discrete approximations in econometrics', *Journal of Econometrics,* **6**, 173–98.
—— (1980*a*), 'Continuous model fitting from discrete data', in *Directions in Time Series,* ed. by D. R. Brillinger and G. C. Tiao (East Lansing, Michigan, Institute of Mathematical Statistics).
—— (1980*b*) 'The efficient estimation of a rational spectral density', *Signal Processing: Theories and Applications,* ed. by M. Kunt and F. de Coulon (Amsterdam, North-Holland).
ROZANOV, Y. A. (1967), *Stationary Random Processes* (San Francisco, Holden-Day).
SARGAN, J. D. (1976), 'Some discrete approximations to continuous time stochastic models', in *Statistical Inference in Continuous Time Economic Models,* ed. by A. R. Bergstrom (Amsterdam, North-Holland).
SIMS, C. A. (1971), 'Discrete approximations to continuous time distributed lag models in econometrics', *Econometrica,* **39**, 545–63.
TITCHMARSH, E. C. (1939) *The Theory of Functions* (London, Oxford University Press).
WALKER, A. M. (1964), 'Asymptotic properties of least-squares estimates of parameters of the spectrum of a stationary non-deterministic time-series', *Journal of the Australian Mathematical Society,* **4**, 363–84.
WHITTLE, P. (1951), *Hypothesis Testing in Time Series Analysis* (Stockholm, Almqvist and Wicksell).
—— (1953), 'The analysis of multiple stationary time series', *Journal of the Royal Statistical Society,* Series B, **15**, 125–39.
—— (1962), 'Gaussian estimation in stationary time series', *Bulletin of the International Statistical Institute,* **39**, 105–29.
WOLD, H. (1938), *A Study in the Analysis of Stationary Time Series* (Stockholm, Almqvist and Wicksell).
WYMER, C. R. (1972), 'Econometric estimation of stochastic differential equation systems', *Econometrica,* **40**, 565–77.
—— (1973), 'A continuous disequilibrium adjustment model of the United Kingdom financial market', in *Econometric Studies of Macro and Monetary Relations,* ed. by A. A. Powell and R. A. Williams (Amsterdam, North-Holland).
ZASUKHIN, V. (1941), 'On the theory of multidimensional stationary processes', *Comptes Rendus (Daklady) de l'Academie des Sciences de l'URSS,* **23**, 435–7.

5

The Estimation of Parameters in Nonstationary Higher-Order Continuous-Time Dynamic Models

1. Introduction

During the decade following the development of the first continuous-time macroeconometric model by Bergstrom and Wymer (1976) [Ch. 10 of this volume] there has been a rapid growth in the use of such models. Continuous-time models have now been developed for most of the leading industrial countries, and their construction and application have been described by various authors in a series of recent contributions.[1] Moreover, some of these models are being regularly revised and used by leading research and forecasting groups (for example, the Reserve Bank of Australia (Jonson, Moses, and Wymer 1977), the World Bank's Division of Global Modelling and Projections (Armington and Wolford 1983), and Wharton Econometric Forecasting Associates (Armington and Wolford 1984)).

Most of this applied work has been very successful as judged by the postsample predictive performance of the models and other criteria. Nevertheless there is scope for further improvement through the use of more sophisticated estimation methods which take advantage of recent developments in computing technology and the great reduction in computing costs that has occurred during the last few years. Nearly all of the continuous-time models that have been developed during the last decade have relied on the application of simultaneous equations estimation methods to approximate discrete models of the type introduced by Bergstrom (1966) [Ch. 3 of this volume], using the computer programs of Wymer (1978). In addition to having a small asymptotic bias[2], estimates obtained by using an approximate discrete model of this type have larger

[1] Armington and Wolford 1983, 1984; Bergstrom 1984*b*; Bergstrom and Wymer 1976; Gandolfo and Padoan 1984; Jonson, Moses, and Wymer 1977; Jonson, McKibbin, and Trevor 1982; Knight and Wymer 1978; Sassanpour and Sheen 1984; Tussio 1981.

[2] Bergstrom 1976; Bergstrom 1984*a*; Sargan 1976; Wymer 1972.

variances than those that could be obtained by taking account of the exact restrictions on the distribution of the discrete data implied by the continuous-time model. Indeed, an important Monte Carlo study by Phillips (1972) shows that the reduction in the variances of the parameter estimates achieved by using the exact discrete model rather than the simultaneous equations approximation is much more important than the elimination of the asymptotic bias.

Although Phillips deals only with the simplest case, a first-order system in which all variables are assumed to be observed at points of time (rather than as integrals), the use of the exact discrete model results in a more than 50 per cent reduction in the root-mean-square error of the estimate for most of the parameters in his model. An even greater increase in precision might be achieved by applying efficient estimation procedures to the exact discrete models derived from higher-order continuous-time dynamic models with mixed stock and flow data. The development of such procedures and the computational algorithm for their implementation is, therefore, of great practical importance and is the main concern of this paper.

There is already a considerable literature on the problem of obtaining asymptotically efficient estimates of the parameters of higher-order continuous-time dynamic models from discrete data. But nearly all of this work is based on the theory of stationary random processes and makes use of the spectral representation of such a process (see, in particular, the contributions of Durbin 1961, Robinson 1976*a, b, c*, and Hansen and Sargent 1981). Although it is realistic to assume that most statistical time series occurring in natural sciences such as meteorology are samples from a stationary random process, this assumption is seldom justified for time series used in econometric work. Indeed, the period covered by such time series is usually chosen so that it commences shortly after major structural change in the economy, such as occurs at the end of a war or at the time of some important change in government policy or economic institutions. For this reason the distribution of the initial observations in the sample may be very different from that of the later observations.

Another reason why the assumption of stationarity may be inappropriate in econometric work is that it excludes systems with unstable roots (i.e. roots with non-negative real parts). The assumption that the system has stable roots will not be required in this paper. But the possibility of unstable roots does raise important problems of statistical inference to which I shall refer later. In Bergstrom (1983) [Ch. 4 of this volume] I proved a general existence and uniqueness theorem for the solution of a higher-order continuous-time dynamic model and, on the basis of this theorem, derived the exact models satisfied by various types of discrete data (stocks, flows, and mixed stocks and flows) generated by the continuous-time model. It was pointed out that these discrete models could be used as a basis for the Gaussian estimation (i.e. estimation by maximizing the Gaussian likelihood, regardless of whether or not the innovations are Gaussian) of the structural parameters of the continuous-time model assuming either stationarity or an initial state vector about which nothing is known except that it is uncorrelated with the subsequent innovations. But a computational algorithm was derived for the former case only. This algorithm involves computing the covariance matrix V of the nT-dimensional vector x of observations (where n is the number of variables in the model and T is the number of observations of each variable) and using the Cholesky factorization of V in a linear transformation of x to a vector z in terms of which the likelihood function is obtained in a very simple form (Bergstrom 1983, Eq. (52)).

It was shown also that the exact discrete model provides a very convenient basis for estimation by maximizing a frequency-domain approximation to the Gaussian likelihood of the type introduced by Whittle (1951, 1953). The use of this type of approximation in estimating the parameters of continuous-time models was proposed by Robinson (1977). But it was shown in (Bergstrom 1983) that it is greatly facilitated by first obtaining the exact discrete model satisfied by the observations, since this yields a compact formula for the discrete spectral density (Bergstrom 1983, Eqs. (58), (63), (80)) as an explicit (matrix valued) function of the structural parameters of the continuous-time model. If the system is stable and very long-time series are available for estimation purposes we can expect to obtain, by this method, estimates which are approximately as good as those obtained by maximizing the exact Gaussian likelihood, even when the initial state vector is fixed. But with time series of the length normally available for econometric work, particularly macroeconomic modelling, estimates obtained by maximizing the exact Gaussian likelihood could be considerably better, especially if the initial observations are far from their steady-state means.

In this paper I shall derive an efficient algorithm for the computation of the exact Gaussian likelihood, assuming a fixed initial state vector, and discuss its application in estimating the parameters of the model. If the initial state vector is fixed then any algorithm which involves the computation of V (the covariance matrix of the observations) will be very costly in terms of both computer time and computer storage space. For, whereas under the assumption of stationarity V will be a block Toeplitz matrix, when the initial state vector is fixed its elements will, apart from the restriction of symmetry, be generally all distinct. In view of this fact we shall derive an algorithm which completely avoids the computation of V. This is achieved by using two successive linear transformations in order to get from the observed vector x to the vector ε occurring in the final form of the likelihood function (see Eq. (36) below). In the first we transform x into an nT-dimensional vector η whose covariance matrix Ω is much easier to work with than V. In addition to being almost a block Toeplitz matrix, Ω has the very convenient property that most of its elements are zeros. The Cholesky factorization of Ω is used in the second linear transformation by which we obtain ε from η.

One reason why the problem of estimating, from discrete data, the structural parameters of a continuous-time model with fixed initial state vector is a much more difficult one than the corresponding discrete-time problem is that in the former case most of the elements of the initial state vector are unobservable. Suppose, for example, that the continuous-time model is a kth-order system in the n-dimensional vector $x(t)$, that it is known to hold for $t \geqslant 0$, and that the variables $x_i(t) (i = 1, \ldots, n)$ are all flow varibles observable at unit intervals of time, in which case the sample comprises the vectors $x_t = \int_{t-1}^{t} x(r)dr$ $(t = 1, \ldots, T)$. Then standard discrete-time estimation procedures might suggest that we should treat $x_1, \ldots, x_k$ as fixed vectors. But this would be incorrect if we wish to make full use of the information in the sample. Instead we should treat as fixed the vectors $x(0)$, $Dx(0), \ldots, D^{k-1}x(0)$, where D is the mean-square differential operator. But none of the elements of these vectors is observable, and they must, therefore, be treated as parameters to be estimated along with the structural parameters of the model. If the model contains some stock variables then the corresponding elements of $x(0)$ will be observable, but the remaining elements of $x(0)$ and all elements of the vectors $Dx(0), \ldots, D^{k-1}x(0)$ must be treated as parameters.

In order to have a single model which is applicable under both the assumption of stationarity and the assumption of an arbitrary initial state vector, the vectors $x(0)$, $Dx(0), \ldots, D^{k-1}x(0)$ were formally treated in (Bergstrom 1983, Assumption 3) as random vectors uncorrelated with the subsequent innovations. But, if nothing is known about the distribution of these vectors, nothing is lost by treating them as nonrandom, and they will be treated as nonrandom in this paper.

The general methods that we shall develop will be applicable to a system of any order with mixed stock and flow variables. The methods are based on the exact discrete model satisfied by the data. The exact discrete models for a second-order system in which all variables are stock variables or all variables are flow variables are derived explicitly in Sections 4 and 5 of (Bergstrom 1983), which also include an outline of the procedure to be followed for systems of order greater than two. The procedure for deriving the exact discrete model in the case in which the model includes both stock and flow variables is outlined in Section 6 of (ibid.). In this paper we shall derive detailed results only for the case of a second-order system in which all the variables are flow variables. But the way in which the results could be extended to more general systems using the methods outlined in (ibid.) will be fairly obvious.

The only other published algorithm for computing the exact Gaussian likelihood for parameters of a higher-order continuous-time dynamic model with a fixed initial state vector and flow data is the Kalman filter algorithm developed by Harvey and Stock (1985). Their algorithm is based on that of Jones (1981), who assumes that all variables are observable at points of time, thus avoiding the complications associated with flow data. The Kalman filter algorithm is applicable in more general circumstances than we shall assume in this paper, for example, when there are missing observations, errors of measurement, and measurements at unequally spaced intervals. But it is strictly comparable with our algorithm in the standard case in which all variables are measurable without error at equally spaced intervals. We shall show that, in this case, our algorithm has computational advantages. In particular, we shall show that it is computationally less costly than the Kalman filter algorithm provided that the sample is sufficiently large to justify the fixed setup cost of obtaining the exact discrete model. Moreover, it should be emphasized that there are many other reasons why practitioners may wish to undertake the costs of finding the exact discrete model and thereby find a comparative advantage in using our algorithm rather than that of Harvey and Stock.

In the next section we shall derive a formula for the exact Gaussian likelihood function and present an algorithm for computing its value for given values of the structural parameters and initial state vectors. In Section 3 we shall propose an iterative estimation procedure in which we alternate between estimating the structural parameters and the initial state vector, taking advantage of an explicit formula for the conditional estimator of the latter. We shall show that the estimates obtained by this procedure converge to the exact Gaussian estimates provided that the likelihood function has a unique local maximum.

The large-sample properties of these estimates will depend on whether or not the model has unstable roots. If there are no unstable roots the model is asymptotically stationary and the classical theory of inference applies. When there are unstable roots much of the classical theory breaks down, as is shown in the parallel literature on statistical inference in unstable discrete-time models (see Basawa and Brockwell 1984; Phillips 1985 and references cited therein). The investigation of the sampling properties of estimates of the parameters of unstable continuous-time models is one of the most important outstanding problems in continuous-time estimation. But such an investigation is beyond the scope of this paper.

2. Computing the Exact Gaussian Likelihood

We shall consider the system

$$d[Dx(t)] = [A_1(\theta)Dx(t) + A_2(\theta)x(t) + b(\theta)]dt + \zeta(dt), \quad t \geq 0, \quad (1)$$

$$x(0) = y_1, \quad Dx(0) = y_2, \tag{2}$$

where $\{x(t), t > 0\}$ is a real n-dimensional continuous-time random process, θ is a p-dimensional vector of unknown structural parameters $[p < n(1 + 2n)]$, A_1 and A_2 are $n \times n$ matrices whose elements are known functions of θ, b is an $n \times 1$ vector whose elements are known functions of θ, y_1 and y_2 are unknown nonrandom $n \times 1$ vectors, D is the mean-square differential operator, and $\zeta(dt)$ is a white noise disturbance vector. With regard to $\zeta(dt)$ we shall assume, more precisely, that $\zeta = [\zeta_1, \ldots, \zeta_n]'$ is a vector of random measures defined on all subsets of the half-line $0 < t < \infty$ with finite Lebesgue measure such that $E[\zeta(dt)] = 0$, $E[\zeta(dt)\zeta'(dt)] = dt\Sigma$, where Σ is an unknown positive definite matrix, and $E[\zeta_i(\Delta_1)\zeta_j(\Delta_2)] = 0$, $(i, j = 1, \ldots, n)$, for any disjoint sets Δ_1 and Δ_2 on the half-line $0 < t < \infty$. (See Bergstrom 1984*a*, p. 1157 for a discussion of random measures and their application to continuous-time stochastic models). We shall assume also that A_2 is nonsingular so that if ζ were a zero vector and the system were stable (an assumption which we shall not require) $x(t)$ would converge to $A_2^{-1}b$ as $t \to \infty$.

As in [2] we shall interpret (1) as meaning that $x(t)$ satisfies the stochastic integral equation

$$Dx(t) - D(0) = \int_0^t [A_1 Dx(r) + A_2 x(r) + b]dr + \int_0^t \zeta(dr) \tag{3}$$

for all $t > 0$, where $\int_0^t \zeta(dr) = \zeta([0, t])$ and the first integral on the right-hand side of (3) is defined in the wide sense (Rozanov 1967, p. 10).

We know (Bergstrom 1983, Theorem 1) that, under the above assumptions, the system (1) (i.e. the system (3)) has a solution $x(t)$, $t \geqslant 0$, subject to the boundary conditions (2) and that this solution is unique in the sense of (Bergstrom 1983, Eq. (21)). We now assume that the solution is observable only as a sequence of integrals (i.e., that all variables are flow variables). In particular, we assume that the sample comprises a set of T n-dimensional observations $x_1., \ldots, x_T$ of the form

$$x_t = \int_{t-1}^t x(r)dr, \; t = 1, \ldots, T. \tag{4}$$

Let the $n \times n$ matrices $[e^{\bar{A}}]_{ij} (i, j = 1, 2)$ be defined by

$$\begin{bmatrix} [e^{\bar{A}}]_{11} & [e^{\bar{A}}]_{12} \\ [e^{\bar{A}}]_{21} & [e^{\bar{A}}]_{22} \end{bmatrix} = \begin{bmatrix} I & 0 \\ 0 & I \end{bmatrix} + \begin{bmatrix} 0 & I \\ A_2 & A_1 \end{bmatrix} + \frac{1}{2}\begin{bmatrix} 0 & I \\ A_2 & A_1 \end{bmatrix}^2 + \frac{1}{3!}\begin{bmatrix} 0 & I \\ A_2 & A_1 \end{bmatrix}^3 + \ldots$$

$$= I + \bar{A} + \frac{1}{2}\bar{A}^2 + \frac{1}{3}\bar{A}^3 + \ldots. \tag{5}$$

In order to exclude certain degenerate cases (see ibid., p. 74 this volume) we shall assume that $[e^{\bar{A}}]_{12}$ is non singular. Then (ibid., Theorem 3) the sample vectors defined by (4) satisfy the system

$$
\begin{aligned}
x_t &= F_1 x_{t-1} + F_2 x_{t-2} + g + \eta_t, \quad t = 3, \ldots, T, \\
E(\eta_t) &= 0, \quad E(\eta_t \eta_t') = \Omega_0, \quad t = 3, \ldots, T, \\
E(\eta_t \eta_{t-1}') &= \Omega_1, \quad t = 4, \ldots, T, \\
E(\eta_t \eta_{t-2}') &= \Omega_2, \quad t = 5, \ldots, T, \\
E(\eta_s \eta_t') &= 0, \quad |s - t| > 2; s, t = 3, \ldots, T,
\end{aligned}
\tag{6}
$$

where F_1, F_2, g, Ω_0, Ω_1, and Ω_2 are defined by

$$F_1 = [e^{\bar{A}}]_{12}[e^{\bar{A}}]_{22}[e^{\bar{A}}]_{12}^{-1} + [e^{\bar{A}}]_{11}, \tag{7}$$

$$F_2 = [e^{\bar{A}}]_{12}[e^{\bar{A}}]_{21} - [e^{\bar{A}}]_{12}[e^{\bar{A}}]_{22}[e^{\bar{A}}]_{12}^{-1}[e^{\bar{A}}]_{11}, \tag{8}$$

$$g = -(I - F_1 - F_2)A_2^{-1}b, \tag{9}$$

$$
\begin{aligned}
\Omega_0 = & \int_0^1 [A_2^{-1}Q(r)]\sum[A_2^{-1}Q(r)]'dr \\
& + \int_0^1 \{A_2^{-1}S(r) + [e^{\bar{A}}]_{12}[e^{r\bar{A}}]_{12} \\
& + WQ(r)\}\sum\{A_2^{-1}S(r) + [e^{\bar{A}}]_{12}[e^{r\bar{A}}]_{12} + WQ(r)\}'dr \\
& + \int_0^1 \{WS(r) + [e^{\bar{A}}]_{12}([e^{r\bar{A}}]_{12} - [e^{\bar{A}}]_{12})\}\Sigma\{WS(r) \\
& + [e^{\bar{A}}]_{12}([e^{r\bar{A}}]_{12} - [e^{\bar{A}}]_{12})\}'dr,
\end{aligned}
\tag{10}
$$

$$
\begin{aligned}
\Omega_1 = & -\int_0^1 \{A_2^{-1}S(r) + [e^{\bar{A}}]_{12}[e^{r\bar{A}}]_{12} + WQ(r)\}\sum\{A_2^{-1}Q(r\}'dr \\
& -\int_0^1 \{WS(r) + [e^{\bar{A}}]_{12}([e^{r\bar{A}}]_{12} - [e^{\bar{A}}]_{12})\}\sum\{A_2^{-1}S(r) \\
& + [e^{\bar{A}}]_{12}[e^{r\bar{A}}]_{12} + WQ(r)\}'dr,
\end{aligned}
\tag{11}
$$

$$\Omega_2 = \int_0^1 \{WS(r) + [e^{\bar{A}}]_{12}([e^{r\bar{A}}]_{12} - [e^{\bar{A}}]_{12})\}\sum[A_2^{-1}Q(r)]'dr, \tag{12}$$

$$
\begin{aligned}
Q(r) &= A_1[e^{r\bar{A}}]_{12} - [e^{r\bar{A}}]_{22} + I, \\
S(r) &= A_1[e^{r\bar{A}}]_{12} - [e^{r\bar{A}}]_{22} - A_1[e^{\bar{A}}]_{12} + [e^{\bar{A}}]_{22}, \\
W &= [e^{\bar{A}}]_{12}[e^{\bar{A}}]_{22}[e^{\bar{A}}]_{12}^{-1}A_2^{-1}.
\end{aligned}
$$

We shall now derive two supplementry equations relating x_1 and x_2 to the initial state vectors and two unobservable random vectors η_1 and η_2.

From Eq. (3) we obtain

$$Dx(1) - Dx(0) = \int_0^1 [A_1 Dx(r) + A_2 x(r) + b]dr + \int_0^1 \zeta(dr)$$

$$= A_1[x(1) - x(0)] + A_2 x_1 + b + \int_0^1 \zeta(dr),$$

and hence

$$x_1 = A_2^{-1}[Dx(1) - Dx(0)] - A_2^{-1}A_1[x(1) - x(0)] - A_2^{-1}b - A_2^{-1}\int_0^1 \zeta(dr). \tag{13}$$

But from [Bergstrom 1983, Eqs. (32), (33)] we obtain

$$x(1) - x(0) = ([e^{\bar{A}}]_{11} - I)x(0) + [e^{\bar{A}}]_{12}Dx(0) + c_1 + \int_0^1 [e^{(1-r)\bar{A}}]_{12}\zeta(dr), \tag{14}$$

$$Dx(1) - Dx(0) = [e^{\bar{A}}]_{12}x(0) + ([e^{\bar{A}}]_{22} - I)Dx(0) + c_2 + \int_0^1 [e^{(1-r)\bar{A}}]_{22}\zeta(dr), \tag{15}$$

where

$$c_1 = [e^{\bar{A}}]_{11}A_2^{-1}b - A_2^{-1}b, \quad c_2 = [e^{\bar{A}}]_{21}A_2^{-1}b.$$

Substituting from (14) and (15) into (13) and using (2) we obtain

$$x_1 = G_{11}y_1 + G_{12}y_2 + g_1 + \eta_1, \tag{16}$$

where

$$G_{11} = A_2^{-1}[e^{\bar{A}}]_{21} - A_2^{-1}A_1([e^{\bar{A}}]_{11} - I), \tag{17}$$

$$G_{12} = A_2^{-1}([e^{\bar{A}}]_{22} - I) - A_2^{-1}A_1[e^{\bar{A}}]_{12}, \tag{18}$$

$$g_1 = A_2^{-1}c_2 - A_2^{-1}A_1c_1 - A_2^{-1}b, \tag{19}$$

$$\eta_1 = -A_2^{-1}\int_0^1 Q(1-r)\zeta(dr). \tag{20}$$

Next, using Bergstrom (1983, Eq. (75)) we obtain

$$x_2 = [e^{\bar{A}}]_{11}x_1 + [e^{\bar{A}}]_{12}[x(1) - x(0)] + c_1 - A_2^{-1}\int_1^2 (A_1[e^{(2-r)\bar{A}}]_{12} - [e^{(2-r)\bar{A}}]_{22} + I)\zeta(dr) + A_2^{-1}\int_0^1 (A_1[e^{(1-r)\bar{A}}]_{12} - [e^{(1-r)\bar{A}}]_{22} - A_1[e^{\bar{A}}]_{12} + [e^{\bar{A}}]_{22})\zeta(dr). \tag{21}$$

Substituting from (14) into (21) and using (2) we obtain

$$x_2 = [e^{\bar{A}}]_{11}x_1 + G_{21}y_1 + G_{22}y_2 + g_2 + \eta_2, \tag{22}$$

where

$$G_{21} = [e^{\tilde{A}}]_{12}([e^{\tilde{A}}]_{11} - I), \tag{23}$$

$$G_{22} = [e^{\tilde{A}}]_{12}^2, \tag{23}$$

$$g_2 = ([e^{\tilde{A}}]_{12} + I)c_1, \tag{25}$$

$$\eta_2 = -\int_1^2 A_2^{-1}Q(2-r)\zeta(dr) + \int_0^1 \{A_2^{-1}S(1-r) + [e^{\tilde{A}}]_{12}[e^{(1-r)\tilde{A}}]_{12}\}\zeta(dr). \tag{26}$$

The vectors η_t, $t = 3,\ldots, T$, in the system (6) are related to the innovation process in the continuous-time model by (see Bergstrom 1983, Eq. (77))

$$\eta_t = -\int_{t-1}^{t}[A_2^{-1}Q(t-r)]\zeta(dr) + \int_{t-2}^{t-1}\{A_2^{-1}S(t-1-r) + [e^{\tilde{A}}]_{12}[e^{(t-1-r)\tilde{A}}]_{12} + WQ(t-1-r)\}\zeta(dr) - \int_{t-3}^{t-2}\{[e^{\tilde{A}}]_{12}([e^{(t-2-r)\tilde{A}}] - [e^{\tilde{A}}]_{12}) + WS(t-2-r)\}\zeta(dr). \tag{27}$$

From (20), (26), and (27) we obtain

$$E(\eta_1\eta_1') = \Omega_{11},$$
$$E(\eta_2\eta_1') = \Omega_{21} = \Omega_{12}',$$
$$E(\eta_3\eta_1') = \Omega_2,$$
$$E(\eta_t\eta_1') = 0,\ t = 4,\ldots, T,$$
$$E(\eta_2\eta_2') = \Omega_{22},$$
$$E(\eta_3\eta_2') = \Omega_{32} = \Omega_{23}',$$
$$E(\eta_4\eta_2') = \Omega_2,$$
$$E(\eta_t\eta_2') = 0,\ t = 5,\ldots, T,$$

where Ω_2 is defined by (12) and Ω_{11}, Ω_{21}, Ω_{22}, and Ω_{32} by

$$\Omega_{11} = \int_0^1 [A_2^{-1}Q(r)]\sum[A_2^{-1}Q(r)]'dr, \tag{28}$$

$$\Omega_{21} = -\int_0^1 \{A_2^{-1}S(r) + [e^{\tilde{A}}]_{12}[e^{r\tilde{A}}]_{12}\}\sum[A_2^{-1}Q(r)]'dr, \tag{29}$$

$$\Omega_{22} = \int_0^1 [A_2^{-1}Q(r)]\sum[A_2^{-1}Q(r)]'dr + \int_0^1 \{A_2^{-1}S(r) + [e^{\tilde{A}}]_{12}[e^{r\tilde{A}}]_{12}\}\sum\{A_2^{-1}S(r) + [e^{\tilde{A}}]_{12}[e^{r\tilde{A}}]_{12}\}'dr, \tag{30}$$

$$\Omega_{32} = -\int_0^1 \{A_2^{-1}S(r) + [e^{\tilde{A}}]_{12}[e^{r\tilde{A}}]_{12} + WQ(r)\}\sum[A_2^{-1}Q(r)]'dr$$

$$-\int_0^1 \{WS(r) + [e^{\tilde{A}}]_{12}([e^{r\tilde{A}}]_{12} - [e^{\tilde{A}}]_{12})\}\sum\{A_2^{-1}S(r)$$

$$+ [e^{\tilde{A}}]_{12}[e^{r\tilde{A}}]_{12}\}dr. \tag{31}$$

The above formulas enable us to evaluate the covariance matrix $\Omega = E(\eta\eta')$ of the nT-dimensional vector $\eta' = [\eta_1', \eta_2', \eta_3', \ldots, \eta_T']$ for any given θ and Σ. Moreover, except for the leading $3n \times 3n$ submatrix, Ω has the form of a block Toeplitz matrix most of whose elements are zero, i.e.

$\Omega =$

$$\begin{bmatrix}
\Omega_{11} & \Omega_{12}' & \Omega_2' & 0 & 0 & 0 & 0 & \cdot & \cdot & \cdot & \cdot & 0 \\
\Omega_{21} & \Omega_{22} & \Omega_{23}' & \Omega_2' & 0 & 0 & 0 & \cdot & \cdot & \cdot & \cdot & 0 \\
\Omega_2 & \Omega_{32} & \Omega_0 & \Omega_1' & \Omega_2' & 0 & 0 & \cdot & \cdot & \cdot & \cdot & 0 \\
0 & \Omega_2 & \Omega_1 & \Omega_0 & \Omega_1' & \Omega_2' & 0 & \cdot & \cdot & \cdot & \cdot & 0 \\
\cdot & \cdot & \cdot & \cdot & \cdot & \cdot & \cdot & \cdot & \cdot & \cdot & \cdot & 0 \\
\cdot & \cdot & \cdot & \cdot & \cdot & \cdot & \cdot & \cdot & \Omega_0 & \Omega_1' & \Omega_2' & 0 \\
\cdot & \cdot & \cdot & \cdot & \cdot & \cdot & \cdot & \cdot & \Omega_1 & \Omega_0 & \Omega_1' & \Omega_2' \\
\cdot & \cdot & \cdot & \cdot & \cdot & \cdot & \cdot & & \Omega_2 & \Omega_1 & \Omega_0 & \Omega_1' \\
0 & 0 & 0 & 0 & 0 & 0 & 0 & & 0 & \Omega_2 & \Omega_1 & \Omega_0
\end{bmatrix}$$

Now let $h' = [g_1', g_2', g', \ldots, g']$, $x' = [x_1', \ldots, x_T']$, $y' = [y_1', y_2']$, and let the $nT \times nT$ matrix F and $nT \times 2T$ matrix G be defined by

$$F = \begin{bmatrix}
I & 0 & 0 & 0 & \cdot & \cdot & \cdot & 0 \\
-[e^{\tilde{A}}]_{11} & I & 0 & 0 & \cdot & \cdot & \cdot & 0 \\
-F_2 & -F_1 & I & 0 & \cdot & \cdot & \cdot & 0 \\
0 & -F_2 & -F_1 & I & \cdot & \cdot & \cdot & 0 \\
\cdot & \cdot & \cdot & \cdot & \cdot & \cdot & \cdot & \cdot \\
\cdot & \cdot & \cdot & \cdot & \cdot & \cdot & \cdot & \cdot \\
0 & 0 & 0 & 0 & \cdot & -F_2 & -F_1 & I
\end{bmatrix},$$

$$G = \begin{bmatrix}
G_{11} & G_{12} \\
G_{21} & G_{22} \\
0 & 0 \\
\vdots & \vdots \\
0 & 0
\end{bmatrix}.$$

Then the system (6), together with the supplementary equations (16) and (22), can be written in the compact form

$$Fx - Gy - h = \eta. \tag{32}$$

Moreover, since $|F| = 1$, the frequency function of x is $\phi(Fx - Gy - h)$, where $\phi(\eta)$ is the frequency function of η.

As in Bergstrom (1983) we shall now parameterize Σ by writing it as $\Sigma(\mu)$, where μ is a vector of parameters. If Σ is unrestricted then μ will have $\frac{1}{2}n(n+1)$ elements. But we could, for example, require Σ to be a diagonal matrix, in which case μ will have n elements, or a block diagonal matrix, in which case the number of elements will be between n and $\frac{1}{2}n(n+1)$. If we now regard y as a vector of supplementary parameters the complete vector of parameters is $[\theta, \mu, y']$.

We shall denote by $[\hat{\theta}, \hat{\mu}, \hat{y}']$ the vector that maximizes the exact Gaussian likelihood and refer to it as the exact Gaussian estimate of $[\theta, \mu, y']$, regardless of whether or not the innovations in the continuous-time model are assumed to be Gaussian. Letting $L(\theta, \mu, y')$ denote minus twice the logarithm of the Gaussian likelihood function, we have

$$\begin{aligned} L(\theta, \mu, y') &= \log|\Omega(\theta, \mu)| + \eta'\Omega^{-1}(\theta, \mu)\eta \\ &= \log|\Omega(\theta, \mu)| + (Fx - Gy - h)'\Omega^{-1}(\theta, \mu)(Fx - Gy - h). \quad (33) \end{aligned}$$

Then $(\hat{\theta}, \hat{\mu}, \hat{y}')$ is the vector that minimizes $L(\theta, \mu, y')$.

In the evaluation of L for a given parameter vector $[\theta, \mu, y']$ the inversion of Ω can be avoided by using the Cholesky factorization

$$\Omega = MM', \quad (34)$$

where M is a real lower triangular matrix with positive elements along the diagonal. All the elements of M above the diagonal and most of the elements below the diagonal will be zeros. Indeed, it is easy to see that the number of nonzero elements in M is approximately $\frac{5}{2}Tn^2$. Moreover, these elements can be computed recursively by using the first method described by Quandt (1984, p. 705). The number of multiplications required in the evaluation of any element will be between 1 and $3n$, depending on its position in M, with the average number of multiplications per element being approximately $\frac{3}{2}n$. The total number of multiplications required in the evaluation of M is, therefore, less than $4Tn^3$.

Now let $\varepsilon = [\varepsilon_1, \ldots, \varepsilon_{nT}]'$ be defined by the linear transformation $\varepsilon = M^{-1}\eta$.[3] Then one can compute the elements of ε recursively from the solution of

$$M\varepsilon = \eta, \quad (35)$$

taking advantage of the fact that no more than $3n$ elements in any row of M are nonzero. Finally, the value of L can be computed from

[3] In the original version of this paper, published in *Econometric Theory* (1985), the vector ε was denoted by z. The change in notation is to avoid confusion with the vector $z(t)$ of exogenous variables introduced in Chapter 6.

$$L = \sum_{i=1}^{nT} (\varepsilon_i^2 + 2 \log m_{ii}), \tag{36}$$

where m_{ii} is the ith diagonal element of M.

A complete list of steps for the computation of L for given values of the structural parameters and given initial state vector, starting with the continuous-time model is as follows.

i. Compute A_1, A_2, b, and Σ, using the specified forms of the functions defining their elements. Most of the elements of these matrices will be equal to zero or one of the structural parameters.
ii. Compute F_1, F_2, g; G_{11}, G_{12}, g_1, G_{21}, G_{22}, g_2 from Eqs. (7)–(9), (17)–(19), (23)–(25).
iii. Compute Ω_0, Ω_1, Ω_2; Ω_{11}, Ω_{21}, Ω_{22}, Ω_{32}, from Eqs. (10)–(12) and (28)–(31). The integrals in these equations can be evaluated by expanding the integrand in powers of r, using (5), and integrating term by term. Since the power series are convergent, even if some of the eigenvalues of $\bar{A}$ have positive real parts, term-by-term integration of these expressions is always valid. The power series should be taken to a sufficient number of terms to achieve the required degree of accuracy, but about ten terms should normally be sufficient.
iv. Compute the elements of M recursively from (34).
v. Compute the elements of η from (32).
vi. Compute the elements of ε recursively from (35).
vii. Compute L from (36).

The only steps of the above algorithm that depend on the sample size are (iv), (v), and (vi), for each of which the computing cost is of order T. We have already shown that the number of multiplications in step (iv) is less than $4Tn^3$. The number of multiplications in steps (v) and (vi) combined is much smaller, being less than $5Tn^2$.

It is of interest to compare the computational cost of using our algorithm with that of the Kalman filter algorithm of Harvey and Stock (1985) to which we have already referred in the Introduction. A precise comparison of these costs depends on the sample size, since a considerable part of the computational cost of our algorithm (steps (i)–(iii)) is independent of the sample size and can be regarded as a setup cost. The computational cost for that part of the algorithm which depends on the sample size is considerably less than the cost for the corresponding part of the algorithm presented in Harvey and Stock (1985). For example, in a second-order system with $n = 10$ the total number of multiplications in steps (iv)–(vi) of our algorithm is considerably less than the number of multiplications in T updatings of the state vector and its covariance matrix in the Kalman filter (Harvey and Stock 1985, Eqs. (3.8a) and (3.8b)). Our algorithm will be less costly than the Kalman filter algorithm, therefore for samples greater than a certain size at which the cost of finding the exact discrete model is offset by the savings in costs that depend on T. Moreover, there will be a comparative advantage in using our algorithm with even smaller samples when the exact discrete model is required for other reasons.

Another advantage of our algorithm as compared with the algorithm of Harvey and Stock is that it does not depend on the assumption that the system has distinct eigenvalues. Although the assumption (made in Harvey and Stock 1985) that the eigenvalues are distinct will normally be satisfied for the final values of the estimated parameters, the likelihood will have to be evaluated several hundred times during the optimization procedure, and it is possible that two or more of the eigenvalues associated with one of these evaluations will be approximately equal. At this stage of the optimization procedure computational difficulties could arise.

3. An Iterative Estimation Procedure

At this stage it should be pointed out that there is an important difference between the estimate $[\hat{\theta}, \hat{\mu}]$ of the structural parameter vector and the estimate $\hat{y}$ of the initial state vector. Whereas $[\hat{\theta}, \hat{\mu}]$ will, under suitable regularity conditions, be a consistent estimator of $[\theta, \mu]$ (sufficient conditions for consistency under the assumption of stationarity are established in (Bergstrom 1983, Section 7), there is no way in which we can obtain a consistent estimate of y. For as the sample size tends to infinity the influence of y on the distribution of most of the observations will tend to become negligible.

This suggests that if the sample is very large there is little to be gained from obtaining the exact Gaussian estimator $\hat{y}$ of the initial state vector. Instead we could obtain an estimate $\tilde{y}$ by some much simpler method and then minimize $L(\theta, \mu, \tilde{y}')$ with respect to $[\theta, \mu]$ using some numerical optimization procedure (e.g. the maximum gradient method) which involves only the repeated evaluation of L for a sequence of vectors of the structural parameter values. A very simple way of obtaining $\tilde{y}$ would be to use the linear extrapolation formulas: $\tilde{y}_1 = x_1 - \frac{1}{2}(x_2 - x_1)$, $\tilde{y}_2 = x_2 - x_1$. An even simpler procedure, which should yield a good approximation to the exact Gaussian estimator when the sample is very large, is to assume $\eta_1 = \eta_2 = 0$, thus avoiding the computation of Ω_{11}, Ω_{21}, Ω_{22}, Ω_{32}, G_{11}, G_{12}, and G_{22}.

With the comparatively small samples normally available for macroeconometric modelling the gain in efficiency achieved by obtaining the exact Gaussian estimator $[\hat{\theta}, \hat{\mu}, \hat{y}']$ rather than one of the approximations that would be obtained by the above procedures may justify the additional computing cost involved. The computation of $[\hat{\theta}, \hat{\mu}, \hat{y}']$ can be greatly facilitated by using an explicit formula for $\hat{y}$ conditional on $[\hat{\theta}, \hat{\mu}]$. We have

$$\left(\frac{\partial}{\partial y} L[\hat{\theta}, \hat{\mu}, \hat{y}]\right)_{y=\hat{y}} = 0,$$

from which, using (33), we obtain

$$\begin{aligned}\hat{y} &= [G'(\hat{\theta}, \hat{\mu})\Omega^{-1}(\hat{\theta}, \hat{\mu})G(\hat{\theta}, \hat{\mu})]^{-1}G'(\hat{\theta}, \hat{\mu})\Omega^{-1} \\ &\quad (\hat{\theta}, \hat{\mu})[F(\hat{\theta}, \hat{\mu})x - h(\hat{\theta}, \hat{\mu})] \\ &= H(\hat{\theta}, \hat{\mu}). \end{aligned} \tag{37}$$

The formula (37) can be used in an iterative procedure the initial steps of which are as follows.

i. Let

$$\tilde{y}^{(1)} = \begin{bmatrix} \tilde{y}_1^{(1)} \\ \tilde{y}_2^{(1)} \end{bmatrix} = \begin{bmatrix} x_1 - \frac{1}{2}(x_2 - x_1) \\ x_2 - x_1 \end{bmatrix}.$$

ii. Obtain $[\tilde{\theta}^{(1)}, \tilde{\mu}^{(1)}]$ by minimizing $L(\theta, \mu, \tilde{y}^{(1)}]$ with respect to $[\theta, \mu]$ using some numerical optimization procedure.
iii. Compute $\tilde{y}^{(2)} = H(\tilde{\theta}^{(1)}, \tilde{\mu}^{(1)})$.
iv. Obtain $[\tilde{\theta}^{(2)}, \tilde{\mu}^{(2)}]$ by minimizing $L(\theta, \mu, \tilde{y}^{(2)}]$ with respect to $[\theta, \mu]$ using a numerical optimization procedure.

If we continue in this way the value of L will decrease at each step and, since $L > 0$, $L(\tilde{\theta}^{(i)}, \tilde{\mu}^{(i)}, \tilde{y}^{(i)})$ must converge to a limit as $i \to \infty$. It follows that, if L has only one local minimum, $[\tilde{\theta}^{(i)}, \tilde{\mu}^{(i)}, \tilde{y}^{(i)}]$ must converge to $[\hat{\theta}, \hat{\mu}, \hat{y}]$ as $i \to \infty$.

It should be noted that in the evaluation of H it is not necessary to invert Ω. For, using the Cholesky factorization (34) we have

$$G'\Omega^{-1}G = Z'Z,$$

where Z can be obtained by solving

$$MZ = G,$$

the elements of each column of Z being computed recursively taking advantage of the fact that no more than $3n$ elements in any row of M are nonzero. In a similar way we can evaluate $G'\Omega^{-1}(Fx - h)$.

Finally, we need a set of starting values of the structural parameters to commence the numerical optimization procedure at step (ii). These could be obtained by using an approximate discrete model of the type discussed in Bergstrom 1966; Sargan 1976; Wymer 1972 together with the computer programs of Wymer (1978).

4. Conclusions

We have derived an efficient algorithm for computing the exact Gaussian likelihood for structural parameters in a higher-order continuous-time dynamic model with a fixed initial state vector. Although the detailed formulas in the algorithm have been derived only for the case of a second-order system with flow data, the general method can be applied to a system of any order with mixed stock and flow data. We have also proposed an iterative estimation procedure in which the structural parameters and initial state vector are estimated alternately and the estimates converge to the exact Gaussian estimates if the likelihood function has only one local maximum.

Although we have been concerned only with a closed model, the algorithm could be extended to an open model by using the methods of Phillips (1974, 1976*a*), who deals only with a first-order system. It is, of course, impossible to obtain, from discrete data, consistent estimates of the parameters of an open continuous-time model without making some assumption about the process generating the continuous-time paths of the exogenous variables (see Bergstrom 1983, p. 150). But both the theoretical results obtained by Phillips (1974, 1976*a*) and his empirical results in (1976*b*) suggest that the asymptotic bias resulting from the application of his method to the exogenous variables occurring in a typical macroeconomic model will be very small.

The application of the methods proposed in this paper can be expected to lead to a considerable increase in the precision of estimates of the parameters of continuous-time econometric models as compared with estimates obtained by the methods currently being used. Moreover, even for the purpose of obtaining the best predictions of future discrete observations there is an important gain from specifying and estimating a continuous-time model rather than fitting a vector autoregressive moving-average model directly to discrete data. Discretely sampled economic time series are usually very well approximated by parsimonious ARMA specifications in discrete time as is well known and well documented. Continuous-time models of the class considered in this paper actually generate such discrete-time specifications and import into them the additional parsimony of restrictions from theoretical sources in the continuous-time formulation. The incorporation of these restrictions can result in a considerable reduction in the mean-square prediction errors of the postsample discrete observations.

References

ARMINGTON, P. and C. WOLFORD (1983), 'PAC-MOD: An econometric model of selected U.S. and global indicators', World Bank (Global Modeling and

Projections Division), Division Working Paper No. 1983–3.

——, ——(1984) 'Exchange rate dynamics and economic policy', Papers of Armington, Wolford and Associates.

BASAWA, I. V. and P. J. BROCKWELL (1984), 'Asymptotic conditional inference for regular nonergodic models with an application to autoregressive processes', *Annals of Statistics*, **12**, 161–71.

BERGSTROM, A. R. (1966), 'Non-recursive models as discrete approximations to systems of stochastic differential equations', *Econometrica*, **34**, 173–82.

——(1983), 'Gaussian estimation of structural parameters in higher order continuous time dynamic models', *Econometrica*, **51**, 117–52.

——(1984*a*), 'Continuous time stochastic models and issues of aggregation over time'. In Z. Griliches and M. D. Intriligator (eds.), *Handbook of Econometrics*, (Amsterdam, North-Holland), ch. 20 and pp. 1145–212.

——(1984*b*), 'Monetary, fiscal and exchange rate policy in a continuous time model of the United Kingdom', in P. Malgrange and P. Muet (eds.), *Contemporary Macroeconomic Modelling* (Oxford, Blackwell), ch. 8 and pp. 183–206.

——and C. R. WYMER (1976), 'A model of disequilibrium neoclassical growth and its application to the United Kingdom, in A. R. Bergstrom (ed.), *Statistical Inference in Continuous Time Economic Models* (Amsterdam, North-Holland), ch. 10 and pp. 267–327.

DURBIN, J. (1961), 'Efficient fitting of linear models for continuous time stationary time series from discrete data', *Bulletin of the International Statistical Institute*, **38**, 273–82.

GANDOLFO, G. and P. C. PADOAN (1984), *A Disequilibrium Model of Real and Financial Accumulation in an Open Economy* (Berlin, Springer).

HANSEN, L. P. and T. J. SARGENT (1981), 'Formulating and estimating continuous time rational expectations models'. Federal Reserve Bank of Minneapolis Research Department Staff Report 75.

HARVEY, A. C. and J. H. STOCK (1985), 'The estimation of higher order continuous time autoregressive models, *Econometric Theory*, **1**, 97–117.

JONES, R. H. (1981), 'Fitting a continuous time autoregression to discrete data.' In D. F. Findley (ed.), *Applied Time Series Analysis II* (New York, Academic).

JONSON, P. D., E. R. MOSES, and C. R. WYMER (1977), 'The RBA76 model of the Australian Economy', in *Conference in Applied Economic Research* (Reserve Bank of Australia).

——W. J. MCKIBBIN, and R. G. TREVOR (1982), 'Exchange rates and capital flows: A sensitivity analysis,' *Canadian Journal of Economics*, **15**, 669–92.

KNIGHT, M. D. and C. R. WYMER (1978), 'A macroeconomic model of the United Kingdom', *IMF Staff Papers*, **25**, 742–78.

PHILLIPS, P. C. B. (1972), 'The structural estimation of a stochastic differential equation system', *Econometrica*, **40**, 1021–41.

——(1974), 'The estimation of some continuous time models', *Econometrica*, **42**, 803–24.

——(1976*a*), 'The estimation of linear stochastic differential equations with exogenous variables', in A. R. Bergstrom (ed.), *Statistical Inference in Continuous Time Economic Models* (Amsterdam, North-Holland) ch. 7 and pp. 135–73.

—— (1976*b*), 'Some computations based on observed data series of the exogenous variable component of continuous systems', in A. R. Bergstrom (ed.), *Statistical Inference in Continuous Time Economic Models* (Amsterdam, North-Holland), ch. 8 and pp. 174–214.

—— (1985), 'Time series regression with unit roots.' Cowles Foundation Discussion Paper No. 740.

QUANDT, R. E. (1984), 'Computational problems and methods.' In Z. Griliches and M. D. Intriligator (eds.), *Handbook of Econometrics* (Amsterdam, North-Holland), ch. 12 and pp. 699–764.

ROBINSON, P. M. (1976*a*), 'Fourier estimation of continuous time models.' In A. R. Bergstrom (ed.), *Statistical Inference in Continuous Time Models* (Amsterdam, North-Holland), ch. 9 and pp. 215–66.

—— (1976*b*), 'The estimation of linear differential equations with constant coefficient', *Econometrica*, **44**, 751–64.

—— (1976*c*), 'Instrumental variables estimation of differential equations', *Econometrica*, **44**, 765–76.

—— (1977), 'The construction and estimation of continuous time models and discrete approximations in econometrics', *Journal of Econometrics*, **6**, 173–98.

ROZANOV, Y. A. (1967), *Stationary Random Processes* (San Francisco, Holden Day)

SARGAN, J. D. (1976), 'Some discrete approximations to continuous time stochastic models', in A. R. Bergstrom (ed.), *Statistical Inference in Continuous Time Economic Models* (Amsterdam, North-Holland) ch. 3 and pp. 27–80.

SASSANPOUR C. and J. SHEEN (1984), 'An empirical analysis of the effect of monetary disequilibrium in open economies', *Journal of Monetary Economics*, **13**, 127–63.

TUSSIO, G. (1981), 'Demand management and exchange rate policy: the Italian experience', *IMF Staff Papers*, **28**, 80–117.

WHITTLE, P. (1951), *Hypothesis Testing in Time Series Analysis* (Stockholm, Almqvist and Wicksell).

—— (1953), 'The analysis of multiple stationary time series', *Journal of the Royal Statistical Society, Series B*, **15**, 125–39.

WYMER, C. R. (1972), 'Econometric estimation of stochastic differential equation systems', *Econometrica*, **40**, 565–77.

—— (1978), *Computer Programs* (International Monetary Fund).

6

The Estimation of Open Higher-Order Continuous Time Dynamic Models with Mixed Stock and Flow Data[1]

1. Introduction

In two recent articles (Bergstrom 1983, 1985) [Chs. 4 and 5 of this volume] I have developed a method of obtaining Gaussian or quasi-maximum likelihood estimates of the parameters of a closed higher-order continuous time dynamic model from discrete data. The theoretical foundations for the method are provided in Bergstrom (1983) with the proof of an existence and uniqueness theorem for the solution of the model under the most general assumptions about the white noise innovation process, and without assuming that the system is stationary or even stable. The assumptions about the innovation process include the case in which the innovations are a mixture of Brownian motion and Poisson processes and allow for more general innovation processes in which the increments are not independent but merely orthogonal.

The article (Bergstrom 1983) shows how to obtain the exact discrete model satisfied by the data and includes a detailed discussion of the use of this model in obtaining the Gaussian estimates of the parameters of the continuous time model when the system is stationary, together with some discussion of the asymptotic sampling properties of the estimates. It is shown, in particular, that the estimates are strongly consistent, even if the innovation process is not Gaussian, provided that the system is strictly stationary and the coefficients of the model are continuous functions of the structural parameters, which are assumed to belong to a closed bounded set.

The second article (Bergstrom 1985) provides a detailed treatment of the problem of obtaining the Gaussian estimates when the system is non-stationary, in which case the initial state vector is assumed to be fixed

[1] This article is based on research funded by the Economic and Social Research Council (ESRC) reference number: H 00 24 2002.

and the system is allowed to have unstable roots. It is argued that in most econometric work the assumption of a fixed initial state vector is more realistic than that of stationarity. An algorithm for computing the estimates is presented and shown to be even more efficient than the Kalman filter algorithm of Harvey and Stock (1985), although the latter algorithm is applicable in more general circumstances, for example, when there are missing observations or errors of measurement.

A serious limitation of the methods developed in Bergstrom 1983; 1985 at least for the purpose of applied econometric work, is that the algorithms, as presented, are applicable only to a closed model. The purpose of this article is to develop the methods further by introducing exogenous variables. This is not a trivial problem. Indeed, there is no way of obtaining consistent estimates of the parameters of an open higher-order continuous time dynamic model from discrete data without making some assumptions about the process generating the continuous time paths of the exogenous variables.

So far, all of the applied econometric work on continuous time models with exogenous variables (see references cited in Bergstrom 1985) has made use of approximate discrete models of the type discussed in the articles of Bergstrom 1966 [Ch. 3 of this volume], Sargan 1976, and Wymer 1972. In these models, integrals over the observation interval of both the endogenous and exogenous variables are replaced by trapezoidal approximations, and the parameters estimated by standard simultaneous equations estimation procedures. Estimates obtained in this way will, of course, be asymptotically biased even if the model is closed. Indeed, Sargan (1976) (see also Bergstrom 1974) shows that the asymptotic bias is of the order δ^2 as the observation period δ tends to zero, both when the model is closed and when it contains exogenous variables, provided that the time paths of the exogenous variables satisfy certain smoothness conditions.

Several more sophisticated methods of estimating open continuous time dynamic models have been proposed in the literature. Each of these methods will yield consistent estimates of the parameters of the model under certain assumptions about the behavior of the exogenous variables and can be expected to yield good estimates under more general conditions. The most obvious method of this sort is to assume that the exogenous variables are themselves generated by a continuous time dynamic model of the same form as the open model, except that it is closed, and extend the open model to include this so that the complete model is closed. A second method is the Fourier method developed by Robinson (1976*a, b, c*). His method yields consistent estimates when both the exogenous variable and the innovations are generated by stationary random processes and the spectral density of the former is zero outside the frequency range $(-\pi, \pi)$. Moreover, it has been generalized to other aliasing assumptions giving a class of estimates for more or less smooth z

(see Robinson 1977). A third method, which we shall follow in this article, is that of Phillips (1974, 1976*a*).

In Phillips' method the time path of each exogenous variable is approximated by a sequence of quadratic functions of time, on overlapping intervals, obtained from the observations by the three-point Lagrange interpolation formula. The method yields consistent estimates when the exogenous variables are polynomials in time of degree not exceeding two. But it is very flexible and can be expected to yield good estimates even when the behavior of the exogenous variables varies greatly as between different parts of the sample period, provided that certain smoothness conditions are satisfied. Phillips shows that, under these smoothness conditions and suitable regularity conditions, the asymptotic bias of estimates obtained by his method is $0(\delta^3)$ as $\delta \rightarrow 0$. Moreover, the computations based on observed data reported in another article (Phillips 1976*b*) indicate that, when the exogenous variables are typical economic time series observed at quarterly intervals, the error of specification in the approximate discrete model obtained by Phillips' method, and used in the estimation procedure, will be much smaller than that in the type of approximate discrete model currently being used in applied econometric work with continuous time models.

Phillips (in 1974, 1976*a*) deals only with a first-order system and assumes that all of the variables are observable at equi-spaced points of time, i.e. there are no flow variables. His main concern, in these articles, is with the derivation of the asymptotic sampling properties of the estimates. These properties can be expected to carry over to estimates obtained by applying his method to higher-order systems provided that certain smoothness and regularity conditions are satisfied. But the application of Phillips' method to higher-order systems does present serious technical algebraic problems, particularly when, as we shall assume in this article, the model includes both stock and flow variables so that some variables are observable at points of time and others only as integrals. These problems arise because of the complexity of the convolution integrals through which the exogenous variables enter the exact discrete model and the difficulty of explicitly evaluating these integrals after replacing the unobservable continuous time paths of the exogenous variables by their quadratic approximations. We shall, in fact, use a method of argument which avoids the explicit evaluation of these integrals and, by making use of results obtained for the closed model, leads to a great simplification of the algebra and the formulas for the coefficient matrices of the exogenous variables in the discrete model.

Although the methods that we shall use are applicable, in principle, to a system of any order we shall deal particularly with a second-order system. We shall derive precise formulas for the most general case in which both the endogenous and exogenous variables are a mixture of stock and flow variables. The results obtained will be more general, therefore, than those presented in Bergstrom 1983, 1985, even for a closed system. Although a method of deriving the exact discrete model satisfied by a sample of mixed stock and flow data generated by a second-order continuous time dynamic model was outlined in Bergstrom (1983, section 6) the precise formulas were obtained only for the special cases in which the data relates to either all stock variables or all flow variables, and the supplementary equations through which the initial state vectors are introduced, in the non-stationary case, were derived in Bergstrom 1985 only for the case in which all variables are flow variables. The formulas for the coefficient matrices in the exact discrete model satisfied by mixed stock and flow data are considerably more complicated than those for either of the simpler cases, even in a first-order system (see Agbeyegbe 1987).

In the next section, we shall derive the exact discrete model satisfied by mixed stock and flow data generated by a closed second-order system (Theorem 2.1) together with the supplementary equations involving the initial state vectors (Theorem 2.2). These results will then be extended to include exogenous variables (Theorems 2.3 and 2.4). In Section 3, we shall deal with the derivation of the likelihood function and the computation of the estimates. The computational algorithms presented in Bergstrom 1985 will require only minor modification to allow, firstly, for the inclusion of exogenous variables in the model, and secondly, for the fact that, with mixed stock and flow data, part of the initial state vector (i.e. the part comprising the levels of the stock variables at $t = 0$) is observable.

2. Exact and Approximate Discrete Models

We shall deal first with the closed model

$$d[Dx(t)] = \{A_1(\theta)Dx(t) + A_2(\theta)x(t)\}dt + \zeta(dt) \qquad (t \geqslant 0) \qquad (1)$$

$$x(0) = y_1, \qquad Dx(0) = y_2 \qquad (2)$$

where $\{x(t), (t > 0)\}$ is a real n-dimensional continuous time random process, θ is a p-dimensional vector of unknown structural parameters ($p \leqslant 2n^2$), A_1 and A_2 are matrices whose elements are known functions of θ, y_1 and y_2 are non-random $n \times 1$ vectors, D is the mean square differential operator, and $\zeta(dt)$ is a white noise disturbance vector which is precisely defined by the following assumption.

Assumption 2.1. $\zeta = [\zeta_1, \ldots, \zeta_n]'$ is a vector of random measures, defined on all subsets of the half line $0 < t < \infty$ with finite Lebesgue, measure, such that $E[\zeta(dt)] = 0$, $E[\zeta(dt)\zeta'(dt)] = (dt)\Sigma$ where Σ is an unknown positive definite matrix and $E[\zeta_i(\Delta_1)\zeta_j(\Delta_2)] = 0$ $(i, j = 1, \ldots, n)$

for any disjoint sets Δ_1 and Δ_2 on the half line $0 < t < \infty$. (See Bergstrom 1984, p. 1157, for a discussion of random measures and their application to continuous time stochastic models.)

We shall interpret equation (1) as meaning that $x(t)$ satisfies the stochastic integral equation

$$Dx(t) - Dx(0) = \int_0^t [A_1 Dx(r) + A_2 x(r)]dr + \int_0^t \zeta(dr) \tag{3}$$

for all $t > 0$, where $\int_0^t \zeta(dr) = \zeta[0, t]$ and the first integral on the right-hand side of equation (3) is defined in the wide sense (Bergstrom 1984, p. 1152).

We shall make the following assumptions about the coefficient matrices of the continuous time system.

Assumption 2.2. $A_2(\theta)$ is nonsingular.

Assumption 2.3. The $n \times n$ matrix $[e^{\bar{A}}]_{12}$ defined by equation (4) is non-singular.

$$\begin{bmatrix} [e^{\bar{A}}]_{11} & [e^{\bar{A}}]_{12} \\ [e^{\bar{A}}]_{21} & [e^{\bar{A}}]_{22} \end{bmatrix} = \begin{bmatrix} I & 0 \\ 0 & I \end{bmatrix} + \begin{bmatrix} 0 & I \\ A_2 & A_1 \end{bmatrix}$$

$$+ \frac{1}{2}\begin{bmatrix} 0 & I \\ A_2 & A_1 \end{bmatrix}^2 + \frac{1}{3!}\begin{bmatrix} 0 & I \\ A_2 & A_1 \end{bmatrix}^3 + \dots$$

$$= I = \bar{A} + \frac{1}{2}\bar{A}^2 + \frac{1}{3!}\bar{A}^3 + \dots. \tag{4}$$

We shall also need the following assumption.

Assumption 2.4. The matrix C_{12} defined in Theorem 2.1 is nonsingular.

Although Assumption 2.2 excludes the case in which the system has one or more zero eigenvalues it does not exclude eigenvalues with positive real parts. We do not, therefore, require the system to be stable. Moreover, the general methods that we shall develop could be extended to cover the case in which the system has zero eigenvalues. But, as was pointed out in Bergstrom 1985, the asymptotic sampling properties of the estimates will depend on whether or not the system has unstable roots (eigenvalues with non-negative real parts). Assumptions 2.3 and 2.4 merely exclude certain degenerate cases (see Bergstrom 1983, p. 74 this volume) which could be dealt with separately.

The system of equation (1) is somewhat less general than that considered in Bergstrom 1983, 1985 since we have excluded the constant vector **b**. This can be allowed for by letting the first element of the vector $z(t)$ of exogenous variables in the open system of equation (26) be one.

We shall assume that the variables in the system include n^s stock variables which are observed at the points of time $0, 1, 2, \ldots, T$ and n^f flow variables which are observed as integrals over the intervals $[0, 1], [1, 2], \ldots, [T-1, T]$, where $n^s + n^f = n$. Let the elements of $x(t)$ be ordered (without loss of generality) so that it can be written in the partitioned form.

$$x(t) = \begin{bmatrix} x^s(t) \\ x^f(t) \end{bmatrix} \tag{5}$$

where $x^s(t)$ is an $n^s \times 1$ vector of stock variables and $x^f(t)$ is a $n^f \times 1$ vector of flow variables. Then the sample comprises the initial stock vector $x^s(0) = y_1^s$ and the vectors $\bar{x}_1, \bar{x}_2, \ldots, \bar{x}_T$, where

$$\bar{x}_t = \begin{bmatrix} x^s(t) - x^s(t-1) \\ \int_{t-1}^{t} x^f(r)dr \end{bmatrix}. \tag{6}$$

We shall now prove a theorem which includes, as special cases, Theorems 2 and 3 of Bergstrom 1983 (when $\mathbf{b} = 0$). In the statement and proof of Theorems 2.1 and 2.2 we shall use the notation

$$A = \begin{bmatrix} A^{ss} & A^{sf} \\ A^{fs} & A^{ff} \end{bmatrix}, \qquad x = \begin{bmatrix} x^s \\ x^f \end{bmatrix}$$

to denote the partitioning of any $n \times n$ matrix A in such a way that A^{ss} is an $n^s \times n^s$ matrix and the partitioning of any $n \times 1$ vector x in such a way that x^s is an $n^s \times 1$ vector.

THEOREM 2.1 *Let* x(t) *be the solution of equation (1), on the interval* [0, T], *subject to the boundary conditions of equation (2). Then, under Assumptions 2.1 to 2.4, the vectors* $\bar{\mathrm{x}}_1, \bar{\mathrm{x}}_2, \ldots, \bar{\mathrm{x}}_{\mathrm{T}}$ *defined by equation (6) satisfy the system*

$$\bar{x}_t = F_1\bar{x}_{t-1} + F_2\bar{x}_{t-2} + \eta_t \qquad (t = 3, \ldots, T). \tag{7}$$

$$\eta_t = \int_{t-1}^{t} K_1(t-r)\zeta(dr) + \int_{t-2}^{t-1} K_2(t-1-r)\zeta(dr)$$
$$+ \int_{t-3}^{t-2} K_3(t-2-r)\zeta(dr) \quad (t = 3, \ldots, T), \tag{8}$$

where

$$F_1 = C_{11} + C_{12}C_{22}C_{12}^{-1},$$
$$F_2 = C_{12}C_{21} - C_{12}C_{22}C_{12}^{-1}C_{11},$$

$$K_1(r) = H_{11}(r),$$
$$K_2(r) = H_{12}(r) - C_{12}C_{22}C_{12}^{-1}H_{11}(r) + C_{12}H_{21}(r),$$
$$K_3(r) = C_{12}H_{22}(r) - C_{12}C_{22}C_{12}^{-1}H_{12}(r),$$

$$C_{11} = \begin{bmatrix} [e^{\bar{A}}]_{22}^{ss} & [e^{\bar{A}}]_{21}^{sf} \\ [e^{\bar{A}}]_{12}^{fs} & [e^{\bar{A}}]_{11}^{ff} \end{bmatrix},$$

$$C_{12} = \begin{bmatrix} [e^{\bar{A}}]_{21}^{ss} & [e^{\bar{A}}]_{22}^{sf} \\ [e^{\bar{A}}]_{11}^{fs} & [e^{\bar{A}}]_{12}^{ff} \end{bmatrix},$$

$$C_{21} = \begin{bmatrix} [e^{\bar{A}}]_{12}^{ss} & [e^{\bar{A}}]_{11}^{sf} \\ [e^{\bar{A}}]_{22}^{fs} & [e^{\bar{A}}]_{21}^{ff} \end{bmatrix},$$

$$C_{22} = \begin{bmatrix} [e^{\bar{A}}]_{11}^{ss} & [e^{\bar{A}}]_{12}^{sf} \\ [e^{\bar{A}}]_{21}^{fs} & [e^{\bar{A}}]_{22}^{ff} \end{bmatrix},$$

$$H_{11}^{ss}(r) = [e^{r\bar{A}}]_{12}^{ss},$$
$$H_{11}^{sf}(r) = [e^{r\bar{A}}]_{12}^{sf},$$
$$H_{11}^{fs}(r) = -P^{fs}[e^{r\bar{A}}]_{12}^{ss} - P^{ff}[e^{r\bar{A}}]_{12}^{fs} + [A_2^{-1}]^{fs}[e^{r\bar{A}}]_{22}^{ss} + [A_2^{-1}]^{ff}[e^{r\bar{A}}]_{22}^{fs} - [A_2^{-1}]^{fs},$$
$$H_{11}^{ff}(r) = -P^{fs}[e^{r\bar{A}}]_{12}^{sf} - P^{ff}[e^{r\bar{A}}]_{12}^{ff} + [A_2^{-1}]^{fs}[e^{r\bar{A}}]_{22}^{sf} + [A_2^{-1}]^{ff}[e^{r\bar{A}}]_{22}^{ff} - [A_2^{-1}]^{ff},$$
$$H_{12}^{ss}(r) = -[e^{r\bar{A}}]_{12}^{ss} + [e^{\bar{A}}]_{12}^{ss},$$
$$H_{12}^{sf}(r) = -[e^{r\bar{A}}]_{12}^{sf} + [e^{\bar{A}}]_{12}^{sf},$$
$$H_{12}^{fs}(r) = P^{fs}[e^{r\bar{A}}]_{12}^{ss} + P^{ff}[e^{r\bar{A}}]_{12}^{fs} - [A_2^{-1}]^{fs}[e^{r\bar{A}}]_{22}^{ss} - [A_2^{-1}]^{ff}[e^{r\bar{A}}]_{22}^{fs} - P^{fs}[e^{\bar{A}}]_{12}^{ss} - P^{ff}[e^{\bar{A}}]_{12}^{fs} + [A_2^{-1}]^{fs}[e^{\bar{A}}]_{22}^{ss} + [A_2^{-1}]^{ff}[e^{\bar{A}}]_{22}^{fs},$$
$$H_{12}^{ff}(r) = P^{fs}[e^{r\bar{A}}]_{12}^{sf} + P^{ff}[e^{r\bar{A}}]_{12}^{ff} - [A_2^{-1}]^{fs}[e^{r\bar{A}}]_{22}^{sf} - [A_2^{-1}]^{ff}[e^{r\bar{A}}]_{22}^{ff} - P^{fs}[e^{\bar{A}}]_{12}^{sf} - P^{ff}[e^{\bar{A}}]_{12}^{ff} + [A_2^{-1}]^{fs}[e^{\bar{A}}]_{22}^{sf} + [A_2^{-1}]^{ff}[e^{\bar{A}}]_{22}^{ff},$$

$$
\begin{aligned}
H_{21}^{ss}(r) &= -P^{ss}[e^{r\bar{A}}]_{12}^{ss} - P^{sf}[e^{r\bar{A}}]_{12}^{fs} + [A_2^{-1}]^{ss}[e^{r\bar{A}}]_{22}^{ss} \\
&\quad + [A_2^{-1}]^{sf}[e^{r\bar{A}}]_{22}^{fs} - [A_2^{-1}]^{ss}, \\
H_{21}^{sf}(r) &= -P^{ss}[e^{r\bar{A}}]_{12}^{sf} - P^{sf}[e^{r\bar{A}}]_{12}^{ff} + [A_2^{-1}]^{ss}[e^{r\bar{A}}]_{22}^{sf} \\
&\quad + [A_2^{-1}]^{sf}[e^{r\bar{A}}]_{22}^{ff} - [A_2^{-1}]^{sf}, \\
H_{21}^{fs}(r) &= [e^{r\bar{A}}]_{12}^{fs}, \\
H_{21}^{ff}(r) &= [e^{r\bar{A}}]_{12}^{ff}, \\
H_{22}^{ss}(r) &= P^{ss}[e^{r\bar{A}}]_{12}^{ss} + P^{sf}[e^{r\bar{A}}]_{12}^{fs} - [A_2^{-1}]^{ss}[e^{r\bar{A}}]_{22}^{ss} \\
&\quad - [A_2^{-1}]^{sf}[e^{r\bar{A}}]_{22}^{fs} - P^{ss}[e^{\bar{A}}]_{12}^{ss} - P^{sf}[e^{\bar{A}}]_{12}^{fs} \\
&\quad + [A_2^{-1}]^{ss}[e^{\bar{A}}]_{22}^{ss} + [A_2^{-1}]^{sf}[e^{\bar{A}}]_{22}^{fs}, \\
H_{22}^{sf}(r) &= P^{ss}[e^{r\bar{A}}]_{12}^{sf} + P^{sf}[e^{r\bar{A}}]_{12}^{ff} - [A_2^{-1}]^{ss}[e^{r\bar{A}}]_{22}^{sf} \\
&\quad - [A_2^{-1}]^{sf}[e^{r\bar{A}}]_{22}^{ff} - P^{ss}[e^{\bar{A}}]_{12}^{sf} - P^{sf}[e^{\bar{A}}]_{12}^{ff} \\
&\quad + [A_2^{-1}]^{ss}[e^{\bar{A}}]_{22}^{sf} + [A_2^{-1}]^{sf}[e^{\bar{A}}]_{22}^{ff}, \\
H_{22}^{fs}(r) &= -[e^{r\bar{A}}]_{12}^{fs} + [e^{\bar{A}}]_{12}^{fs}, \\
H_{22}^{ff}(r) &= -[e^{r\bar{A}}]_{12}^{ff} + [e^{\bar{A}}]_{12}^{ff}, \\
P &= A_2^{-1}A_1.
\end{aligned}
$$

Proof. We start with equations (75) and (76) of Bergstrom 1983 which can be written as equations (9) and (10) respectively.

$$
\begin{aligned}
\int_{t-1}^{t} x(r)dr &= [e^{\bar{A}}]_{11}\int_{t-2}^{t-1} x(r)dr + [e^{\bar{A}}]_{12}\{x(t-1) - x(t-2)\} \\
&\quad - A_2^{-1}\int_{t-1}^{t}\{A_1[e^{(t-r)\bar{A}}]_{12} - [e^{(t-r)\bar{A}}]_{22} + I\}\zeta(dr) \\
&\quad + A_2^{-1}\int_{t-2}^{t-1}\{A_1[e^{(t-1-r)\bar{A}}]_{12} - [e^{(t-1-r)\bar{A}}]_{22} \\
&\quad - A_1[e^{\bar{A}}]_{12} + [e^{\bar{A}}]_{22}\}\zeta(dr).
\end{aligned}
\tag{9}
$$

$$
\begin{aligned}
x(t) - x(t-1) &= [e^{\bar{A}}]_{21}\int_{t-2}^{t-1} x(r)dr + [e^{\bar{A}}]_{22}\{x(t-1) - x(t-2)\} \\
&\quad + \int_{t-1}^{t}[e^{(t-r)\bar{A}}]_{12}\zeta(dr) \\
&\quad - \int_{t-2}^{t-1}\{[e^{(t-1-r)\bar{A}}]_{12} - [e^{\bar{A}}]_{12}\}\zeta(dr).
\end{aligned}
\tag{10}
$$

Let x_t^* be defined by

$$x_t^* = \begin{bmatrix} \int_{t-1}^{t} x^s(r)dr \\ x^f(t) - x^f(t-1) \end{bmatrix}. \tag{11}$$

Then combining the first n^s equations of equation (10) with the last n^f equations of equation (9) we obtain

$$\bar{x}_t = C_{11}\bar{x}_{t-1} + C_{12}x_{t-1}^* + \int_{t-1}^{t} H_{11}(t-r)\zeta(dr) + \int_{t-2}^{t-1} H_{12}(t-1-r)\zeta(dr), \tag{12}$$

and combining the first n^s equations of equation (9) with the last n^f equations of equation (10) we obtain

$$x_t^* = C_{21}\bar{x}_{t-1} + C_{22}x_{t-1}^* + \int_{t-1}^{t} H_{21}(t-r)\zeta(dr) + \int_{t-2}^{t-1} H_{22}(t-1-r)\zeta(dr). \tag{13}$$

From equation (12) we obtain

$$x_{t-1}^* = C_{12}^{-1}\bar{x}_t - C_{12}^{-1}C_{11}\bar{x}_{t-1} - C_{12}^{-1}\int_{t-1}^{t} H_{11}(t-r)\zeta(dr) - C_{12}^{-1}\int_{t-2}^{t-1} H_{12}(t-1-r)\zeta(dr), \tag{14}$$

$$x_{t-2}^* = C_{12}^{-1}\bar{x}_{t-1} - C_{12}^{-1}C_{11}\bar{x}_{t-2} - C_{12}^{-1}\int_{t-2}^{t-1} H_{11}(t-1-r)\zeta(dr) - C_{12}^{-1}\int_{t-3}^{t-2} H_{12}(t-2-r)\zeta(dr), \tag{15}$$

and from equation (13)

$$x_{t-1}^* = C_{21}\bar{x}_{t-2} + C_{22}x_{t-2}^* + \int_{t-2}^{t-1} H_{21}(t-1-r)\zeta(dr) + \int_{t-3}^{t-2} H_{22}(t-2-r)\zeta(dr). \tag{16}$$

Substituting from equations (14) and (15), for x_{t-1}^* and x_{t-2}^* respectively, into equation (16) and premultiplying by C_{12} we obtain equations (7) and (8). ■

The next theorem, which includes as a special case, equations (16) and (22) of Bergstrom 1985, relates $\bar{x}_1$ and $\bar{x}_2$ to the initial state vectors.

THEOREM 2.2. *Let* x(t) *be the solution of equation (1) on the interval* [0, T] *subject to the boundary conditions of equation (2). Then, under Assumptions 2.1 to 2.4, the vectors* $\bar{x}_1$ *and* $\bar{x}_2$ *satisfy the equations.*

$$\bar{x}_1 = G_{11}y_1 + G_{12}y_2 + \eta_1, \tag{17}$$

$$\bar{x}_2 = C_{11}\bar{x}_1 + G_{21}y_1 + G_{22}y_2 + \eta_2, \tag{18}$$

$$\eta_1 = \int_0^1 K_{11}(1-r)\zeta(dr), \tag{19}$$

$$\eta_2 = \int_0^1 K_{21}(1-r)\zeta(dr) + \int_1^2 K_{22}(2-r)\zeta(dr). \tag{20}$$

where

$$G_{11}^{ss} = [e^{\bar{A}}]_{11}^{ss} - I,$$
$$G_{11}^{sf} = [e^{\bar{A}}]_{11}^{sf},$$
$$G_{11}^{fs} = [A_2^{-1}]^{fs}[e^{\bar{A}}]_{21}^{ss} + [A_2^{-1}]^{ff}[e^{\bar{A}}]_{21}^{fs} - P^{fs}([e^{\bar{A}}]_{11}^{ss} - I) - P^{ff}[e^{\bar{A}}]_{11}^{fs},$$
$$G_{11}^{ff} = [A_2^{-1}]^{fs}[e^{\bar{A}}]_{21}^{sf} + [A_2^{-1}]^{ff}[e^{\bar{A}}]_{21}^{ff} - P^{fs}[e^{\bar{A}}]_{11}^{sf} - P^{ff}([e^{\bar{A}}]_{11}^{ff} - I),$$
$$G_{12}^{ss} = [e^{\bar{A}}]_{12}^{ss},$$
$$G_{12}^{sf} = [e^{\bar{A}}]_{12}^{sf}$$
$$G_{12}^{fs} = [A_2^{-1}]^{fs}([e^{\bar{A}}]_{22}^{ss} - I) + [A_2^{-1}]^{ff}[e^{\bar{A}}]_{22}^{fs} - P^{fs}[e^{\bar{A}}]_{12}^{ss} - P^{ff}[e^{\bar{A}}]_{12}^{fs},$$
$$G_{12}^{ff} = [A_2^{-1}]^{fs}[e^{\bar{A}}]_{22}^{sf} + [A_2^{-1}]^{ff}([e^{\bar{A}}]_{22}^{ff} - I) - P^{fs}[e^{\bar{A}}]_{12}^{sf} - P^{ff}[e^{\bar{A}}]_{12}^{ff},$$
$$G_{21} = C_{12}G_{11}^{*},$$
$$G_{22} = C_{12}G_{12}^{*}$$
$$[G_{11}^{*}]^{ss} = [A_2^{-1}]^{ss}[e^{\bar{A}}]_{21}^{ss} + [A_2^{-1}]^{sf}[e^{\bar{A}}]_{21}^{fs} - P^{ss}([e^{\bar{A}}]_{11}^{ss} - I) - P^{sf}[e^{\bar{A}}]_{11}^{fs},$$
$$[G_{11}^{*}]^{sf} = [A_2^{-1}]^{ss}[e^{\bar{A}}]_{21}^{sf} + [A_2^{-1}]^{sf}[e^{\bar{A}}]_{21}^{ff} - P^{ss}[e^{\bar{A}}]_{11}^{sf} - P^{sf}([e^{\bar{A}}]_{11}^{ff} - I),$$
$$[G_{11}^{*}]^{fs} = [e^{\bar{A}}]_{11}^{fs},$$
$$[G_{11}^{*}]^{ff} = [e^{\bar{A}}]_{11}^{ff} - I,$$
$$[G_{12}^{*}]^{ss} = [A_2^{-1}]^{ss}([e^{\bar{A}}]_{22}^{ss} - I) + [A_2^{-1}]^{sf}[e^{\bar{A}}]_{22}^{fs} - P^{ss}[e^{\bar{A}}]_{12}^{ss} - P^{sf}[e^{\bar{A}}]_{12}^{fs},$$
$$[G_{12}^{*}]^{sf} = [A_2^{-1}]^{ss}[e^{\bar{A}}]_{22}^{sf} + [A_2^{-1}]^{sf}([e^{\bar{A}}]_{22}^{ff} - I) - P^{ss}[e^{\bar{A}}]_{12}^{sf} - P^{sf}[e^{\bar{A}}]_{12}^{ff},$$
$$[G_{12}^{*}]^{fs} = [e^{\bar{A}}]_{12}^{fs},$$
$$[G_{12}^{*}]^{ff} = [e^{\bar{A}}]_{12}^{ff},$$
$$K_{11}^{ss}(r) = [e^{r\bar{A}}]_{12}^{ss},$$
$$K_{11}^{sf}(r) = [e^{r\bar{A}}]_{12}^{sf},$$

$$K_{11}^{fs}(r) = [A_2^{-1}]^{fs}[e^{r\bar{A}}]_{22}^{ss} + [A_2^{-1}]^{ff}[e^{r\bar{A}}]_{22}^{fs} - P^{fs}[e^{r\bar{A}}]_{12}^{ss} - P^{ff}[e^{r\bar{A}}]_{12}^{fs} - [A_2^{-1}]^{fs},$$

$$K_{11}^{ff}(r) = [A_2^{-1}]^{fs}[e^{r\bar{A}}]_{22}^{sf} + [A_2^{-1}]^{ff}[e^{r\bar{A}}]_{22}^{ff} - P^{fs}[e^{r\bar{A}}]_{12}^{sf} - P^{ff}[e^{r\bar{A}}]_{12}^{ff} - [A_2^{-1}]^{ff},$$

$$K_{21}(r) = H_{12}(r) + C_{12}K_{11}^{*}(r),$$

$$K_{22}(r) = H_{11}(r),$$

$$[K_{11}^{*}(r)]^{ss} = [A_2^{-1}]^{ss}[e^{r\bar{A}}]_{22}^{ss} + [A_2^{-1}]^{sf}[e^{r\bar{A}}]_{22}^{fs} - P^{ss}[e^{r\bar{A}}]_{12}^{ss} - P^{sf}[e^{r\bar{A}}]_{12}^{fs} - [A_2^{-1}]^{ss},$$

$$[K_{11}^{*}(r)]^{sf} = [A_2^{-1}]^{ss}[e^{r\bar{A}}]_{22}^{sf} + [A_2^{-1}]^{sf}[e^{r\bar{A}}]_{22}^{ff} - P^{ss}[e^{r\bar{A}}]_{12}^{sf} - P^{sf}[e^{r\bar{A}}]_{12}^{ff} - [A_2^{-1}]^{sf},$$

$$[K_{11}^{*}(r)]^{fs} = [e^{r\bar{A}}]_{12}^{fs},$$

$$[K_{11}^{*}(r)]^{ff} = [e^{r\bar{A}}]_{12}^{ff},$$

Proof. From equation (1) we obtain

$$Dx(1) - Dx(0) = A_1\{x(1) - x(0)\} + A_2\int_0^1 x(r)dr + \int_0^1 \zeta(dr), \quad (21)$$

and hence

$$\int_0^1 x(r)dr = A_2^{-1}\{Dx(1) - Dx(0)\} - P\{x(1) - x(0)\} - \int_0^1 A_2^{-1}\zeta(dr). \quad (22)$$

From equations (32) and (33) of Bergstrom 1983, using the initial conditions of equation (2), we obtain equations (23) and (24), respectively.

$$x(1) - x(0) = \{[e^{\bar{A}}]_{11} - I\}y_1 + [e^{\bar{A}}]_{12}y_2 + \int_0^1 [e^{(1-r)\bar{A}}]_{12}\zeta(dr). \quad (23)$$

$$Dx(1) - Dx(0) = [e^{\bar{A}}]_{21}y_1 + \{[e^{\bar{A}}]_{22} - I\}y_2 + \int_0^1 [e^{(1-r)\bar{A}}]_{22}\zeta(dr). \quad (24)$$

We now obtain equations (17) and (19) by combining the first n^s equations of equation (23) with the last n^f equations of equation (22), after first substituting for $x(1) - x(0)$ and $Dx(1) - Dx(0)$ from equations (23) and (24), respectively into equation (22).

Similarly, we obtain

$$x_1^* = G_{11}^* y_1 + G_{12}^* y_2 + \int_0^1 K_{11}^*(1-r)\zeta(dr) \quad (25)$$

by combining the first n^s equations of equation (22) (after substituting for $x(1) - x(0)$ and $Dx(1) - Dx(0)$) with the last n^f equations of equation (23). We then obtain equations (18) and (20) by putting $t = 2$ in equation (12) and substituting for x_1^* from equation (25). ∎

We turn now to the open model

$$d[Dx(t)] = \{A_1(\theta)Dx(t) + A_2(\theta)x(t) + B(\theta)z(t)\}dt + \zeta(dt) \qquad (t \geq 0),$$

with

$$x(0) = y_1,\ Dx(0) = y_2, \tag{26}$$

where $z(t)$ is an $m \times 1$ vector of non-random functions (exogenous variables), $B(\theta)$ is an $n \times m$ matrix of known functions of θ, and all other terms are defined as for the closed system of equation (1). We interpret equation (26) as meaning that $x(t)$ satisfies the stochastic integral equation

$$Dx(t) - Dx(0) = \int_0^t [A_1Dx(r) + A_2x(r) + Bz(r)]dr + \int_0^t \zeta(dr) \tag{27}$$

for all $t > 0$. This reduces to equation (3) when $z(t)$ is a zero vector.

The first element of $z(t)$ will normally be 1, and in this case the first column of B will be the vector of constants in the system. The vector $z(t)$ could also include t (and possibly t^2) to allow for trends arising from such factors as technical progress which are not, directly, observable. (See, for example, the model of Bergstrom and Wymer (1976).) But, generally, there will be no continuous record of the complete vector $z(t)$ over the sample period.

We shall assume that the exogenous variables include m^s stock variables which are observed at points of time $0, 1, 2, \ldots, T$ and m^f flow variables which are observed as integrals over the intervals $[0, 1]$, $[1, 2]$, $\ldots, [T - 1, T]$ where $m^s + m^f = m$. Let the elements of $z(t)$ be ordered (without loss of generality) so that it can be written in the form

$$z(t) = \begin{bmatrix} z^s(t) \\ z^f(t) \end{bmatrix} \tag{28}$$

where $z^s(t)$ is an $m^s \times 1$ vector of stock variables and $z^f(t)$ is an $m^f \times 1$ vector of flow variables. Then we can define a sequence of observable vectors $\bar{z}_1, \ldots, \bar{z}_T$ by

$$\bar{z}_t = \begin{bmatrix} \frac{1}{2}\{z^s(t) + z^s(t-1)\} \\ \int_{t-1}^{t} z^f(r)dr \end{bmatrix} \qquad (t = 1, \ldots, T). \tag{29}$$

These vectors will be used in Theorems 2.3 and 2.4 which provide the basis for the estimation procedure discussed in Section 3.

The exact discrete analogue of the open system of equation (26) is

$$\bar{x}_t = F_1\bar{x}_{t-1} + F_2\bar{x}_{t-2} + \int_{t-1}^{t} K_1(t-r)Bz(r)dr + \int_{t-2}^{t-1} K_2(t-1-r)Bz(r)dr$$
$$+ \int_{t-3}^{t-2} K_3(t-2-r)Bz(r)dr + \eta_t \qquad (t = 3, \ldots, T) \tag{30}$$

which reduces to equation (7) when $z(t)$ is a zero vector. Equation (30) cannot be used, directly, as a basis for estimation since the convolution integrals on the right-hand side are unobservable. But, if over an interval $[t'-3, t']$ the elements of $z(t)$ are polynomials in t of degree not exceeding two, then the vectors of coefficients of these polynomials can be expressed in terms of $\bar{z}_{t'}$, $\bar{z}_{t'-1}$, and $\bar{z}_{t'-2}$, so that the integrals on the right-hand side of equation (30) can be expressed in terms of the observations. In this way we can obtain a discrete model which is exact if the elements of $z(t)$ are polynomials of degree not exceeding two and, otherwise, has an error depending on the smoothness properties of $z(t)$. Estimates of the parameters of the continuous time model can then be obtained by applying the Gaussian or quasi-maximum likelihood estimation procedure to this approximate discrete model.

This is the method proposed and discussed by Phillips (1974; 1976*a*), particularly in relation to a first-order system with point observations. The application of the method to a second-order system with mixed stock and flow data is much more difficult, however, because of the complexity of the convolution integrals on the right-hand side of equation (30), even when the elements of $z(t)$ are quadratic functions of t. These integrals can, of course, be approximated to any required degree of accuracy by expanding the vectors $K_1(t-r)z(r)$, $K_2(t-1-r)z(r)$, and $K_3(t-2-r)z(r)$ as power series in r and integrating term by term. In the proof of Theorem 2.3 we shall use a much simpler procedure which avoids the explicit evaluation of these integrals and makes use of the results obtained in Theorem 2.1 for the closed system. We shall obtain the coefficient matrices of the exogenous vectors in a comparatively simple form in which the submatrices of e^A enter only through the matrices F_1 and F_2 which must be evaluated to obtain the exact discrete model, even if there are no exogenous variables.

In the statement and proof of Theorems 2.3 and 2.4 we shall use the notation

$$A = \begin{bmatrix} A^{ss} & A^{sf} \\ A^{fs} & A^{ff} \end{bmatrix} = \begin{bmatrix} A^s & A^f \end{bmatrix}$$

and

$$x = \begin{bmatrix} x^s \\ x^f \end{bmatrix},$$

to denote the partitioning of any $n \times m$ matrix A in such a way that A^{ss} is an $n^s \times m^s$ matrix, and the partitioning of any $m \times 1$ vector x in such a way that x^s is an $m^s \times 1$ vector.

THEOREM 2.3. *Let* x(t) *be the solution of equation (26) on the interval* [0, T] *subject to the boundary conditions of equation (2), and assume that, over the subinterval* [t' − 3, t'] *of* [0, T], *the elements of* z(t) *are polynomials in* t *of degree not exceeding two. Then, under Assumptions 2.1 to 2.4, the vectors* $\bar{x}_{t'}$, $\bar{x}_{t'-1}$, $\bar{x}_{t'-2}$, $\bar{z}_{t'}$, $\bar{z}_{t'-1}$, $\bar{z}_{t'-2}$ *defined by equations (6) and (29) satisfy the equation*

$$\bar{x}_{t'} = F_1\bar{x}_{t'-1} + F_2\bar{x}_{t'-2} + E_0\bar{z}_{t'} + E_1\bar{z}_{t'-1} + E_2\bar{z}_{t'-2} + \eta_t \qquad (31)$$

where F_1, F_2 *and* η_t *are defined as in Theorem 2.1 and* E_0, E_1, *and* E_2 *are given by*

$$E_0 = F_1L_{10} + F_2L_{20} - L_{00},$$

$$E_1 = F_1L_{11} + F_2L_{21} - L_{01},$$

$$E_2 = F_1L_{12} + F_2L_{22} - L_{02},$$

$$L_{00}^{ss} = [(\tfrac{3}{2}I - P)Q]^{ss},$$

$$L_{00}^{sf} = [(\tfrac{3}{2}I - P)Q]^{sf},$$

$$L_{00}^{fs} = [(\tfrac{11}{12}I + A_2^{-1} - \tfrac{3}{2}P + P^2)Q]^{fs},$$

$$L_{00}^{ff} = [(I + A_2^{-1} - \tfrac{3}{2}P + P^2)Q]^{ff},$$

$$L_{01}^{ss} = 2[(P - I)Q]^{ss},$$

$$L_{01}^{sf} = 2[(P - I)Q]^{sf},$$

$$L_{01}^{fs} = [(\tfrac{1}{6}I - 2A_2^{-1} + 2P - 2P^2)Q]^{fs},$$

$$L_{01}^{ff} = [(-2A_2^{-1} + 2P - 2P^2)Q]^{ff},$$

$$L_{02}^{ss} = [(\tfrac{1}{2}I - P)Q]^{ss},$$

$$L_{02}^{sf} = [(\tfrac{1}{2}I - P)Q]^{sf},$$

$$L_{02}^{fs} = [(-\tfrac{1}{12}I + A_2^{-1} - \tfrac{1}{2}P + P^2)Q]^{fs},$$

$$L_{02}^{ff} = [(A_2^{-1} - \tfrac{1}{2}P + P^2)Q]^{ff},$$

$$L_{10}^{ss} = [(\tfrac{1}{2}I - P)Q]^{ss},$$

$$L_{10}^{sf} = [(\tfrac{1}{2}I - P)Q]^{sf},$$

$$L_{10}^{fs} = [(-\tfrac{1}{12}I + A_2^{-1} - \tfrac{1}{2}P + P^2)Q]^{fs},$$

$$L_{10}^{ff} = [(A_2^{-1} - \tfrac{1}{2}P + P^2)Q]^{ff},$$

$$L_{11}^{ss} = 2[PQ]^{ss},$$

$$L_{11}^{sf} = 2[PQ]^{sf},$$

$$L_{11}^{fs} = [(\tfrac{7}{6}I - 2A_2^{-1} - 2P^2)Q]^{fs},$$

$$L_{11}^{ff} = [(I - 2A_2^{-1} - 2P^2)Q]^{ff},$$

$$L_{12}^{ss} = -[(\tfrac{1}{2}I + P)Q]^{ss},$$

$$L_{12}^{sf} = -[(\tfrac{1}{2}I + P)Q]^{sf},$$

$$L_{12}^{fs} = [(-\tfrac{1}{12}I + A_2^{-1} + \tfrac{1}{2}P + P^2)Q]^{fs},$$
$$L_{12}^{ff} = [(A_2^{-1} + \tfrac{1}{2}P + P^2)Q]^{ff},$$
$$L_{20}^{ss} = -[(\tfrac{1}{2}I + P)Q]^{ss},$$
$$L_{20}^{sf} = -[(\tfrac{1}{2}I + P)Q]^{sf},$$
$$L_{20}^{fs} = [(-\tfrac{1}{12}I + A_2^{-1} + \tfrac{1}{2}P + P^2)Q]^{fs},$$
$$L_{20}^{ff} = [(A_2^{-1} + \tfrac{1}{2}P + P^2)Q]^{ff},$$
$$L_{21}^{ss} = 2[(I + P)Q]^{ss},$$
$$L_{21}^{sf} = 2[(I + P)Q]^{sf},$$
$$L_{21}^{fs} = [(\tfrac{1}{6}I - 2A_2^{-1} - 2P - 2P^2)Q]^{fs},$$
$$L_{21}^{ff} = -2[A_2^{-1} + P + P^2)Q]^{ff},$$
$$L_{22}^{ss} = -[(\tfrac{3}{2}I + P)Q]^{ss},$$
$$L_{22}^{sf} = -[(\tfrac{3}{2}I + P)Q]^{sf},$$
$$L_{22}^{fs} = [(\tfrac{11}{12}I + A_2^{-1} + \tfrac{3}{2}P + P^2)Q]^{fs},$$
$$L_{22}^{ff} = [(I + A_2^{-1} + \tfrac{3}{2}P + P^2)Q]^{ff},$$
$$Q = A_2^{-1}B.$$

Proof. Let

$$z(t) = a + b(t' - t) + c(t' - t)^2 \qquad (t' - 3 \leq t \leq t'), \tag{32}$$

where a, b, and c are $n \times 1$ vectors of constants. Then, by substituting equation (32) into equation (29) and letting t assume the values t', $t' - 1$, and $t' - 2$ successively, we obtain a system of equations which can be solved to yield

$$\begin{aligned}
a^s &= \tfrac{7}{4}\bar{z}_{t'}^s - \bar{z}_{t'-1}^s + \tfrac{1}{4}\bar{z}_{t'-2}^s \\
a^f &= \tfrac{11}{6}\bar{z}_{t'}^s - \tfrac{7}{6}\bar{z}_{t'-1}^s + \tfrac{1}{3}\bar{z}_{t'-2}^s \\
b^s &= -2\bar{z}_{t'}^s + 3\bar{z}_{t'-1}^s - \bar{z}_{t'-2}^s \\
b^f &= -2\bar{z}_{t'}^f + 3\bar{z}_{t'-1}^f - \bar{z}_{t'-2}^f \\
c^s &= \tfrac{1}{2}\bar{z}_{t'}^s - \bar{z}_{t'-1}^s + \tfrac{1}{2}\bar{z}_{t'-2}^s \\
c^f &= \tfrac{1}{2}\bar{z}_{t'}^f - \bar{z}_{t'-1}^f + \tfrac{1}{2}\bar{z}_{t'-2}^f
\end{aligned} \tag{33}$$

Now let $u(t)$ be defined by

$$u(t) = x(t) + Qa + PQb + 2(A_2^{-1} + P^2)Qc + (Qb + 2PQc)(t' - t)$$
$$+ Qc(t' - t)^2 \qquad (t' - 3 \leq t \leq t'). \tag{34}$$

Then it can be verified, using equations (27) and (32), that

$$Du(t) - Du(t'-3) = \int_{t'-3}^{t} \{A_1 Du(r) + A_2 u(r)\} dr + \int_{t'-3}^{t} \zeta(dr) \qquad (t'-3 \leqslant t \leqslant t'). \qquad (35)$$

It follows from Theorem 2.1, therefore, that if $\bar{u}_t$ is defined by

$$\bar{u}_t = \begin{bmatrix} u^s(t) - u^s(t-1) \\ \int_{t-1}^{t} u^f(r) dr \end{bmatrix} \qquad (36)$$

then

$$\bar{u}_{t'} = F_1 \bar{u}_{t'-1} + F_2 \bar{u}_{t'-2} + \eta_{t'}. \qquad (37)$$

Moreover, from equations (34) and (36) we obtain (after substituting for a, b, and c from equation (33) into equation (34))

$$\begin{aligned} \bar{u}_{t'} &= \bar{x}_{t'} + L_{00}\bar{z}_{t'} + L_{01}\bar{z}_{t'-1} + L_{02}\bar{z}_{t'-2}, \\ \bar{u}_{t'-1} &= \bar{x}_{t'-1} + L_{10}\bar{z}_{t'} + L_{11}\bar{z}_{t'-1} + L_{12}\bar{z}_{t'-2}, \\ \bar{u}_{t'-2} &= \bar{x}_{t'-2} + L_{20}\bar{z}_{t'} + L_{21}\bar{z}_{t'-1} + L_{22}\bar{z}_{t'-2}. \end{aligned} \qquad (38)$$

For example, $\bar{u}^f_{t'-2}$ is obtained by integrating the last n^f equations of equation (34) over the interval $[t'-3, t'-2]$ and replacing a^s, a^f, b^s, b^f, c^s, and c^f by the expressions given in equation (33). Then by substituting for $\bar{u}_t$, $\bar{u}_{t'-1}$, and $\bar{u}_{t'-2}$ from equation (38) into equation (37) we obtain equation (31). ■

Finally, we shall obtain a pair of supplementary equations relating $\bar{x}_1$ and $\bar{x}_2$ to $\bar{z}_1$, $\bar{z}_2$, $\bar{z}_3$ and the initial state vectors.

THEOREM 2.4. *Let* x(t) *be the solution of equation (26) on the interval* [0, T] *subject to the boundary conditions of equations (2), and assume that, over the interval* [0, 3] *the elements of* z(t) *are polynomials in* t *of degree not exceeding two. Then, under Assumptions 2.1 to 2.4, the vectors* $\bar{x}_1$, $\bar{x}_2$, $\bar{z}_1$, $\bar{z}_2$, *and* $\bar{z}_3$ *defined by equations (6) and (29) satisfy the equations.*

$$\bar{x}_1 = G_{11}y_1 + G_{12}y_2 + E_{11}\bar{z}_1 + E_{12}\bar{z}_2 + E_{13}\bar{z}_3 + \eta_1, \qquad (39)$$

$$\bar{x}_2 = C_{11}\bar{x}_1 + G_{21}y_1 + G_{22}y_2 + E_{21}\bar{z}_1 + E_{22}\bar{z}_2 + E_{23}\bar{z}_3 + \eta_2, \qquad (40)$$

where C_{11}, G_{11}, G_{12}, G_{21}, G_{22}, η_1, *and* η_2 *are defined as in Theorem 2.2 and* E_{11}, E_{12}, E_{13}, E_{21}, E_{22}, *and* E_{23} *are given by*

$$E_{11} = G_{11}N_{11} + G_{12}N_{21} - L_{22},$$

$$E_{12} = G_{11}N_{12} + G_{12}N_{22} - L_{21},$$

$$E_{13} = G_{11}N_{13} + G_{12}N_{23} - L_{20},$$
$$E_{21} = G_{21}N_{11} + G_{22}N_{21} - L_{12} + C_{11}L_{22},$$
$$E_{22} = G_{21}N_{12} + G_{22}N_{22} - L_{11} + C_{11}L_{21},$$
$$E_{23} = G_{21}N_{13} + G_{22}N_{23} - L_{10} + C_{11}L_{20},$$
$$N_{11}^s = [(\tfrac{7}{4}I + A_2^{-1} + 2P + P^2)Q]^s,$$
$$N_{11}^f = [(\tfrac{11}{6}I + A_2^{-1} + 2P + P^2)Q]^f,$$
$$N_{12}^s = -[(I + 2A_2^{-1} + 3P + 2P^2)Q]^s,$$
$$N_{12}^f = -[(\tfrac{7}{6}I + 2A_2^{-1} + 3P + 2P^2)Q]^f,$$
$$N_{13}^s = [(\tfrac{1}{4}I + A_2^{-1} + P + P^2)Q]^s,$$
$$N_{13}^f = [(\tfrac{1}{3}I + A_2^{-1} + P + P^2)Q]^f,$$
$$N_{21} = -(2I + P)Q,$$
$$N_{22} = (3I + 2P)Q,$$
$$N_{23} = -(I + P)Q,$$

Proof. Let

$$z(t) = a + b(3 - t) + c(3 - t)^2 \qquad (0 \leq t \leq 3), \tag{41}$$

where a, b, and c are $n \times 1$ vectors of constants. Then, putting $t' = 3$ in equation (33), we have

$$\begin{aligned} a^s &= \tfrac{7}{4}\bar{z}_3^s - \bar{z}_2^s + \tfrac{1}{4}\bar{z}_1^s, \\ a^f &= \tfrac{11}{6}\bar{z}_3^f - \tfrac{7}{6}\bar{z}_2^f + \tfrac{1}{3}\bar{z}_1^f, \\ b^s &= -2\bar{z}_3^s + 3\bar{z}_2^s - \bar{z}_1^s, \\ b^f &= -2\bar{z}_3^f + 3\bar{z}_2^f - \bar{z}_1^f, \\ c^s &= \tfrac{1}{2}\bar{z}_3^s - \bar{z}_2^s + \tfrac{1}{2}\bar{z}_1^s, \\ c^f &= \tfrac{1}{2}\bar{z}_3^f - \bar{z}_2^f + \tfrac{1}{2}\bar{z}_1^f. \end{aligned} \tag{42}$$

Now let $u(t)$ be defined by

$$\begin{aligned} u(t) = x(t) &+ Qa + PQb + 2(A_2^{-1} + P^2)Qc \\ &+ (Qb + 2PQc)(3 - t) + Qc(3 - t)^2 \qquad (0 \leq t \leq 3). \end{aligned} \tag{43}$$

Then, putting $t' = 3$ in equation (35) we have

$$Du(t) - Du(0) = \int_0^t \{A_1 Du(r) + A_2 u(r)\}dr + \int_0^t \zeta(dr) \qquad (0 \leq t \leq 3). \tag{44}$$

It follows from Theorem 2.2, therefore, that if $\bar{u}_t$ is defined by equation (36) then

$$\bar{u}_1 = G_{11}u(0) + G_{12}Du(0) + \eta_1, \tag{45}$$

$$\bar{u}_2 = C_{11}\bar{u}_1 + G_{21}u(0) + G_{22}Du(0) + \eta_2. \tag{46}$$

Moreover, from equation (38) we obtain

$$\bar{u}_1 = \bar{x}_1 + L_{20}\bar{z}_3 + L_{21}\bar{z}_2 + L_{22}\bar{z}_1, \tag{47}$$

$$\bar{u}_2 = \bar{x}_2 + L_{10}\bar{z}_3 + L_{11}\bar{z}_2 + L_{12}\bar{z}_1, \tag{48}$$

and from equations (42) and (43)

$$u(0) = N_{11}\bar{z}_1 + N_{12}\bar{z}_2 + N_{13}\bar{z}_3 + y_1, \tag{49}$$

$$Du(0) = N_{21}\bar{z}_1 + N_{22}\bar{z}_2 + N_{23}\bar{z}_3 + y_2. \tag{50}$$

Then substituting from equations (47)–(50) into equations (45) and (46) we obtain equations (39) and (40) respectively. ■

If, over the whole interval $[0, T]$, the elements of $z(t)$ are polynomials in t of degree not exceeding two, then equation (31) will be satisfied exactly for $t = 1, \ldots, T$, and Theorems 2.3 and 2.4 will provide the basis for the computation of the exact Gaussian likelihood. This will be the case if the continuous time model is closed but incorporates linear and quadratic trends, in which case $z'(t) = [1, t, t^2]$.

Under more general conditions the system of equations (31), (39), and (40) is an approximate discrete model whose errors depend on the smoothness properties of $z(t)$. It is important to notice that these errors do not depend on the accuracy with which each element of $z(t)$ can be approximated by a single quadratic function over the whole interval $[0, T]$, but only on the accuracy with which it can be approximated by a sequence of quadratic functions over the overlapping intervals $[0, 3], [1, 4], [2, 5], \ldots, [T-3, T]$. This is because $[E_0, E_1, E_2]$ depends only on the matrix $[A_1, A_2, B]$ of coefficients of the continuous time model and not on the matrix $[a, b, c]$ of coefficients of the polynomials approximating $z(t)$ over any subinterval. The errors in the approximate model of equations (31), (39), and (40) could be very small even when $[a, b, c]$ varies greatly as between different subintervals of $[0, T]$. Indeed, it is easy to show by an extension of the argument of Phillips (1976*a*, p. 139), that, if $z(t)$ is a vector of thrice continuously differentiable functions of t, then the errors in the approximate discrete model will be of order δ^4 as the observation period δ tends to zero.

Phillips (1976*a*) also shows that, under suitable regularity and smoothness conditions, the asymptotic bias of the parameter estimates obtained by applying the Gaussian or quasi-maximum likelihood procedure to his approximate discrete model is $0(\delta^3)$ as $\delta \to \infty$. A similar result could undoubtedly be proved for the more complicated model discussed in this article.

The general method of this section could be used for obtaining the exact discrete model for mixed stock and flow data generated by a continuous time dynamic model of order greater than two, on the assumption that the paths of the exogenous variables are quadratic functions of time. Under more general conditions the model would be an approximate discrete model whose errors would depend on the accuracy with which the paths of the exogenous variables could be approximated by quadratic functions on overlapping intervals of length depending on the order of the system. This fact suggests that, instead of using quadratic interpolation of the exogenous variables, we should use polynomial interpolation with the order of the polynomial depending on the order of the system. This could be done by the general methods of this section, although it would obviously result in more complicated formulas.

3. The Likelihood Function and Computation of the Estimates

In Bergstrom 1985 I presented, in detail, an algorithm for computing the exact Gaussian estimates of the parameters of a closed second-order continuous time dynamic model with flow data. We shall now discuss the modifications of that algorithm that are required to allow, firstly, for the inclusion of exogenous variables in the model and, secondly, for the fact that, with mixed stock and flow data, part of the initial state vector (i.e. the part comprising the levels of the stock variables at $t = 0$) is observable. We start by deriving the exact Gaussian likelihood function under the assumption that the elements of $z(t)$ are polynomials in t of degree not exceeding two over the whole interval $[0, T]$.

For computational purposes it will be convenient to write the vector y_1 in the partitioned form

$$y_1 = \begin{bmatrix} y_1^s \\ y_1^f \end{bmatrix}$$

where y_1^s is the $n^s \times 1$ vector of initial values of the stock variables and y_1^f the $n^f \times 1$ vector of initial values of the flow variables. Equations (39) and (40) can then be written

$$\bar{x}_1 = G_{11}^s y_1^s + G_{11}^f y_1^f + G_{12} y_2 + E_{11}\bar{z}_1 + E_{12}\bar{z}_2 + E_{13}\bar{z}_3 + \eta_1, \tag{51}$$

$$\bar{x}_2 = C_{11}\bar{x}_1 + G_{21}^s y_1^s + G_{21}^f y_1^f + G_{22} y_2 + E_{21}\bar{z}_1 + E_{22}\bar{z}_2 + E_{23}\bar{z}_3 + \eta_2. \tag{52}$$

Now let the $(n^f + n) \times 1$ vector y and the $nT \times 1$ vectors x, η, and h be defined by

$$y = \begin{bmatrix} y_1^f \\ y_2 \end{bmatrix},$$

$$x' = [\bar{x}'_1, \bar{x}'_2, \ldots, \bar{x}'_T],$$

$$\eta' = [\eta'_1, \eta'_2, \ldots, \eta'_T],$$

$$h' = [h'_1, h'_2, \ldots, h'_T],$$

where

$$h_1 = G_{11}^s y_1^s + E_{11}\bar{z}_1 + E_{12}\bar{z}_2 + E_{13}\bar{z}_3,$$

$$h_2 = G_{21}^s y_1^s + E_{21}\bar{z}_1 + E_{22}\bar{z}_2 + E_{23}\bar{z}_3,$$

$$h_t = E_0\bar{z}_t + E_1\bar{z}_{t-1} + E_2\bar{z}_{t-2} \qquad (t = 3, \ldots, T), \tag{53}$$

and let the $nT \times nT$ matrix F and the $nT \times (n^f + n)$ matrix G be defined by

$$F = \begin{bmatrix} I & 0 & 0 & 0 & . & . & . & 0 \\ -C_{11} & I & 0 & 0 & . & . & . & 0 \\ -F_2 & -F_1 & I & 0 & . & . & . & 0 \\ 0 & -F_2 & -F_1 & I & . & . & . & 0 \\ \vdots & \vdots & \vdots & \vdots & \vdots & \vdots & \vdots & \vdots \\ 0 & 0 & 0 & 0 & . & -F_2 & -F_1 & I \end{bmatrix}, \tag{54}$$

$$G = \begin{bmatrix} G_{11}^f & G_{12} \\ G_{21}^f & G_{22} \\ 0 & 0 \\ \vdots & \vdots \\ 0 & 0 \end{bmatrix}. \tag{55}$$

Then the system of equations (31), (39), and (40) can be written as

$$Fx - Gy - h = \eta. \tag{56}$$

Moreover, we have

$$E(\eta\eta') = \Omega$$

where

$$\Omega =$$

$$\begin{bmatrix} \Omega_{11} & \Omega_{12} & \Omega_{13} & 0 & 0 & 0 & 0 & . & . & . & . & 0 \\ \Omega_{21} & \Omega_{22} & \Omega_{23} & \Omega_{24} & 0 & 0 & 0 & . & . & . & . & 0 \\ \Omega_{31} & \Omega_{32} & \Omega_0 & \Omega'_1 & \Omega'_2 & 0 & 0 & . & . & . & . & 0 \\ 0 & \Omega_{42} & \Omega_1 & \Omega_0 & \Omega'_1 & \Omega'_2 & 0 & . & . & . & . & 0 \\ . & . & . & . & . & . & . & . & . & . & . & . \\ . & . & . & . & . & . & . & . & \Omega_0 & \Omega'_1 & \Omega'_2 & 0 \\ . & . & . & . & . & . & . & . & \Omega_1 & \Omega_0 & \Omega'_1 & \Omega'_2 \\ . & . & . & . & . & . & . & . & \Omega_2 & \Omega_1 & \Omega_0 & \Omega'_1 \\ 0 & 0 & 0 & 0 & 0 & 0 & 0 & 0 & 0 & \Omega_2 & \Omega_1 & \Omega_0 \end{bmatrix}, \tag{57}$$

$$\Omega_0 = \int_0^1 K_1(r)\Sigma K_1'(r)dr + \int_0^1 K_2(r)\Sigma K_2'(r)dr + \int_0^1 K_3(r)\Sigma K_3'(r)dr,$$

$$\Omega_1 = \int_0^1 K_2(r)\Sigma K_1'(r)dr + \int_0^1 K_3(r)\Sigma K_2'(r)dr,$$

$$\Omega_2 = \int_0^1 K_3(r)\Sigma K_1'(r)dr,$$

$$\Omega_{11} = \int_0^1 K_{11}(r)\Sigma K_{11}'(r)dr,$$

$$\Omega_{22} = \int_0^1 K_{21}(r)\Sigma K_{21}'(r)dr + \int_0^1 K_{22}(r)\Sigma K_{22}'(r)dr,$$

$$\Omega_{21} = \Omega_{12}' = \int_0^1 K_{21}(r)\Sigma K_{11}'(r)dr,$$

$$\Omega_{31} = \Omega_{13}' = \int_0^1 K_3(r)\Sigma K_{11}'(r)dr,$$

$$\Omega_{32} = \Omega_{23}' = \int_0^1 K_2(r)\Sigma K_{22}'(r)dr + \int_0^1 K_3(r)\Sigma K_{21}'(r)dr,$$

$$\Omega_{42} = \Omega_{24}' = \int_0^1 K_3(r)\Sigma K_{22}'(r)dr.$$

We note, incidentally, that when all the variables are flow variables $\Omega_{31} = \Omega_{42} = \Omega_2$ as is implicit in the formula for Ω given in Bergstrom 1985.

As in Bergstrom 1983, 1985 we shall parametrize Σ by writing it as $\Sigma(\mu)$ where μ is a vector of parameters. The complete vector of parameters (including the unobservable part of the initial state vector) is then $[\theta, \mu, y']$.

Now let $L(\theta, \mu, y']$ denote minus twice the logarithm of the Gaussian likelihood function. Then, since $|F| = 1$, we have

$$\begin{aligned} L(\theta, \mu, y') &= \log|\Omega(\theta, \mu)| + \eta'\Omega^{-1}(\theta, \mu)\eta \\ &= \log|\Omega(\theta, \mu)| + \{F(\theta)x - G(\theta)y - h(\theta)\}'\Omega^{-1}(\theta, \mu) \\ &\quad \{F(\theta)x - G(\theta)y - h(\theta)\}. \end{aligned} \tag{58}$$

Equation (58) has the same form as that given in Bergstrom 1985, eq. (33), but differs from it in that the elements of F, G, and h are now more complicated functions of θ and that h also involves the observations of the exogenous variables, which can be regarded as given numbers.

We can express L in an even simpler form, which is very convenient for computational purposes, by using the Cholesky factorization

$$\Omega = MM', \tag{59}$$

where M is a real lower triangular matrix with positive elements along the diagonal. All of the elements of M above the diagonal and most of the elements below the diagonal are zero, and, as is shown in Bergstrom 1985, the total number of multiplications required in the evaluation of M is less than $4Tn^3$. We then have

$$L = \sum_{i=1}^{nT} (\varepsilon_i^2 + 2 \log m_{ii}), \tag{60}$$

where m_{ii} is the ith diagonal element of M and $\varepsilon = [\varepsilon_1, \ldots, \varepsilon_{nT}]'$ is a vector whose elements can be evaluated recursively from

$$M\varepsilon = \eta.^2 \tag{61}$$

A computational algorithm for evaluating L for given values of the structural parameters is set out in Bergstrom (1985, p. 112 this vol.) In order to apply the algorithm we must first evaluate $e^{\bar{A}}$. Although $e^{\bar{A}}$ is well defined by equation (4), even when $\bar{A}$ has unstable roots, the truncation of this series does not provide a very satisfactory basis for computation. The best, general, and current method for calculating matrix exponentials is the Padé rational approximation method (see Golub and van Loan 1983, pp. 396–402). Having evaluated $e^{\bar{A}}$ we can evaluate F, G, h, and Ω as defined, in this article, by equations (54), (55), (53), and (57) respectively. We can then carry out the last four steps of the algorithm presented in Bergstrom (1985, p. 112 this vol.), in which equations (34), (32), (35), and (36) of that article correspond to equations (59), (56), (61), and (60) respectively of this article.

We can also use the iterative procedure proposed in Bergstrom (1985, p. 114 this vol.) for obtaining the Gaussian estimate $[\hat{\theta}, \hat{\mu}, \hat{y}']$ of $[\theta, \mu, y']$. In this procedure we alternate between minimizing L with respect to $[\theta, \mu]$, for given y, using some numerical optimization procedure, and minimizing L with respect to y, for given $[\theta, \mu]$ using equation (37) of (ibid.). If L has only one local minimum then the sequence of estimates obtained in this iterative procedure must converge to $[\hat{\theta}, \hat{\mu}, \hat{y}]$. The numerical optimization procedure by which we minimize L with respect to $[\theta, \mu]$ for given y will require the evaluation of the partial derivatives of L. These could be approximated from the changes in L resulting from small changes in the parameters or, alternatively, could be obtained by a modification of the procedure for computing L. Evaluations of L could also be used in a grid search to obtain starting values for the iterative procedure.

It might be thought that, if we are not interested in estimating y, the results of Theorems 2.2 and 2.4 can be dispensed with by forming the likelihood function conditional on $\bar{x}_1$ and $\bar{x}_2$ for the purpose of estimating

[2] In the original version of this paper, published in *Econometric Theory* (1986), the vector ε was denoted by z. The change in notation is to avoid confusion with the vector $z(t)$ of exogenous variables.

$[\theta, \mu]$. But this is not so. Since η_3 and η_4 are not independent of $\bar{x}_1$ and $\bar{x}_2$ the conditional distribution of $\bar{x}_3, \ldots, \bar{x}_T$ for given $\bar{x}_1$ and $\bar{x}_2$ cannot be obtained from Theorems 2.1 and 2.3 by simply treating $\bar{x}_1$ and $\bar{x}_2$ as if they were constant vectors. The vector y is included, in an essential way, as part of the parameter vector of the conditional distribution. The only way of dispensing with the results of Theorems 2.2 and 2.4 is to put $\eta_1 = \eta_2 = 0$ in equation (58) (which, of course, does not give the conditional distribution of $\bar{x}_3, \ldots, \bar{x}_T$ relative to the assumption $\eta_1 = \eta_2 = 0$) before minimizing with respect to $[\theta, \mu]$. As was pointed out in Bergstrom (1985, p. 113 this vol.), this procedure avoids the calculation of $\Omega_{11}, \Omega_{12}, \Omega_{13}, \Omega_{22}, \Omega_{23}, \Omega_{24}$, G_{11}, G_{12}, G_{21}, and G_{22}, and it can be expected to yield a good approximation to the exact Gaussian estimator when the sample is large.

4. Conclusions

We have developed a method of estimating the parameters of an open higher-order continuous time dynamic model with mixed stock and flow data. The method is applicable, in principle, to a system of any order, and the precise formulas for its implementation have been derived for the most general second-order case in which both the endogenous and exogenous variables are a mixture of stock and flow variables.

The method yields exact maximum likelihood estimates when the innovations are Gaussian and the exogenous variables are polynomials in time of degree not exceeding two, and it can be expected to yield very good estimates under much more general conditions. Under similar smoothness and regularity conditions to those assumed by Phillips (1976*a*), in his study of an open first-order system, our estimates can be expected to have an asymptotic bias of order δ^3 as the observation period δ tends to zero, as compared with an asymptotic bias of order δ^2 for estimates obtained from the type of approximate discrete model currently used in applied econometric work with continuous time models. Another advantage of our method is that it can be expected to yield estimates with considerably lower variances than those obtained from such approximate discrete models. The Monte Carlo study of Phillips (1972) discussed in my article (1985) suggests that the application of our method could yield estimates whose root-mean-square errors are, typically, less than 50 per cent of those of estimates obtained by the methods in current use.

The developments in computing technology that have occurred during the last few years are such that it would now be feasible to apply the methods of this article to a small macroeconometric model such as a second-order version of the continuous time model of the United Kingdom developed by Bergstrom and Wymer (1976). Moreover, I believe that the advantages of using these methods would easily outweigh the additional computing cost as compared with that of using approximate discrete models of the type that have been used so far.

References

AGBEYEGBE, T. D. (1987), 'The exact discrete analogue to a closed linear first order continuous time system with mixed sample', *Econometric Theory,* **3**, 143–9.

BERGSTROM, A. R. (1966), 'Non-recursive models as discrete approximations to systems of stochastic differential equations', *Econometrica,* **34**, 173–82.

—— (1983), 'Gaussian estimation of structural parameters in higher-order continuous time dynamic models', *Econometrica,* **51**, 117–52.

—— (1984), 'Continuous time stochastic models and issues of aggregation over time'. In Z. Griliches and M. D. Intriligator (eds.), *Handbook of Econometrics* (Amsterdam, North Holland), ch. 20 and pp. 1145–212.

—— (1985), 'The estimation of parameters in nonstationary higher-order continuous-time dynamic models', *Econometric Theory,* **1**, 369–85.

—— and C. R. WYMER (1976), 'A model of disequilibrium neoclassical growth and its application to the United Kingdom.' In A. R. Bergstrom (ed.), *Statistical Inference in Continuous Time Economic Models* (Amsterdam, North Holland), ch. 10 and pp. 267–327.

GOLUB, G. H. and C. F. VAN LOAN (1983), *Matrix Computations* (Baltimore, John Hopkins University Press).

HARVEY, A. C. and J. H. STOCK (1985), 'The estimation of higher-order continuous time autoregressive models', *Econometric Theory,* **1**, 97–117.

PHILLIPS, P. C. B. (1972), 'The structural estimation of a stochastic differential equation system', *Econometrica,* **40**, 1021–41.

—— (1974), 'The estimation of some continuous time models', *Econometrica,* **42**, 803–24.

—— (1976*a*), 'The estimation of linear stochastic differential equations with exogenous variables. In A. R. Bergstrom (ed.), *Statistical Inference in Continuous Time Economic Models,* (Amsterdam, North Holland), ch. 7 and pp. 135–73.

—— (1976*b*), 'Some computations based on observed data series of the exogenous variable component of continuous systems', in A. R. Bergstrom (ed.), *Statistical Inference in Continuous Time Economic Models,* (Amsterdam, North Holland), ch. 8 and pp. 174–214.

ROBINSON, P. M. (1976*a*), 'Fourier estimàtion of continuous time models'. In A. R. Bergstrom (ed.), *Statistical Inference in Continuous Time Economic Models,* (Amsterdam, North Holland), ch. 9 and pp. 215–66.

—— (1976*b*), 'The estimation of linear differential equations with constant coefficients', *Econometrica,* **44**, 751–64.

—— (1976*c*), 'Instrumental variables estimation of differential equations', *Econometrica,* **44**, 765–76.

—— (1977), 'The construction and estimation of continuous time models and discrete approximations in econometrics', *Journal of Econometrics,* **6**, 173–98.

SARGAN, J. D. (1976), 'Some discrete approximations to continuous time stochastic models', in A. R. Bergstrom (ed.), *Statistical Inference in Continuous Time Economic Models,* (Amsterdam, North Holland), ch. 3 and pp. 27–80.

WYMER, C. R. (1972), 'Econometric estimation of stochastic differential equation systems', *Econometrica,* **40**, 565–77.

7

Hypothesis Testing in Continuous Time Econometric Models[1]

1. Introduction

THE preceding three chapters have been concerned with the development of methods of obtaining exact Gaussian estimates of the parameters of higher-order continuous time dynamic models from discrete data. These methods can be expected to yield much more precise estimates of the parameters than methods that rely on approximate discrete models, and more accurate forecasts (see Ch. 8) of the post sample discrete observations. Another advantage of the new methods is that they provide the basis for exact asymptotic tests of the specification of a continuous time model, and of hypotheses concerning its parameters. The main purpose of this chapter is to develop practical testing procedures based on the new methods.

The testing procedures will be developed within the framework of the VARMA (vector autoregressive moving average) representation of the exact discrete model satisfied by the observations. This is a very convenient framework, since it enables us to rely, to a considerable extent, on the standard procedures applicable to models that are directly formulated in discrete time and greatly facilitates the comparison of continuous time and discrete time models. The fact that the discrete data generated by a closed continuous time model satisfy a VARMA model was proved in Chapter 4, under very general assumptions concerning the innovations, and without assuming that the system is stationary or even stable. Detailed formulae for the autoregressive coefficient matrices and the covariance matrix of the moving average terms were derived in Chapters 5 and 6 for various special cases. For the purpose of this chapter a more detailed analysis of the VARMA representation of the exact discrete model will be required. This will be presented in Section 2, where we shall derive some new results

[1] This chapter is based on research funded by the Economic and Social Research Council (ESRC), reference number RB0023 2215.

(Theorem 1) which are of considerable practical importance for the implementation of the testing procedures developed in Section 3.

Throughout Sections 2 and 3 we shall confine our attention to a closed model, since our aim is to obtain exact asymptotic tests, and these can be obtained for an open model only by making specific assumptions about the unobservable continuous time paths of the exogenous variables. The treatment of exogenous variables will be discussed in Section 4.

2. The VARMA Representation of the Exact Discrete Model

In order to simplify the exposition we shall deal, specifically, with a second-order system, the way in which the arguments and formulae can be extended to a system of any order being indicated at the end of the section. The assumed continuous time model is

$$\mathrm{d}[Dx(t)] = \{A_1(\theta)Dx(t) + A_2(\theta)x(t)\}\mathrm{d}t + \zeta(\mathrm{d}t) \qquad (\mathrm{t} \geq 0), \qquad (1)$$

$$x(0) = y_1,\ Dx(0) = y_2, \qquad (2)$$

where $\{x(t), (t \geq 0)\}$ is real n-dimensional continuous time random process, θ is a p-dimensional vector of unknown structural parameters $(p \leq 2n^2)$, A_1 and A_2 are matrices whose elements are known functions of θ, y_1, and y_2 and non-random $n \times 1$ vectors, D is the mean square differential operator, and $\zeta(\mathrm{d}t)$ is a white noise innovation vector which is precisely defined by the following assumption.

Assumption 1. $\zeta = [\zeta_1, \ldots, \zeta_n]$ is a vector of random measures defined on all subsets of the half line $0 < t < \infty$ with finite Lebesgue measure, such that $E[\zeta(\mathrm{d}t)] = 0$, $E[\zeta(\mathrm{d}t)\zeta'(\mathrm{d}t)] = \mathrm{d}t\Sigma(\mu)$, where Σ is a positive definite matrix whose elements are known functions of a q-dimensional vector μ of unknown parameters $(q \leq n(n+1)/2)$, and $E[\zeta_i(\Delta_1)\zeta_j(\Delta_2)] = 0$ $(i, j = 1, \ldots, n)$ for any disjoint sets Δ_1 and Δ_2 on the half line $0 < t < \infty$. (See Bergstrom 1984, p. 1157 for a discussion of random measures and their application to continuous time stochastic models.)

We shall also require Assumptions 2, 3, and 4, which are identical with Assumptions 2.2, 2.3, and 2.4 respectively of Chapter 6. (See Ch. 6 for a discussion of those assumptions and the precise interpretation of equation (1).)

Assumption 2. $A_2(\theta)$ is nonsingular.

Assumption 3. The matrix $[e^{\bar{A}}]_{12}$ defined by equation (4) of Chapter 6 is nonsingular.

Assumption 4. The matrix C_{12} defined in Theorem 2.1 of Chapter 6 is nonsingular.

As in Chapter 6 we shall assume that the variables in the model include

n^s stock variables, which are observed at points of time $0, 1, 2, \ldots, T$, and n^f flow variables which are observed as integrals over the intervals $[0, 1]$ $[1, 2], \ldots, [T-1, T]$, where $n^s + n^f = n$. Let the elements of the vector $x(t)$ be ordered (without loss of generality) so that it can be written in the partitioned form

$$x(t) = \begin{bmatrix} x^s(t) \\ x^f(t) \end{bmatrix}$$

where $x^s(t)$ is an $n^s \times 1$ vector of stock variables and $x^f(t)$ is an $n^f \times 1$ vector of flow variables. Then the sample comprises of the initial stock vector $x^s(0) = y_1^s$ and the vectors $\bar{x}_1, \bar{x}_2, \ldots, \bar{x}_T$, where

$$\bar{x}_t = \begin{bmatrix} x^s(t) - x^s(t-1) \\ \int_{t-1}^{t} x^f(r)\mathrm{d}r \end{bmatrix}. \tag{3}$$

The unknown parameter vector is the $(p + q + n + n^f)$-dimensional vector $[\theta, \mu, y']$, where

$$y = \begin{bmatrix} y_1^f \\ y_2 \end{bmatrix},$$

i.e. y is the unobservable part of the initial state vector.

It has been shown in Chapter 6 that the $nT \times 1$ vector $\bar{x}$ defined by

$$\bar{x} = \begin{bmatrix} \bar{x}_1 \\ \bar{x}_2 \\ \vdots \\ \bar{x}_T \end{bmatrix}$$

satisfies the system

$$F\bar{x} - Gy - h = \eta, \tag{4}$$

where

$$F = \begin{bmatrix} I & 0 & 0 & 0 & . & . & . & 0 \\ -C_{11} & I & 0 & 0 & . & . & . & 0 \\ -F_2 & -F_1 & I & 0 & . & . & . & 0 \\ 0 & -F_2 & -F_1 & I & . & . & . & 0 \\ . & . & . & . & . & . & . & . \\ . & . & . & . & . & . & . & . \\ . & . & . & . & . & . & . & . \\ 0 & 0 & 0 & 0 & . & -F_2 & -F_1 & I \end{bmatrix}, \tag{5}$$

$$G = \begin{bmatrix} G^f_{11} & G_{12} \\ G^f_{21} & G_{22} \\ 0 & 0 \\ \cdot & \cdot \\ \cdot & \cdot \\ \cdot & \cdot \\ 0 & 0 \end{bmatrix}, \tag{6}$$

$$h = \begin{bmatrix} G^s_{11} y^s_1 \\ G^s_{21} y^s_1 \\ 0 \\ \cdot \\ \cdot \\ \cdot \\ 0 \end{bmatrix}, \tag{7}$$

$$E(\eta) = 0,\ E(\eta\eta') = \Omega,$$

$$\Omega =$$

$$\begin{bmatrix}
\Omega_{11} & \Omega_{12} & \Omega_{13} & 0 & 0 & 0 & 0 & \cdot & \cdot & \cdot & \cdot & 0 \\
\Omega_{21} & \Omega_{22} & \Omega_{23} & \Omega_{24} & 0 & 0 & 0 & \cdot & \cdot & \cdot & \cdot & 0 \\
\Omega_{31} & \Omega_{32} & \Omega_{0} & \Omega'_{1} & \Omega'_{2} & 0 & 0 & \cdot & \cdot & \cdot & \cdot & 0 \\
0 & \Omega_{42} & \Omega_{1} & \Omega_{0} & \Omega'_{1} & \Omega'_{2} & 0 & \cdot & \cdot & \cdot & \cdot & 0 \\
\cdot & \cdot & \cdot & \cdot & \cdot & \cdot & \cdot & \cdot & \cdot & \cdot & \cdot & \cdot \\
\cdot & \cdot & \cdot & \cdot & \cdot & \cdot & \cdot & \cdot & \cdot & \cdot & \cdot & \cdot \\
\cdot & \cdot & \cdot & \cdot & \cdot & \cdot & \cdot & \cdot & \Omega_{0} & \Omega'_{1} & \Omega'_{2} & 0 \\
\cdot & \cdot & \cdot & \cdot & \cdot & \cdot & \cdot & \cdot & \Omega_{1} & \Omega_{0} & \Omega'_{1} & \Omega'_{2} \\
\cdot & \cdot & \cdot & \cdot & \cdot & \cdot & \cdot & \cdot & \Omega_{2} & \Omega_{1} & \Omega_{0} & \Omega'_{1} \\
0 & 0 & 0 & 0 & 0 & 0 & 0 & \cdot & 0 & \Omega_{2} & \Omega_{1} & \Omega_{0}
\end{bmatrix}, \tag{8}$$

C_{11}, F_1, F_2, $G_{11} = [G^s_{11}, G^f_{11}]$, G_{12}, $G_{21}, = [G^s_{21}, G^f_{21}]$, and G_{22} are $n \times n$ matrix functions of θ, and $\Omega_{11}, \Omega_{12}, \Omega_{13}, \Omega_{21}, \Omega_{22}, \Omega_{23}, \Omega_{24}, \Omega_{31}, \Omega_{32}, \Omega_{42}$, Ω_0, Ω_1, and Ω_2 are $n \times n$ matrix functions of $[\theta, \mu]$. The explicit form of these functions is given in Chapter 6.

Now let M be the real lower triangular matrix, with positive elements along the diagonal such that

$$MM' = \Omega, \tag{9}$$

and let ε be defined by

$$M_\varepsilon = \eta, \tag{10}$$

so that $E(\varepsilon) = 0$ and $E(\varepsilon\varepsilon') = I$. Then from equations (4) and (10) we obtain the system

$$F\bar{x} - Gy - h = M\varepsilon \tag{11}$$

which is the VARMA representation of the exact model written in its most compact form.

If we now define the n-dimensional vectors ε_t $(t = 1, \ldots, T)$ by the partition

$$\varepsilon' = [\varepsilon_1', \varepsilon_2', \ldots, \varepsilon_T'],$$

and the $n \times n$ matrices M_{st} $(s, t = 1, \ldots, T)$ by the partition

$$M = \begin{bmatrix} M_{11} & 0 & 0 & 0 & . & . & . & . & 0 \\ M_{21} & M_{22} & 0 & 0 & . & . & . & . & 0 \\ M_{31} & M_{32} & M_{33} & 0 & . & . & . & . & 0 \\ 0 & M_{42} & M_{43} & M_{44} & . & . & . & . & 0 \\ . & . & . & . & . & . & . & . & . \\ . & . & . & . & . & . & . & . & . \\ . & . & . & . & . & . & . & . & . \\ 0 & 0 & 0 & 0 & . & 0 & M_{T,T-2} & M_{T,T-1} & M_{TT} \end{bmatrix},$$

we obtain, from equations (5), (6), (7), (11), and (12), the system

$$\begin{aligned} \bar{x}_1 - G_{11}(\theta)y_1 - G_{12}(\theta)y_2 &= M_{11}(\theta, \mu)\varepsilon_1, \\ \bar{x}_2 - C_{11}(\theta)\bar{x}_1 - G_{21}(\theta)y_1 - G_{22}(\theta)y_2 &= M_{21}(\theta, \mu)\varepsilon_1 + M_{22}(\theta, \mu)\varepsilon_2, \\ \bar{x}_t - F_1(\theta)\bar{x}_{t-1} - F_2(\theta)\bar{x}_{t-2} &= M_{tt}(\theta, \mu)\varepsilon_t + M_{t,t-1}(\theta, \mu)\varepsilon_{t-1} \\ &\quad + M_{t,t-2}(\theta, \mu)\varepsilon_{t-2} \quad (t = 3, \ldots, T), \\ E(\varepsilon_t) = 0\,(t = 1, \ldots, T),\ & E(\varepsilon_t\varepsilon_t') = I\,(t = 1, \ldots, T), \\ E(\varepsilon_s\varepsilon_t') &= 0\,(s \neq t; s, t = 1, \ldots, T). \end{aligned} \tag{13}$$

The system (13) is the VARMA representation of the exact discrete model written in its expanded form. It is more complicated in several ways than a standard VARMA model formulated directly in discrete time.

The main complication is that, when the initial state vector in the continuous time model is fixed (as is assumed by equation (2)), the moving average coefficient matrices M_{tt}, $M_{t,t-1}$, and $M_{t,t-2}$ are changing over time, whereas they are invariably assumed to be constant matrices when a VARMA model is formulated directly in discrete time. But, since the elements of these matrices are all functions of a comparatively small number of parameters (i.e. functions of $[\theta, \mu]$), the continuous time model implies complicated nonlinear restrictions on the VARMA model, both across time and across equations within each time period. Moreover, further restrictions are implied by the fact that both the autoregressive and moving average coefficient matrices depend on θ. These restrictions have important implications for hypothesis testing as we shall show in Section 3.

For given values of the parameters of the continuous time model, the matrices $M_{11}, M_{21}, M_{22}, M_{tt}, M_{t,t-1}, M_{t,t-2}$ ($t = 3, \ldots, T$) can be computed recursively using the sequence of formulae (14) to (25), which are obtained from equations (8), (9), and (12).

$$M_{11}M'_{11} = \Omega'_{11} \tag{14}$$

$$M_{21} = \Omega_{21}[M'_{11}]^{-1}, \tag{15}$$

$$M_{22}M'_{22} = \Omega_{22} - M_{21}M'_{21}, \tag{16}$$

$$M_{31} = \Omega_{31}[M'_{11}]^{-1}, \tag{17}$$

$$M_{32} = [\Omega_{32} - M_{31}M'_{21}][M'_{22}]^{-1}, \tag{18}$$

$$M_{33}M'_{33} = \Omega_0 - M_{31}M'_{31} - M_{32}M'_{32}, \tag{19}$$

$$M_{42} = \Omega_{42}[M'_{22}]^{-1} \tag{20}$$

$$M_{43} = [\Omega_1 - M_{42}M'_{32}][M'_{33}]^{-1}, \tag{21}$$

$$M_{44}M'_{44} = \Omega_0 - M_{42}M'_{42} - M_{43}M'_{43}, \tag{22}$$

and, for $t = 5, \ldots, T$,

$$M_{t,t-2} = \Omega_2[M'_{t-2,t-2}]^{-1}, \tag{23}$$

$$M_{t,t-1} = [\Omega_1 - M_{t,t-2}M'_{t-1,t-2}][M'_{t-1,t-1}]^{-1}, \tag{24}$$

$$M_{tt}M'_{tt} = \Omega_0 - M_{t,t-2}M'_{t,t-2} - M_{t,t-1}M'_{t-1,t-1}. \tag{25}$$

Indeed, the recursive application of these formulae provides a computationally efficient procedure for the Cholesky factorization of the $nT \times nT$ matrix Ω (represented by equation (9) which is an important step in the estimation procedure developed in Chapter 6. It automatically takes account of the fact that Ω is a very sparse matrix and avoids the waste of computing time that would result from the direct application of a standard Cholesky factorization program to this matrix. All of the computations in the formulae (14) to (25) involve only standard operations on $n \times n$ matrices including Cholesky factorizations in equations (14), (16), (19), (22), and (25).

The computation is further simplified by the fact that the sequence of $n \times 3n$ matrices $\bar{M}_t = [M_{tt}, M_{t,t-1}, M_{t,t-2}]$ ($t = 3, 4, \ldots$) converges very rapidly to a constant limit matrix. Indeed, in a study by Nowman (1990*a*), in which the computer program described in Nowman (1990*b*) was applied to a 3 equation, 15 parameter trade cycle model, the above matrix had converged for 7 significant figures by $t = 12$, in spite of the fact that some of the eigenvalues of the system were very close to zero (the real part of one pair of complex eigenvalues being -0.01). In practical applications, therefore, the matrices $\bar{M}_t$ ($t = 3, 4, \ldots$) need to be computed only up to about $t = 12$ (say for the first 3 years of quarterly observations) and for $t = 13, \ldots, T$, put equal to the value for $t = 12$.

The system of equations (23), (24), and (25) is a system of nonlinear second-order difference equations in the elements of the $n \times 3n$ matrix $\bar{M}_t$. It follows from equation (25) that the solution of this system must remain on the surface of the $3n^2$-dimensional hypersphere with its centre at the origin and radius equal to $(\operatorname{tr} \Omega_0)^{1/2}$. The solution must, therefore, converge either to a limit point, which will be an asymptotically stable solution to the system, or to a limit cycle. We shall now prove the existence of an asymptotically stable solution, i.e. an $n \times 3n$ matrix $\bar{M}$ such that, if for some integer $t' \geq 5$ $M_{t'}$ is sufficiently close to $\bar{M}$, then $M_t \to \bar{M}$ as $t \to \infty$. For this purpose we shall use an argument which is equivalent to the transformation of the second-order system into a first-order system of higher dimension.

Since we are concerned, here, with asymptotic theory, there is nothing lost by assuming that $T = 2T^*$ where T^* is a positive integer. We can then partition Ω and M into $2n \times 2n$ matrices as follows:

$$\Omega = \begin{bmatrix} W_{11} & W_{12} & 0 & 0 & 0 & . & . & . & 0 \\ W_{21} & W_0 & W_1' & 0 & 0 & . & . & . & 0 \\ 0 & W_1 & W_0 & W_1' & 0 & . & . & . & 0 \\ . & . & . & . & . & . & . & . & . \\ . & . & . & . & . & . & . & . & . \\ . & . & . & . & . & . & . & . & . \\ 0 & 0 & 0 & 0 & 0 & . & 0 & W_1 & W_0 \end{bmatrix}, \tag{26}$$

$$M = \begin{bmatrix} U_{11} & 0 & 0 & 0 & 0 & . & . & . & 0 \\ U_{21} & U_{22} & 0 & 0 & 0 & . & . & . & 0 \\ 0 & U_{32} & U_{33} & 0 & 0 & . & . & . & 0 \\ . & . & . & . & . & . & . & . & . \\ . & . & . & . & . & . & . & . & . \\ . & . & . & . & . & . & . & . & . \\ 0 & 0 & 0 & 0 & 0 & . & . & U_{T^*,T^*-1} & U_{T^*T^*} \end{bmatrix}, \tag{27}$$

where

$$W_{11} = \begin{bmatrix} \Omega_{11} & \Omega_{12} \\ \Omega_{21} & \Omega_{22} \end{bmatrix},$$

$$W_{12} = \begin{bmatrix} \Omega_{13} & 0 \\ \Omega_{23} & \Omega_{24} \end{bmatrix},$$

$$W_{21} = \begin{bmatrix} \Omega_{31} & \Omega_{32} \\ 0 & \Omega_{42} \end{bmatrix},$$

$$W_0 = \begin{bmatrix} \Omega_0 & \Omega_1' \\ \Omega_1 & \Omega_0 \end{bmatrix},$$

$$W_1 = \begin{bmatrix} \Omega_2 & \Omega_1 \\ 0 & \Omega_2 \end{bmatrix},$$

and, for $\tau = 1, \ldots, T^*$,

$$U_{\tau\tau} = \begin{bmatrix} M_{2\tau-1,2\tau-1} & 0 \\ M_{2\tau,2\tau-1} & M_{2\tau,2\tau} \end{bmatrix}, \tag{28}$$

$$U_{\tau,\tau-1} = \begin{bmatrix} M_{2\tau-1,2\tau-3} & M_{2\tau-1,2\tau-2} \\ 0 & M_{2\tau,2\tau-2} \end{bmatrix}, \tag{29}$$

Then, by substituting from equations (26) and (27) into equation (9), we obtain, for $\tau = 3, \ldots, T^*$, the equations

$$U_{\tau,\tau-1}U'_{\tau,\tau-1} + U_{\tau\tau}U'_{\tau\tau} = W_0, \tag{30}$$

$$U_{\tau,\tau-1}U'_{\tau-1,\tau-1} = W_1. \tag{31}$$

From equation (31) we obtain

$$U_{\tau,\tau-1} = W_1[U'_{\tau-1,\tau-1}]^{-1}, \tag{32}$$

and then, from equations (30) and (32),

$$U_{\tau\tau}U'_{\tau\tau} + W_1[U'_{\tau-1,\tau-1}]^{-1}[U_{\tau-1,\tau-1}]^{-1}W_1' = W_0. \tag{33}$$

The system of equations (32) and (33) is a system of nonlinear first-order difference equations in the elements of the $2n \times 4n$ matrix $\bar{U}_\tau = [U_{\tau\tau}, U_{\tau,\tau-1}]$. We shall now find an asymptotically stable solution of this system.

Let the $2n \times 2n$ matrix function $f(\lambda)$ be defined, for all real λ, by

$$f(\lambda) = W_0 + W_1 e^{-i\lambda} + W_1' e^{i\lambda}. \tag{34}$$

By using the formulae for Ω_0, Ω_1, and Ω_2 given immediately below equation (57) of Chapter 6, it can be shown that $f(\lambda)$ is a positive definite matrix function. It then follows from Rosanov (1967, Theorem 10.1, p. 47) that $f(\lambda)$ can be factorized as

$$f(\lambda) = [\Gamma_0 + e^{-i\lambda}\Gamma_1][\Gamma_0' + e^{i\lambda}\Gamma_1'],$$

where Γ_0 and Γ_1 are $2n \times 2n$ matrices such that $g(z) = |\Gamma_0 + z\Gamma_1|$ has no zeros within the closed unit disc. Moreover, since the right-hand sides of equations (34) and (35) are identically equal, we have

$$\Gamma_0\Gamma_0' + \Gamma_1\Gamma_1' = W_0, \tag{36}$$

$$\Gamma_1\Gamma_0' = W_1, \tag{37}$$

$$\Gamma_0\Gamma_1' = W_1', \tag{38}$$

and from equations (36) and (37) we obtain

$$\Gamma_0\Gamma_0' + W_1[\Gamma_0']^{-1}[\Gamma_0]^{-1}W_1'] = W_0. \tag{39}$$

Now let U_0 be a real lower triangular matrix, with positive elements along the diagonal, such that

$$U_0U_0' = \Gamma_0\Gamma_0'. \tag{40}$$

Then, from equations (39) and (40), we have

$$U_0U_0' + W_1[U_0']^{-1}[U_0^{-1}]W_1' = W_0. \tag{41}$$

Let U_1 be the $2n \times 2n$ matrix defined by

$$U_1 = W_1[U_0']^{-1}. \tag{42}$$

Then equations (41) and (42) imply that a stationary solution of the difference equation system comprising equations (32) and (33) is given by

$$\bar{U}_\tau = \bar{U} = [U_0, U_1]. \tag{43}$$

We shall now show that this is an asymptotically stable solution.

From equations (33) and (41) we obtain

$$\begin{aligned} U_{\tau+1,\tau+1}U_{\tau+1,\tau+1}' - U_0U_0' &= W_1\{[U_0']^{-1}[U_0]^{-1} - [U_{\tau\tau}']^{-1}[U_{\tau\tau}]^{-1}\}W_1' \\ &= W_1[U_{\tau\tau}']^{-1}[U_{\tau\tau}]^{-1}\{U_{\tau\tau}U_{\tau\tau}' - U_0U_0'\}[U_0']^{-1}[U_0]^{-1}W_1'. \end{aligned} \tag{44}$$

The fact that $|\Gamma_0 + z\Gamma_1|$ has no zeros in the closed unit disc, together with equations (37) and (40), implies that $|U_0U_0' + zW_1|$ has no zeros in the closed unit disc and, hence, that all the eigenvalues of $[U_0']^{-1}[U_0]^{-1}W_1'$ (which are identical with the eigenvalues of $[U_0']^{-1}[U_0]^{-1}W_1$) are in the open unit disc. Moreover, if $U_{\tau\tau}$ is sufficiently close to U_0, then all the eigenvalues of $W_1[U_{\tau\tau}']^{-1}[U_{\tau\tau}]^{-1}$ are in the open unit disc. Equation (44) implies, therefore, that, if for some value of τ $U_{\tau\tau}$ is sufficiently close to U_0, then $U_{\tau\tau} \rightarrow U_0$ as $\tau \rightarrow \infty$. It then follows from equations (32) and (42) that $U_{\tau,\tau-1} \rightarrow U_1$ as $t \rightarrow \infty$ and, hence, that the stationary solution $\bar{U}$ (see equation (43)) of the first-order system comprising equations (32) and (33) is asymptotically stable.

We now return to the consideration of the second-order system in $\bar{M}_t = [M_{tt}, M_{t,t-1}, M_{t,t-2}]$, i.e. the system comprising equations (23), (24), and (25). It follows from equations (28) and (29) that this system has an asymptotically stable solution

$$\bar{M}_t = \bar{M} = [M_0, M_1, M_2], \tag{45}$$

where M_0, M_1, and M_2 are obtained from the partitions

$$U_0 = \begin{bmatrix} M_0 & 0 \\ M_1 & M_0 \end{bmatrix}, \tag{46}$$

$$U_1 = \begin{bmatrix} M_2 & M_1 \\ 0 & M_2 \end{bmatrix}. \qquad (47)$$

The results obtained in this section can be summed up in the following theorem.

THEOREM 1: (*a*) *Under Assumptions* 1 *to* 4, *the observed vectors* $\bar{x}_t$ ($t = 1, 2, \ldots$) *defined by equation* (3) *satisfy the vector autoregressive moving average model* (13), *in which the matrices* $M_{tt}(\theta, \mu)$ *are obtained recursively from the system comprising equations* (14) *to* (25).

(*b*) *The difference equation system comprising equations* (23), (24), *and* (25) *has an asymptotically stable stationary solution given by equations* (45), (46), *and* (47), *where* U_0 *and* U_1 *are defined by equations* (40) *and* (42).

Finally, it should be noticed that all of the above results can be extended to a system of any order. If, instead of the second-order system (2), we have a kth order continuous time system, then the discrete observations will satisfy a vector autoregressive moving average model which is of order k in both its autoregressive and moving average parts. Moreover, the analysis of the behaviour of the moving average coefficient matrices $M_{t,t-r}$ ($r = 1, \ldots, k$) as $t \rightarrow \infty$ is then identical with that for the second-order system, except that the matrices W_0 and W_1 will now be $kn \times kn$ matrices partitioned into submatrices of order $n \times n$.

3. Hypothesis Testing in a Closed Model

We turn now to the problem of testing hypotheses relating to a closed continuous time model of the type discussed in the preceding section. We shall deal first with the testing of the hypothesis that a sample of discrete data has been generated by a specified continuous time model of that form, and then with the testing of hypotheses that can be represented by sets of restrictions on the parameters of the model. For convenience of exposition, and to facilitate the use of the results obtained in Section 2, we shall deal specifically with the second-order model comprising equations (1) and (2) together with Assumptions 1 to 4. The proposed procedures can, however, be applied to a similar model of any order.

The results obtained in Section 2 suggest the following three-stage testing strategy.

1. Test the hypothesis that the sample has been generated by a VARMA (2.2) model (i.e. a vector autoregressive moving average model which is of the second order in both its autoregressive and moving average parts) against the hypothesis that it has been generated by a VARMA model of higher order.

2. Assuming that the tests in Stage 1 do not result in the rejection of the VARMA (2.2) model, test the hypothesis that the autoregressive and moving average coefficient matrices of this model satisfy the restrictions implied by the specified continuous time model.

3. Assuming that the test in Stage 2 does not result in the rejection of the restrictions on the VARMA model, test the hypotheses represented by sets of restrictions on the parameters of the continuous time model.

Since procedures for testing the order of an unrestricted VARMA model are well known, we shall not discuss the first stage. It is worth noting, however, that this stage is unlikely to be worth undertaking except for very small models or when exceptionally large samples are available. It is very unlikely that we could reject the hypothesis that a sample of less than 100 observations on 10 or more economic variables has been generated by an unrestricted VARMA (2.2) model, since with 10 variables such a model has 455 parameters.

In order to rigorously justify the tests proposed for stages 2 and 3, it is necessary to introduce further assumptions.

Assumption 5. The true value of the parameter vector $[\theta, \mu]$ is an interior point of a closed bounded set Θ in R^{p+q}.

Assumption 6. For any vector $[\theta, \mu]$ belonging to the set Θ, the zeros of the polynomial $\phi(\lambda) = |\lambda^2 I + \lambda A_1(\theta) + A_2(\theta)|$ have negative real parts (i.e. the system (1) is stable), and the matrix $\Sigma(\mu)$ is positive definite.

Assumption 7. In a neighbourhood of the true parameter vector $[\theta_0, \mu_0]$, the elements of the matrices $A_1(\theta)$ and $A_2(\theta)$ are twice continuously differentiable functions of θ, and the elements of $\Sigma(\mu)$ are twice continuously differentiable functions of μ.

Assumption 8. The matrix $(1/T)E[\partial^2 L/\partial(\theta, \mu)'\partial(\theta, \mu)]$ evaluated in Theorem 2 tends to a nonsingular matrix as $T \to \infty$.

Assumption 9. The innovation vector $\zeta(dt)$ in the system (1) is Gaussian, i.e. $\int_0^t \zeta(dr)$ is Brownian motion.

Stage 2 of the testing strategy is complicated by the fact that the matrices M_{tt}, $M_{t,t-1}$, and $M_{t,t-2}$ in the system (13) are changing over time and, consequently, this system is not exactly nested in a VARMA model with constant moving average coefficient matrices. But the matrices M_{tt}, $M_{t,t-1}$, and $M_{t,t-2}$ converge very rapidly to the matrices M_0, M_1, and M_2 given by equations (46) and (47), and for values of t greater than about 12 can be regarded, for practical purposes, as equal to these matrices. The simplest procedure, therefore, is to test the restricted VARMA model

$$\bar{x}_t - F_1(\theta)\bar{x}_{t-1} - F_2(\theta)\bar{x}_{t-2} = M_0(\theta, \mu)\varepsilon_t + M_1(\theta, \mu)\varepsilon_{t-1} + M_2(\theta, \mu)\varepsilon_{t-2} \quad (t = t_1, \ldots, T), \tag{48}$$

against the unrestricted VARMA model

$$\bar{x}_t - F_1\bar{x}_{t-1} - F_2\bar{x}_{t-2} = M_0\varepsilon_t + M_1\varepsilon_{t-1} + M_2\varepsilon_{t-2} \quad (t = t_1, \ldots, T), \tag{49}$$

where t_1 is an integer depending on the degree of accuracy to which the calculations are being made. The value of t_1 should be chosen so that for $t \geq t_1$ the matrices $M_{tt}(\theta, \mu)$, $M_{t,t-1}(\theta, \mu)$, and $M_{t,t-2}(\theta, \mu)$ are (to the required degree of accuracy) equal to the matrices $M_0(\theta, \mu)$, $M_1(\theta, \mu)$, $M_2(\theta, \mu)$ respectively for values of the vector $[\theta, \mu]$ in a neighbourhood the Gaussian estimator $[\hat{\theta}, \hat{\mu}]$.

It is important to notice that we are not assuming that the system (48) can be treated as stationary for $t \geq t_1$. The vectors $\bar{x}_{t_1-1}$ and $\bar{x}_{t_1-2}$ depend on the initial state vector $[y_1', y_2']$ defined by equation (2), and their distributions could differ considerably from the limiting distribution of $\bar{x}_t$ as $t \to \infty$, in spite of the fact that M_{tt}, $M_{t,t-1}$, and $M_{t,t-2}$ have converged (to the required degree of accuracy) to their limits when $t = t_1$. This is because the matrices M_{tt}, $M_{t,t-1}$, and $M_{t,t-2}$ depend only on the parameter vector $[\theta, \mu]$ and not on the initial state vector $[y_1', y_2']$. When testing the restricted model (48) against the unrestricted model (49) it is best, therefore, to treat the observed vectors $\bar{x}_{t_1-1}$ and $\bar{x}_{t_1-2}$ as non-random. We can also treat the unobserved vectors ε_{t_1-1} and ε_{t_1-2} as non-random by putting them equal to the estimates $\hat{\varepsilon}_{t_1-1}$ and $\hat{\varepsilon}_{t_1-2}$ obtained in the application of the Gaussian estimation procedure (described in Chapter 6) to the complete sample. The vectors $\bar{x}_{t_1-1}$, $\bar{x}_{t_1-2}$ ε_{t_1-1} and ε_{t_1-2} are, of course, random. Under Assumptions 1 to 9, however, they are independent of the vectors $\varepsilon_{t_1}, \varepsilon_{t_1+2}, \ldots, \varepsilon_T$. The formal treatment of the vectors $\bar{x}_{t_1-1}, \bar{x}_{t_1-2}, \varepsilon_{t_1-1}$, and ε_{t_1-2} as non-random is equivalent, therefore, to basing the test on the conditional distribution of $\bar{x}_{t_1}, \bar{x}_{t_1+2}, \ldots, \bar{x}_T$ relative to the assumption that $\bar{x}_{t_1-1}$ and $\bar{x}_{t_1-2}$ are equal to their observed values and that $\varepsilon_{t_1-1} = \hat{\varepsilon}_{t_1-1}$ and $\varepsilon_{t_1-2} = \hat{\varepsilon}_{t_1-2}$.

Let $L_1(\theta, \mu)$ denote minus twice the logarithm of the Gaussian likelihood function (less a constant) obtained from the system (48) and $L_2(F_1, F_2, M_0, M_1, M_2)$ denote minus twice the logarithm of the Gaussian likelihood function (less a constant) obtained from the system (49), the vectors $\bar{x}_{t_1-1}$, $\bar{x}_{t_1-2}$, ε_{t_1-1}, and ε_{t_1-2} being treated, in each case, as non-random with $\varepsilon_{t_1-1} = \hat{\varepsilon}_{t_1-1}$ and $\varepsilon_{t_1-2} = \hat{\varepsilon}_{t_1-2}$. Then

$$L_1(\theta, \mu) = 2(T - t_1 + 1) \log |M_0(\theta, \mu)| + \sum_{t=t_1}^{T} \varepsilon_t' \varepsilon_t, \tag{50}$$

where the vectors $\varepsilon_{t_1}, \ldots, \varepsilon_T$ are regarded as functions of $[\theta, \mu]$ which can be obtained recursively from the system

$$\varepsilon_t = M_0^{-1}(\theta, \mu)[\bar{x}_t - F_1(\theta)\bar{x}_{t-1} - F_2(\theta)\bar{x}_{t-2} - M_1(\theta, \mu)\varepsilon_{t-1} - M_2(\theta, \mu)\varepsilon_{t-2}]$$
$$(t = t_1, \ldots, T). \tag{51}$$

Similarly,

$$L_2(F_2, F_2, M_0, M_1, M_2) = 2(T - t_1 + 1) \log |M_0| + \sum_{t=t_1}^{T} \varepsilon_t' \varepsilon_t, \tag{52}$$

where the vectors $\varepsilon_{t_1}, \ldots, \varepsilon_T$ are regarded as functions of $[F_1, F_2, F_2, M_0, M_1, M_2]$ which can be obtained recursively from the system

$$\varepsilon_t = M_0^{-1}[\bar{x}_t - F_1\bar{x}_{t-1} - F_2\bar{x}_{t-2} - M_1\varepsilon_{t-1} - M_2\varepsilon_{t-2}] \quad (t = t_1, \ldots, T). \tag{53}$$

Now let $\bar{L}_1$ denote the minimum of L_1 with respect to $[\theta, \mu]$ and $\bar{L}_2$ the minimum of L_2 with respect to $[F_1, F_2, M_0, M_1, M_2]$. Then by using standard asymptotic theory (see Hannan and Deistler 1988, ch. 4, and Amemiya 1985, pp. 142–4) it can be shown that, if the data are generated by the continuous time model comprising equations (1) and (2) then, under Assumptions 1 to 9, the distribution of $\bar{L}_1 - \bar{L}_2$ tends to a chi-square distribution with $(9n^2 + n)/2 - (p + q)$ degrees of freedom as $T - t_1 \rightarrow \infty$ and $t_1 \rightarrow \infty$. In practical applications there is, of course, nothing to be gained by choosing t_1 larger than is required for the convergence of M_{tt}, $M_{t,t-1}$, and $M_{t,t-2}$ to the required degree of accuracy.

We turn now to the third stage of the testing strategy. We assume that the hypothesis to be tested can be represented by a set of restrictions

$$h(\theta, \mu) = 0 \tag{54}$$

where h is an $r \times 1$ vector of differentiable functions. The most convenient way of testing such a hypothesis is to apply a Wald test. For this purpose we require a formula for the asymptotic covariance matrix of the Gaussian estimator $[\hat{\theta}, \hat{\mu}]$, i.e. the estimator obtained by minimizing the function L defined by equation (55). The formula for the asymptotic covariance matrix of $[\hat{\theta}, \hat{\mu}]$ depends crucially on Assumption 9, which implies that $[\hat{\theta}, \hat{\mu}]$ is the true maximum likelihood estimator rather than merely a pseudo-maximum likelihood estimator.

If Assumptions 1 to 9 are satisfied, $[\hat{\theta}, \hat{\mu}]$ is asymptotically normally distributed with the covariance matrix $2\{E[\partial^2 L/\partial(\theta, \mu)'\partial(\theta, \mu)]\}^{-1}$. If Assumptions 1 to 8 are satisfied but Assumption 9 is replaced by a weaker assumption, then $[\hat{\theta}, \hat{\mu}]$ may still be asymptotically normally distributed, but its asymptotic covariance matrix will have a more complicated form which depends on the higher moments of the innovation vector. The reason for this is that the moving average coefficient matrices M_{tt}, $M_{t,t-1}$, and $M_{t,t-2}$ as well as the autoregressive coefficient matrices F_1 and F_2 depend on the parameter vector θ. (See Robinson 1988 for a general discussion of this problem.)

In order to derive a formula for the expected Hessian $E[\partial^2 L/\partial(\theta, \mu)'\partial(\theta, \mu)]$ it is convenient to write L in the form

$$L(\theta, \mu) = \log|V(\theta, \mu)| + tr\{[V^{-1}(\theta, \mu)][x - m(\theta)][x - m(\theta)]'\} \tag{55}$$

where

$$m(\theta) = F^{-1}[Gy + h], \tag{56}$$

$$V(\theta, \mu) = [F]^{-1}\Omega[F']^{-1}. \tag{57}$$

The expression on the right-hand side of equation (55) is identical with that on the right-hand side of equation (58) in Chapter 6 as can be seen by substituting from equations (56) and (57) into (55) and using the equation $|F| = 1$. Equation (55) can also be obtained directly, since it follows from equation (4) that $m(\theta)$ and $V(\theta, \mu)$ are the mean and covariance matrix, respectively, of $\bar{x}$. Formulae expressing $E[\partial^2 L/\partial(\theta, \mu)'\partial(\theta, \mu)]$ in terms of the parameters of the model are given by the following theorem.

THEOREM 2: *Under Assumptions 1 to 4, the elements of* $E[\partial^2 L/\partial(\theta, \mu)'\partial(\theta, \mu)]$ *are given by equations* (58), (59), *and* (60).

$$E[\partial^2 L/\partial\theta_i, \partial\theta_j] = \operatorname{tr}\left\{V^{-1}\left[\frac{\partial}{\partial\theta_i}V\right]V^{-1}\left[\frac{\partial}{\partial\theta_j}V\right]\right\} + 2\operatorname{tr}\left\{V^{-1}\left[\frac{\partial}{\partial\theta_i}m\right]\left[\frac{\partial}{\partial\theta_j}m'\right]\right\}, \tag{58}$$

$$E[\partial^2 L/\partial\theta_i, \partial\mu_j] = \operatorname{tr}\left\{V^{-1}\left[\frac{\partial}{\partial\theta_i}V\right]V^{-1}\left[\frac{\partial}{\partial\mu_j}V\right]\right\}, \tag{59}$$

$$E[\partial^2 L/\partial\mu_i, \partial\mu_j] = \operatorname{tr}\left\{V^{-1}\left[\frac{\partial}{\partial\mu_i}V\right]V^{-1}\left[\frac{\partial}{\partial\mu_j}V\right]\right\}. \tag{60}$$

PROOF: Differentiating equation (55) with respect to θ_i we obtain

$$\frac{\partial L}{\partial\theta_i} = \operatorname{tr}\left\{V^{-1}\left[\frac{\partial}{\partial\theta_i}V\right][I - V^{-1}(x - m)(x - m)']\right\} - 2\operatorname{tr}\left\{V^{-1}\left[\frac{\partial}{\partial\theta_i}m\right](x - m)'\right\}. \tag{61}$$

Then differentiating equation (61) with respect to θ_j, we obtain

$$\frac{\partial^2 L}{\partial\theta_i\partial\theta_j} = -\operatorname{tr}\left\{V^{-1}\left[\frac{\partial}{\partial\theta_j}V\right]V^{-1}\left[\frac{\partial}{\partial\theta_i}V\right][I - V^{-1}(x - m)(x - m)']\right\}$$
$$+ \operatorname{tr}\left\{V^{-1}\left[\frac{\partial^2}{\partial\theta_i\partial\theta_j}V\right][I - V^{-1}(x - m)(x - m)']\right\}$$
$$+ \operatorname{tr}\left\{V^{-1}\left[\frac{\partial}{\partial\theta_i}V\right]V^{-1}\left[\frac{\partial}{\partial\theta_j}V\right]V^{-1}(x - m)(x - m)'\right\}$$

$$+ 2\mathrm{tr}\left\{V^{-1}\left[\frac{\partial}{\partial\theta_i}V\right]V^{-1}\left[\frac{\partial}{\partial\theta_j}m\right](x-m)'\right]\right\}$$

$$+ 2\mathrm{tr}\left\{V^{-1}\left[\frac{\partial}{\partial\theta_j}V\right]V^{-1}\left[\frac{\partial}{\partial\theta_i}m\right](x-m)'\right]\right\}$$

$$- 2\mathrm{tr}\left\{V^{-1}\left[\frac{\partial^2}{\partial\theta_i\partial\theta_j}m\right](x-m)'\right]\right\}$$

$$+ 2\mathrm{tr}\left\{V^{-1}\left[\frac{\partial}{\partial\theta_i}m\right]\left[\frac{\partial}{\partial\theta_j}m\right]'. \qquad (62)$$

Equation (58) is obtained by applying the operator E to equation (62) and taking account of the relations $E(x - m) = 0$ and $E(x - m)(x - m)' = V$. In a similar way we obtain equations (59) and (60). *End of Proof*

It is not necessary to compute F^{-1} in order to obtain the matrix V occurring in the formulae (58), (59), and (60). The easiest way of computing V is to use the formulae

$$V = RR',$$

where R is an $nT \times nT$ lower triangular matrix whose $n \times n$ submatrices can be computed by the recursive solution of the system

$$FR = M,$$

taking advantage of the fact that F is a sparse matrix (see equation (5)). Moreover, the easiest way of computing V^{-1}, which also occurs in the formulae (58), (59), and (60), is to use the formula

$$V^{-1} = [R']^{-1}[R]^{-1},$$

where R^{-1} is an $nT \times nT$ lower triangular matrix whose $n \times n$ submatrices can be computed by the recursive solution of the system

$$MR^{-1} = F,$$

taking account of the fact that M is a sparse matrix (see equation (12)).

In order to apply the above results in a Wald test of the hypothesis represented by equation (54) we use the statistic

$$W = 2\left| h'(\theta,\mu)\left[\frac{\partial h}{\partial(\theta,\mu)}\left\{E\left[\frac{\partial^2 L}{\partial(\theta,\mu)'\partial(\theta,\mu)}\right]\right\}^{-1}\frac{\partial h'}{\partial(\theta,\mu)'}\right]^{-1} h(\theta,\mu)\right|_{\mu=\hat{\mu}}^{\theta=\hat{\theta}}.$$

If equation (54) is satisfied then, under Assumptions 1 to 9, the distribution of W tends to a chi-square distribution with r degrees of freedom as $T \to \infty$.

In this section we have discussed a three-stage strategy which involves the testing of a sequence of successively more restrictive hypotheses against specified alternatives. Further tests of dynamic specification, which make no assumptions about alternative hypotheses, can be made by applying test statistics of the type introduced by Box and Pierce (1970) to the residual vector $\hat{\varepsilon}$ obtained from the estimated exact discrete model (11). This vector will have been computed in the course of obtaining the Gaussian estimate $[\hat{\theta}, \hat{\mu}]$ by the procedure described in Chapter 6. Since, under Assumptions 1 to 9, ε is an $nT \times 1$ vector of independent normally distributed random variables each with mean 0 and variance 1, the test statistic can be formulated in terms of the autocovariances of $\hat{\varepsilon}_t$ rather than their serial correlation coefficients. Moreover, since each element of ε_t depends on the innovations in all of the equations of the continuous time model it is appropriate to use a single test statistic for the complete system. A simple test can be based on the statistic S defined by

$$S = \frac{1}{n(T-l)} \sum_{r=1}^{l} \left(\sum_{t=l+1}^{T} \hat{\varepsilon}_t' \hat{\varepsilon}_{t-r} \right)^2. \tag{63}$$

If l and $T-l$ are sufficiently large the distribution of S will be approximately a chi-square distribution with l degrees of freedom.

4. The Treatment of Exogenous Variables

The general procedures developed in Section 3 can be applied to an open model by extending the exact discrete model to include exogenous variables, using the results of Theorems 2.3 and 2.4 of Chapter 6. The resulting model is, in fact, a pseudo-exact discrete model of the type introduced by Phillips (1974, 1976*a*); i.e. it will be exact if the time paths of the exogenous variables are polynomials in t of degree not exceeding two. Instead of the closed continuous time system (1) we now assume the open system

$$d[Dx(t)] = \{A_1(\theta)Dx(t) + A_2(\theta)x(t) + B(\theta)z(t)\}dt + \zeta(dt) \qquad (t \geq 0), \tag{64}$$

where $z(t)$ is an $m \times 1$ vector of exogenous variables which are non-random functions of t. We assume that $z(t)$ includes an $m^s \times 1$ vector $z^s(t)$ of stock variables and an $m^f \times 1$ vector $z^f(t)$ of flow variables and that its elements are ordered (without loss of generality) so that

$$z(t) = \begin{bmatrix} z^s(t) \\ z^f(t) \end{bmatrix}.$$

We can then define a sequence of observable vectors $\bar{z}_1, \ldots, \bar{z}_T$ by

$$\bar{z}_t = \begin{bmatrix} \frac{1}{2}\{z^s(t) + z^s(t-1)\} \\ \int_{t-1}^{t} z^f(r)dr \end{bmatrix} \qquad (t = 1, \ldots, T).$$

If the elements of $z(t)$ are polynomials in t of degree not exceeding two, then the observations $x^s(0), \bar{x}_1, \ldots, \bar{x}_T, \bar{z}_1, \ldots, \bar{z}_T$ satisfy an exact discrete model which has the VARMAX representation

$$\bar{x}_1 - G_{11}(\theta)y_1 - G_{12}(\theta)y_2 - E_{11}(\theta)\bar{z}_1 - E_{12}(\theta)\bar{z}_2 - E_{13}(\theta)\bar{z}_3$$
$$= M_{11}(\theta, \mu)\varepsilon_1,$$
$$\bar{x}_2 - C_{11}(\theta)\bar{x}_1 - G_{21}(\theta)y_1 - G_{22}(\theta)y_2 - E_{21}(\theta)\bar{z}_1 - E_{22}(\theta)\bar{z}_2 - E_{23}(\theta)\bar{z}_3,$$
$$= M_{21}(\theta, \mu)\varepsilon_1 + M_{22}(\theta, \mu)\varepsilon_2,$$
$$\bar{x}_t - F_1(\theta)\bar{x}_{t-1} - F_2(\theta)\bar{x}_{t-2} - E_0(\theta)\bar{z}_t - E_1(\theta)\bar{z}_{t-1} - E_2(\theta)\bar{z}_{t-2}$$
$$= M_{tt}(\theta, \mu)\varepsilon_t + M_{t't-1}(\theta, \mu)\varepsilon_{t-1} + M_{t,t-2}(\theta, \mu)\varepsilon_{t-2}$$
$$(t = 3, \ldots, T). \quad (65)$$

$$E(\varepsilon_t) = 0 \qquad (t = 1, \ldots, T), \qquad E(\varepsilon_t\varepsilon_t') = I \qquad (t = 1, \ldots, T)$$
$$E(\varepsilon_s\varepsilon_t') = 0 \qquad (s \neq t;\, s, t = 1, \ldots, T).$$

The explicit form of the matrix functions E_{11}, E_{12}, E_{13}, E_{21}, E_{22}, E_{23}, E_0, E_1, and E_2 in the system (65) are given by Theorems 2.3 and 2.4 of Chapter 6, while all the other matrix functions in the system (including the moving average coefficient matrices on the right-hand side) are identical with those in the VARMA system (13) in Section 2 of this Chapter.

It is important to notice that all of the coefficient matrices in the VARMAX model (65) (including the coefficient matrices of the exogenous variables) depend only on the parameters of the continuous time model. They do not depend on the parameters of the quadratic functions of t defining the paths of the exogenous variables. For this reason the model (65) will be a good approximation even if the behaviour of the exogenous variable changes greatly as between different parts of the sample period. Its accuracy does not depend on the accuracy with which each element of $z(t)$ can be approximated by a single quadratic function over the whole sample period $[0, T]$, but only on the accuracy with which each element can be approximated by a sequence of quadratic functions over the overlapping intervals $[0, 3], [1, 4], [2, 5], \ldots, [T-3, T]$. Indeed, if $z(t)$ is a vector of arbitrary thrice differentiable functions of t, then the errors in the model (65) will be of the order δ^4 as the unit observation period δ tends to zero.

The hypothesis that the discrete data have been generated by the continuous time model (64) can be tested by using a likelihood ratio test of the restricted VARMAX model

$$\bar{x}_t - F_1(\theta)\bar{x}_{t-1} - F_2(\theta)\bar{x}_{t-2} - E_0(\theta)\bar{z}_t - E_1(\theta)\bar{z}_{t-1} - E_2(\theta)\bar{z}_{t-2}$$
$$= M_0(\theta, \mu)\varepsilon_t + M_1(\theta, \mu)\varepsilon_{t-1} + M_2(\theta, \mu)\varepsilon_{t-2} \qquad (t = t_1, \ldots, T) \tag{66}$$

against the unrestricted VARMAX model

$$\bar{x}_t - F_1\bar{x}_{t-1} - F_2\bar{x}_{t-2} - E_0\bar{z}_t - E_1\bar{z}_{t-1} - E_2\bar{z}_{t-2}$$
$$= M_0\varepsilon_t + M_1\varepsilon_{t-1} + M_2\varepsilon_{t-2} \qquad (t = t_1, \ldots, T) \tag{67}$$

where the matrix functions $M_0(\theta, \mu)$, $M_1(\theta, \mu)$, and $M_2(\theta, \mu)$ in the model (66) are identical with those in the VARMA model (48), and t_1 is the value of t by which $M_{tt}(\theta, \mu)$, $M_{t,t-1}(\theta, \mu)$, and $M_{t,t-2}(\theta, \mu)$ have converged to $M_0(\theta, \mu)$, $M_1(\theta, \mu)$, and $M_2(\theta, \mu)$, respectively, to the required degree of accuracy. The test statistic is $\bar{L}_1 - \bar{L}_2$ where $\bar{L}_1$ is the minimum, with respect $[\theta, \mu]$, of the function L_1 defined by equation (50) and $\bar{L}_2$ is the minimum, with respect to $[F_1, F_2, E_0, E_1, E_2, M_0, M_1, M_2]$, of the function L_2 defined by equation (52). The vectors $\varepsilon_1, \ldots, \varepsilon_T$ in the function L_1 are now regarded as functions of $[\theta, \mu]$ obtained, recursively, from the system

$$\varepsilon_t = M_0^{-1}(\theta, \mu)[\bar{x}_t - F_1(\theta)\bar{x}_{t-1} - F_2(\theta)\bar{x}_{t-2} - E_0(\theta)\bar{z}_t - E_1(\theta)\bar{z}_{t-1}$$
$$- E_2(\theta)\bar{z}_{t-2} - M_1(\theta, \mu)\varepsilon_{t-1} - M_2(\theta, \mu)\varepsilon_{t-2}] \qquad (t = t_1, \ldots, T) \tag{68}$$

while the vectors $\varepsilon_1, \ldots, \varepsilon_T$ in the function L_2 are regarded as functions of $[F_1, F_2, E_0, E_1, E_2, M_0, M_1, M_2]$ obtained recursively, from the system

$$\varepsilon_t = \mathrm{M}_0^{-1}[\bar{x}_t - F_1\bar{x}_{t-1} - F_2\bar{x}_{t-2} - E_0\bar{z}_t - E_1\bar{z}_{t-1} - E_2\bar{z}_{t-2}$$
$$- M_1\varepsilon_{t-1} - M_2\varepsilon_{t-2}] \qquad (t = t_1, \ldots, T). \tag{69}$$

If the model (66) were exactly satisfied by the data and the $\bar{z}_t$ $(t = t_1, \ldots, T)$ satisfied suitable regularity conditions (see Hannan and Deistler 1988, p. 102), then the distribution of $\bar{L}_1 - \bar{L}_2$ would tend to a chi-square distribution with $(15n^2 + n)/2 - (p + q)$ degrees of freedom as $T \to \infty$.

Because of the errors resulting from the quadratic approximation of $z(t)$ used in the construction of the model (66), the likelihood ratio test described above will not be an exact asymptotic test of the hypothesis that the data have been generated by the continuous time model (64). But the empirical study of Phillips (1976*b*) suggests that, with exogenous variables as smooth as those commonly used in continuous time macroeconometric models and quarterly observations, the errors resulting from the quadratic approximations are likely to be fairly small, indeed smaller than those resulting from errors in the data.

Hypotheses relating to parameters of the continuous time model can be tested by a Wald test, as described in Section 3, using the formulae given by Theorem 2. The vector h, which occurs in the formula for $m(\theta)$ given by equation (56) will now include exogenous variables and will be as defined by equation (53) of Chapter 6 rather than equation (7) of this chapter.

Finally, we can apply a more general test of dynamic specification, which makes no assumptions about alternative hypotheses, by using the test statistic S defined by equation (63). The vectors $\hat{\varepsilon}_t$ $(t = l + 1, \ldots, T)$ occurring in this formula will now be the estimates of ε_t obtained by applying the Gaussian estimation procedure described in Chapter 6 to the open model using the complete sample.

5. Conclusion

Our main concern in this chapter has been the development of practical procedures for testing hypotheses relating to a continuous time econometric model, using a sample of discrete stock and flow data. We have considered tests of both the general hypothesis that the discrete data have been generated by a continuous time model of a specified form and tests of more restrictive hypotheses relating to the parameters of the specified continuous time model.

The tests for a closed model have been developed within the framework of the VARMA representation of the exact discrete model, and one of the main contributions of the chapter has been to provide a more detailed discussion of this representation than has previously appeared in the literature. The result of this discussion are summed up in Theorem 1.

The problem created by the inclusion of exogenous variables and the fact that the process generating their continuous time paths is unknown have been discussed in Section 4. Here we have developed tests within the framework of a pseudo-exact discrete model of the type used in Chapter 6, i.e. a model which would be exactly satisfied by the discrete data if the unobservable continuous time paths of the exogenous variables were polynomials in t of degree not exceeding two. Although the resulting tests for an open model are not exact asymptotic tests, as are those for a closed model, they can be expected to be fairly accurate when the exogenous variables are as smooth as those occurring in a typical continuous time macroeconometric model.

References

AMEMIYA, T. (1985), *Advanced Econometrics* (Oxford, Blackwell).

BERGSTROM, A. R. (1984), 'Continuous time stochastic models and issues of aggregation over time', in Z. Griliches and M. D. Intriligator (eds.), *Handbook of Econometrics* (Amsterdam, North-Holland).

BOX, G. E. P. and D. A. PIERCE (1970), 'Distribution of residual autocorrelations in autoregressive-integrated moving average time series models', *Journal of the American Statistical Association,* **64**, 1509–26.

HANNAN, E. J. and M. DEISTLER (1988), *The Statistical Theory of Linear Systems* (New York, Wiley).

NOWMAN, K. B. (1990*a*), 'Finite sample properties of the Gaussian estimation of an open higher order continuous time dynamic model with mixed stock and flow data', unpublished paper, University of Essex.

—— (1990*b*), 'Computer program manual for computing the Gaussian estimates of an open second order continuous time dynamic model with mixed stock and flow data', unpublished paper, University of Essex.

PHILLIPS, P. C. B. (1974), 'The estimation of some continuous time models', *Econometrica,* **42**, 803–24.

—— (1976*a*), 'The estimation of linear stochastic differential equations with exogenous variables', in A. R. Bergstrom (ed.), *Statistical Inference in Continuous Time Economic Models* (Amsterdam, North-Holland).

—— (1976*b*), 'Some computations based on observed data series of the exogenous variable component in continuous systems', in A. R. Bergstrom (ed.), *Statistical Inference in Continuous Time Economic Models* (Amsterdam, North-Holland).

ROBINSON, P. M. (1988), 'Using Gaussian estimators robustly', *Oxford Bulletin of Economics and Statistics,* **50**, 97–106.

ROZANOV, Y. A. (1967), *Stationary Random Processes* (San Francisco, Holden-Day).

8

Optimal Forecasting of Discrete Stock and Flow Data Generated by a Higher Order Continuous Time System*

1. Introduction

A MAIN source of difficulty in macroeconomic modelling is the fact that the unit observation period for most macroeconomic variables is much longer than the intervals between the microeconomic decisions that they reflect. Although such variables as aggregate consumption, exports, imports and the gross domestic product are measured only at quarterly intervals (in earlier years only at annual intervals), they are the outcome of millions of decisions taken by different individuals at different points of time. For most such variables there will be thousands of small changes at random intervals of time in a single day, and the changes can occur at any time. A realistic macroeconomic model which takes account of the microeconomic decision processes must, therefore, be formulated in continuous time.

As a tool for forecasting, a continuous time model has the obvious advantage that it can be used to generate forecasts of the continuous time paths of the variables. Although the accuracy of such forecasts can never be exactly checked, they can be of use to businessmen and policy makers and could facilitate the use of optimal control techniques in economic policy formulation (Bergstrom 1987). But, even if our sole aim is to obtain the best forecasts of the post-sample discrete observations, which is all that we need for checking the predictive power of the model, there is a strong argument for formulating the model in continuous time. For it is only through a continuous time model that we can, accurately, take account of the restrictions implied by economic theory and other *a priori* knowledge. Such *a priori* information can be of great importance in increasing the accuracy of the forecasts.

* This paper is based on research funded by the Economic and Social Science Research Council (ESRC); Reference No. RB0023 2215.

By far the most common type of *a priori* restriction in econometric models is the assumption that certain elements (usually most of the elements) in the matrices of coefficients of the variables are zero. When the model is formulated as a system of differential equations this type of restriction can be interpreted as meaning that the variables can be arranged in a causal chain in which each variable responds, directly, to the stimulus provided by a subset of the other variables in the model. More precisely, it assumes that the change in any variable during a small interval of time depends on the levels, during that interval, of only a subset of the other variables in the model. The assumption that the variables can be arranged in causal chains of this type need not depend on any elaborate or controversial economic theory, but only on our knowledge of the information available to the various economic agents at particular points of time. It is obvious, for example, that aggregate consumers' expenditure on a particular day can be influenced by the levels (on that day) of only those variables which are known to consumers, particularly personal income, personal assets, and prices, and not by the levels (on that day) of such variables as exports, imports, and investment. This is very strong information which can be used to reduce the variances of the parameter estimates and the forecasts based on these estimates. But it can be used, efficiently, only if the model is formulated in continuous time. For, because of the interaction of all variables during the unit observation period (say a quarter), the conditional expectation of the value of each variable in quarter t, conditional on information up to the end of quarter $t-1$, will be a function of the values in period $t-1$ (and generally periods $t-2, t-3, \ldots$) of all the variables in the model, not just a subset of them.

The difficulty of incorporating *a priori* information in a macroeconomic model that is formulated in discrete time is compounded by the fact that macroeconomic variables include both stock and flow variables, and these are measured in different ways. Stock variables such as the money supply, the level of inventories, and the stock of fixed capital are measured at points of time (e.g. at the end of each quarter), while flow variables such as consumption, investment, and the gross domestic product are measured as integrals over periods of time (e.g. integrals over successive quarters). Even in the simple case in which the continuous time model is a first order system of linear stochastic differential equations with constant coefficients and white noise innovations, the exact discrete model satisfied by a sample of flow data, or mixed stock and flow data, is a vector autoregressive moving average process (Bergstrom 1984; Agbeyegbe 1987). Moreover, although most of the coefficients in the differential equation system may be restricted to zero, all of the coefficients in both the autoregressive and moving average parts the exact discrete model will be non-zero and they will be complicated transcendental functions of the non-zero coefficients in the differential equation system. If the underlying continuous time model is a higher order system of differential equations with white noise innovations, then the exact discrete model satisfied by the data is a higher order vector autoregressive moving average model whose coefficient matrices are even more complicated functions of the parameters of the continuous time model (Bergstrom 1983).

The difficulty of taking account of *a priori* knowledge in discrete time econometric models has lead some econometricians to advocate forecasting from unrestricted vector autoregressive models or unrestricted vector autoregressive moving average models. In particular Sims (1980) has advocated the use of higher order vector autoregressive models with either no restrictions on the coefficient matrices or very few restrictions obtained in a Bayesian manner. This strategy, which has become widely known as the VAR methodology, might be justified as a way of avoiding the difficult problem of estimating the parameters of a continuous time model from discrete data. A high order vector autoregressive model could be regarded as an approximation to the vector autoregressive moving average model satisfied by data generated by a stochastic differential equation system. The main disadvantage of this procedure is that the number of coefficients to be estimated is likely to be so large relative the sample size that the estimates will have very large variances and forecasts based on these estimates could be very inaccurate. The potential gain in efficiency (as measured by the smallness of the root mean square errors of the forecasts) from using a forecasting procedure based on a continuous time model could be very important.

During the last few years there has been a rapid growth in the use of continuous time methods in macroeconomic modelling. Complete continuous time models (several versions for each country) have been developed for the United Kingdom (Bergstrom and Wymer 1976; Knight and Wymer 1978; Bergstrom 1984), Italy (Tullio 1981; Gandolfo and Padoan 1984; 1987), Germany (Kirkpatrick 1987), and Australia (Jonson, Moses, and Wymer 1977; Jonson, McKibbin, and Trevor 1982), and partial models for most of the other leading industrial countries. These models have been used, with considerable success, for both forecasting and policy analysis. But there is scope for further improvement through the use of more sophisticated estimation methods, which avoid the use of approximations and take advantage of the enormous developments in computing technology that have occurred during the last decade.

In a series of recent articles (Bergstrom 1983; 1985; 1986) [Chs. 4, 5, and 6 of this volume] I have developed a method of obtaining exact Gaussian (quasi-maximum likelihood) estimates of the parameters of a higher order continuous time system from discrete stock and flow data. The method yields exact maximum likelihood estimates when the innovations are Gaussian and, either the model is closed, or the exogenous variables are polynomials in time of degree not exceeding two. Moreover, it can be expected to yield very good estimates under much more general conditions. It also has the advantage that it is applicable to models that are non-stationary, in which case the initial state vector is fixed and the system is allowed to have roots with non-negative real parts. For reasons discussed in Bergstrom 1985 I believe that the assumption of a fixed initial state vector is more realistic in most econometric work than the assumption of stationarity. Finally, the new method is highly efficient computationally. Indeed, it has been shown (Bergstrom 1985) to be even more efficient than the Kalman filter algorithm of Harvey and Stock (1985), which is the only other published algorithm for obtaining exact Gaussian estimates of a non-stationary higher order continuous time system from mixed stock and flow data.

The purpose of this paper is to show how, by a simple extension of the new estimation procedure, we can obtain optimal forecasts of the post-sample discrete observations generated by the system. The forecasts are optimal in the sense that, when the model is closed and the innovations are Gaussian, they are exact maximum likelihood estimates of the conditional expectations of the post sample discrete observations conditional on all the information in the sample. The forecasting algorithm can be applied to either an open or closed model. If the model is open we must, of course, make some assumption about the post sample behaviour of the exogenous variables. Like the estimation procedure it is also highly efficient computationally. Indeed it relies on simple extensions, which can be computed recursively, of certain matrices and vectors that will have already been computed up to the end of the sample period as part of the estimation procedure.

In Section 2 the estimation procedure will be, briefly, reviewed with references to Bergstrom 1986 for the detailed formulae for its implementation. The forecasting algorithm and a theorem relating to the optimality of the forecasts will be presented in Section 3. For simplicity of exposition we shall deal in Sections 2 and 3 with a closed model. The treatment of exogenous variables will be discussed, briefly, in Section 4. Although the general method of estimation and forecasting is applicable to a system of any order we shall deal specifically with a second order system. The second order system is likely to be of great practical importance since it allows for a more realistic specification of the dynamic adjustment processes in the economy than a first order system and it is likely to be the highest order system for which we can obtain reliable parameter estimates with samples of the size normally available for macroeconomic modelling.

2. The Model and its Estimation

We shall deal, in this section, with the estimation of the parameters of the closed second order system

$$d[D\mathbf{x}(t)] = \{\mathbf{A}_1(\boldsymbol{\theta})D\mathbf{x}(t) + \mathbf{A}_2(\boldsymbol{\theta})\mathbf{x}(t)\}dt + \zeta(dt), \tag{1}$$

$$\mathbf{x}(0) = \mathbf{y}_1, \qquad D\mathbf{x}(0) = \mathbf{y}_2, \tag{2}$$

where $\{\mathbf{x}(t), (t > 0)\}$ is a real n-dimensional continuous time random process, $\boldsymbol{\theta}$ is a p-dimensional vector of unknown structural parameters $(p \leq 2n^2)$, $\mathbf{A}_1$ and $\mathbf{A}_2$ are matrices whose elements are known functions of $\boldsymbol{\theta}$, $\mathbf{y}_1$ and $\mathbf{y}_2$ are non-random $n \times 1$ vectors, D is the mean square differential operator, and $\zeta(dt)$ is a white noise innovation vector which is precisely defined by the following assumption.

Assumption 1

$\zeta' = [\zeta_1, \ldots, \zeta_n]$ is a vector of random measures defined on all subsets of the half line $0 < t < \infty$ with finite Lebesgue measure, such that $E[\zeta(dt)] = 0$, $E(\zeta(dt)\zeta'(dt)] = dt\boldsymbol{\Sigma}$ where $\boldsymbol{\Sigma}$ is an unknown positive definite matrix and $E[\zeta_i(\Delta_1)\zeta_j(\Delta_2)] = 0 (i, j = 1, \ldots, n)$ for any disjoint sets Δ_1 and Δ_2 on the half line $0 < t < \infty$.

We interpret equation (1) as meaning that

$$D\mathbf{x}(t) - D\mathbf{x}(0) = \int_0^t [\mathbf{A}_1 D\mathbf{x}(r) + \mathbf{A}_2\mathbf{x}(r)]dr + \int_0^1 \zeta(dr), \tag{3}$$

for all $t > 0$, where

$$\int_0^t \zeta(r)dr = \zeta[0, t],$$

and the first integral on the right hand side of equation (3) is the integral of a random process as defined in Rosanov 1967, p. 67. See Bergstrom 1983 for a proof of the existence and uniqueness of the solution of the system (1), subject to the boundary conditions given by equation (2), under Assumption 1, which is much weaker than the assumption that the innovations are generated by Brownian motion.

We shall assume that the variables in the system include n^s stock variables which are measured at points of time $0, 1, 2, \ldots, T$ and n^f flow variables which are measured as integrals over the intervals $[0, 1]$, $[1, 2]$, $\ldots$, $[T-1, T]$, where $n^s + n^f = n$. Let the elements of $\mathbf{x}(t)$ be ordered (without loss of generality) so that it can be written in the partitioned form

$$\mathbf{x}(t) = \begin{bmatrix} \mathbf{x}^s(t) \\ \mathbf{x}^f(t) \end{bmatrix}, \tag{4}$$

where $\mathbf{x}^s(t)$ is an $n^s \times 1$ vector of stock variables and $\mathbf{x}^f(t)$ is an $n^f \times 1$ vector of flow variables. Then the sample comprises the initial stock vector $\mathbf{x}^s(0) = \mathbf{y}_1^s$ and the vectors $\bar{\mathbf{x}}_1, \bar{\mathbf{x}}_2, \ldots, \bar{\mathbf{x}}_T$, where

$$\bar{\mathbf{x}}_t = \begin{bmatrix} \mathbf{x}^{\mathrm{s}}(t) - \mathbf{x}^{\mathrm{s}}(t-1) \\ \int_{t-1}^{t} \mathbf{x}^{\mathrm{f}}(r)\mathrm{d}r \end{bmatrix}. \tag{5}$$

The estimation procedure makes use of an exact discrete model, which is a system of stochastic difference equations [equations (6)–(11)] satisfied by the sample. The formulae for the coefficient matrices of this system are given in Bergstrom (1986), Theorems 2.1 and 2.2, and are in a form which requires the following assumptions.

Assumption 2

The matrix $\mathbf{A}_2$ is non-singular.

Assumption 3

The matrices $[e^{\bar{A}}]_{12}$ and $\mathbf{C}_{12}$ defined in Bergstrom (1986, pp. 123 and 125 this vol.) are non-singular.

Although Assumption 2 excludes the case in which the characteristic equation of the system (1) has zero roots, it does not exclude the case in which some roots have positive real parts and does not, therefore, require the system to be stable. Moreover, the zero roots case could be treated separately by the general methods of Bergstrom (1986). Assumption 3 merely excludes certain degenerate cases (Bergstrom 1983, p. 74 this vol.) which could also be treated separately. The exact discrete model is given by the following theorem.

Theorem 1

Let $\mathbf{x}(t)$ be the solution of the system (1) on the interval $[0, T]$ subject to the boundary condition given by equation (2). Then under Assumptions 1–3, the vectors $\bar{\mathbf{x}}_1, \bar{\mathbf{x}}_2, \ldots, \bar{\mathbf{x}}_T$ defined by equation (5) satisfy the system

$$\bar{\mathbf{x}}_1 = \mathbf{G}_{11}\mathbf{y}_1 + \mathbf{G}_{12}\mathbf{y}_2 + \boldsymbol{\eta}_1, \tag{6}$$

$$\bar{\mathbf{x}}_2 = \mathbf{C}_{11}\bar{\mathbf{x}}_1 + \mathbf{G}_{21}\mathbf{y}_1 + \mathbf{G}_{22}\mathbf{y}_2 + \boldsymbol{\eta}_2, \tag{7}$$

$$\bar{\mathbf{x}}_t = \mathbf{F}_1\bar{\mathbf{x}}_{t-1} + \mathbf{F}_2\bar{\mathbf{x}}_{t-2} + \boldsymbol{\eta}_t \quad (t = 3, \ldots, T), \tag{8}$$

$$\boldsymbol{\eta}_1 = \int_0^1 \mathbf{K}_{11}(1-r)\zeta(\mathrm{d}r), \tag{9}$$

$$\boldsymbol{\eta}_2 = \int_0^1 \mathbf{K}_{21}(1-r)\zeta(\mathrm{d}r) + \int_1^2 \mathbf{K}_{22}(2-r)\zeta(\mathrm{d}r), \tag{10}$$

$$\boldsymbol{\eta}_t = \int_{t-1}^{t} \mathbf{K}_1(t-r)\zeta(\mathrm{d}r) + \int_{t-2}^{t-1} \mathbf{K}_2(t-1-r)\zeta(\mathrm{d}r)$$

$$+ \int_{t-3}^{t-2} \mathbf{K}_3\{t-2-r)\zeta(\mathrm{d}r) \quad (t = 3, \ldots, T), \tag{11}$$

where $\mathbf{G}_{11}$, $\mathbf{G}_{12}$, $\mathbf{C}_{11}$, $\mathbf{C}_{21}$, $\mathbf{G}_{21}$, $\mathbf{G}_{22}$, $\mathbf{F}_1$, and $\mathbf{F}_1$ are constant matrices and $\mathbf{K}_{11}, \mathbf{K}_{21}, \mathbf{K}_{22}, \mathbf{K}_1, \mathbf{K}_2$, and $\mathbf{K}_3$ are matrix value functions.

Proof. See Bergstrom (1986, Theorems 2.1 and 2.2) which also gives explicit formulae for $\mathbf{G}_{11}$, $\mathbf{G}_{12}$, $\mathbf{C}_{11}$, $\mathbf{G}_{21}$, $\mathbf{G}_{22}$, $\mathbf{F}_1$, $\mathbf{F}_2$, $\mathbf{K}_{11}(r)$, $\mathbf{K}_{21}(r)$, $\mathbf{K}_{22}(r)$, $\mathbf{K}_1(r)$, $\mathbf{K}_2(r)$, and $\mathbf{K}_3(r)$ in terms of $\mathbf{A}_1$ and $\mathbf{A}_2$.

For the purpose of estimation it is convenient to parametrize $\boldsymbol{\Sigma}$ by writing it as $\boldsymbol{\Sigma}(\boldsymbol{\mu})$ where $\boldsymbol{\mu}$ is a vector of parameters. If there are no restrictions on $\boldsymbol{\Sigma}$ (except for the requirement that it is a symmetric positive definite matrix) then $\boldsymbol{\mu}$ will have $n(n+1)/2$ elements. But we could, for example, require $\boldsymbol{\Sigma}$ to be a diagonal matrix, in which case $\boldsymbol{\mu}$ would have only n elements. In addition to $\boldsymbol{\theta}$ and $\boldsymbol{\mu}$ we must estimate the part of the initial state vector that is unobservable. This includes $\mathbf{y}_1^{\mathrm{f}}$ the vector of initial values of the flow variables and $\mathbf{y}_2$ the vector of initial values of the mean square derivatives of all the variables. The complete vector of parameters to be estimated is, therefore, $[\boldsymbol{\theta}, \boldsymbol{\mu}, \mathbf{y}']$ where

$$\mathbf{y} = \begin{bmatrix} \mathbf{y}_1^{\mathrm{f}} \\ \mathbf{y}_2 \end{bmatrix}. \tag{12}$$

It is important to notice that in order to obtain the exact Gaussian estimate of $[\boldsymbol{\theta}, \boldsymbol{\mu}]$ it is essential to estimate $\mathbf{y}$. For, because $\boldsymbol{\eta}_3$ and $\boldsymbol{\eta}_4$ are not independent of $\bar{\mathbf{x}}_1$ and $\bar{\mathbf{x}}_2$, the conditional distribution of $\bar{\mathbf{x}}_3, \bar{\mathbf{x}}_4, \ldots, \bar{\mathbf{x}}_T$, conditional on $\bar{\mathbf{x}}_1$ and $\bar{\mathbf{x}}_2$, depends in an essential way on $\mathbf{y}$ and cannot be obtained from equations (8) and (11) alone. It should be noticed, also, that whereas $\mathbf{y}_1^{\mathrm{f}}$ is part of the parameter vector, $\mathbf{y}_1^{\mathrm{s}}$ is part of the sample. It is convenient, therefore, to partition $\mathbf{G}_{11}$ and $\mathbf{G}_{21}$ conformably with $\mathbf{y}$ and write equations (6) and (7) in the form

$$\bar{\mathbf{x}}_1 = \mathbf{G}_{11}^{\mathrm{s}}\mathbf{y}_1^{\mathrm{s}} + \mathbf{G}_{11}^{\mathrm{f}}\mathbf{y}_1^{\mathrm{f}} + \mathbf{G}_{12}\mathbf{y}_2 + \boldsymbol{\eta}_1 \tag{13}$$

$$\bar{\mathbf{x}}_2 = \mathbf{C}_{11}\bar{\mathbf{x}}_1 + \mathbf{G}_{21}^{\mathrm{s}}\mathbf{y}_1^{\mathrm{s}} + \mathbf{G}_{21}^{\mathrm{f}}\mathbf{y}_1^{\mathrm{f}} + \mathbf{G}_{22}\mathbf{y}_2 + \boldsymbol{\eta}_2. \tag{14}$$

The exact discrete model satisfied by the sample can now be written in a compact form. Let the $nT \times 1$ vectors $\mathbf{x}$, $\boldsymbol{\eta}$, and $\mathbf{h}$ be defined

$$\bar{\mathbf{x}}' = [\bar{\mathbf{x}}_1', \bar{\mathbf{x}}_2', \ldots, \bar{\mathbf{x}}_T'],$$

$$\boldsymbol{\eta}' = [\boldsymbol{\eta}_1', \boldsymbol{\eta}_2', \ldots, \boldsymbol{\eta}_T'],$$

$$\mathbf{h}' = [\mathbf{h}_1', \mathbf{h}_2', \mathbf{0}, \ldots, \mathbf{0}],$$

where

$$\mathbf{h}_1 = \mathbf{G}_{11}^s \mathbf{y}_1^s, \tag{15}$$

$$\mathbf{h}_2 = \mathbf{G}_{21}^s \mathbf{y}_1^s, \tag{16}$$

and let the $nT \times nT$ matrix $\mathbf{F}$ and the $nT \times (n^f + n)$ matrix $\mathbf{G}$ be defined by

$$\mathbf{F} = \begin{bmatrix} \mathbf{I} & \mathbf{0} & \mathbf{0} & \mathbf{0} & . & . & . & \mathbf{0} \\ -\mathbf{C}_{11} & \mathbf{I} & \mathbf{0} & \mathbf{0} & . & . & . & \mathbf{0} \\ -\mathbf{F}_2 & -\mathbf{F}_1 & \mathbf{I} & \mathbf{0} & . & . & . & \mathbf{0} \\ \mathbf{0} & -\mathbf{F}_2 & -\mathbf{F}_1 & \mathbf{I} & . & . & . & \mathbf{0} \\ . & . & . & . & . & . & . & . \\ . & . & . & . & . & . & . & . \\ . & . & . & . & . & . & . & . \\ \mathbf{0} & \mathbf{0} & \mathbf{0} & \mathbf{0} & . & -\mathbf{F}_2 & -\mathbf{F}_1 & \mathbf{I} \end{bmatrix}, \tag{17}$$

$$\mathbf{G} = \begin{bmatrix} \mathbf{G}_{11}^f & \mathbf{G}_{12} \\ \mathbf{G}_{21}^f & \mathbf{G}_{22} \\ \mathbf{0} & \mathbf{0} \\ . & . \\ . & . \\ . & . \\ \mathbf{0} & \mathbf{0} \end{bmatrix}, \tag{18}$$

The system of equations (8), (13), and (14) can be written as

$$\mathbf{F}\bar{\mathbf{x}} - \mathbf{G}\mathbf{y} - \mathbf{h} = \boldsymbol{\eta}. \tag{19}$$

The covariance matrix $\mathbf{\Omega}$ of the vector $\boldsymbol{\eta}$ can be obtained from equations (9), (10), and (11) and written in the partitioned form

$$\mathbf{\Omega} = \begin{bmatrix} \mathbf{\Omega}_{11} & \mathbf{\Omega}_{12} & \mathbf{\Omega}_{13} & \mathbf{0} & \mathbf{0} & \mathbf{0} & \mathbf{0} & . & . & . & . & \mathbf{0} \\ \mathbf{\Omega}_{21} & \mathbf{\Omega}_{22} & \mathbf{\Omega}_{23} & \mathbf{\Omega}_{24} & \mathbf{0} & \mathbf{0} & \mathbf{0} & . & . & . & . & \mathbf{0} \\ \mathbf{\Omega}_{31} & \mathbf{\Omega}_{32} & \mathbf{\Omega}_{0} & \mathbf{\Omega}_1' & \mathbf{\Omega}_2' & \mathbf{0} & \mathbf{0} & . & . & . & . & \mathbf{0} \\ \mathbf{0} & \mathbf{\Omega}_{42} & \mathbf{\Omega}_{1} & \mathbf{\Omega}_0 & \mathbf{\Omega}_1' & \mathbf{\Omega}_2' & \mathbf{0} & . & . & . & . & \mathbf{0} \\ . & . & . & . & . & . & . & . & . & . & . & . \\ . & . & . & . & . & . & . & . & \mathbf{\Omega}_0 & \mathbf{\Omega}_1' & \mathbf{\Omega}_2' & \mathbf{0} \\ . & . & . & . & . & . & . & . & \mathbf{\Omega}_1 & \mathbf{\Omega}_0 & \mathbf{\Omega}_1' & \mathbf{\Omega}_2' \\ . & . & . & . & . & . & . & . & \mathbf{\Omega}_2 & \mathbf{\Omega}_1 & \mathbf{\Omega}_0 & \mathbf{\Omega}_1' \\ \mathbf{0} & \mathbf{0} & \mathbf{0} & \mathbf{0} & \mathbf{0} & \mathbf{0} & \mathbf{0} & . & \mathbf{0} & \mathbf{\Omega}_2 & \mathbf{\Omega}_1 & \mathbf{\Omega}_0 \end{bmatrix}. \tag{20}$$

It should be noticed that, except for the submatrices in the upper left-hand corner, $\mathbf{\Omega}$ is a block Toeplitz matrix most of whose elements are zero. The non-zero elements of $\mathbf{\Omega}$ are functions of $[\boldsymbol{\theta}, \boldsymbol{\mu}]$ and can be computed from the formulae given in Bergstrom (1986, p. 139 this vol.)

Now let $L(\boldsymbol{\theta}, \boldsymbol{\mu}, \mathbf{y}')$ denote minus twice the logarithm of the Gaussian likelihood function. Then, since $|F| = 1$, we have

$$\begin{aligned} L(\boldsymbol{\theta}, \boldsymbol{\mu}, \mathbf{y}') &= \log|\boldsymbol{\Omega}(\boldsymbol{\theta}, \boldsymbol{\mu})| + \boldsymbol{\eta}'\boldsymbol{\Omega}^{-1}(\boldsymbol{\theta}, \boldsymbol{\mu})\boldsymbol{\eta} \\ &= \log|\boldsymbol{\Omega}(\boldsymbol{\theta}, \boldsymbol{\mu})| + \{\mathbf{F}(\boldsymbol{\theta})\bar{\mathbf{x}} - \mathbf{G}(\boldsymbol{\theta})\mathbf{y} \\ &\quad - \mathbf{h}(\boldsymbol{\theta})\}'\boldsymbol{\Omega}^{-1}(\boldsymbol{\theta}, \boldsymbol{\mu})\{\mathbf{F}(\boldsymbol{\theta})\bar{\mathbf{x}} - \mathbf{G}(\boldsymbol{\theta})\mathbf{y} - \mathbf{h}(\boldsymbol{\theta})\}. \end{aligned} \tag{21}$$

We can express L in an even simpler form, which is very convenient for computational purposes, by using the Cholesky factorization

$$\boldsymbol{\Omega} = \mathbf{M}\mathbf{M}', \tag{22}$$

where $\mathbf{M}$ is a real lower triangular matrix with positive elements along the diagonal. All of the elements of $\mathbf{M}$ above the diagonal and most of the elements below the diagonal are zero and the total number of multiplications required in the evaluation of $\mathbf{M}$ is less than $4Tn^3$ (Bergstrom 1985). Let $\varepsilon' = [\varepsilon_1, \varepsilon_2, \ldots, \varepsilon_{nT}]$ be the vector satisfying

$$\mathbf{M}\boldsymbol{\varepsilon} = \boldsymbol{\eta}.^1 \tag{23}$$

Because $\mathbf{M}$ is a lower triangular matrix the elements of ε can be computed recursively starting with ε_1. We can then compute L from the following formula:

$$L = \sum_{i=1}^{nT} (\varepsilon_i^2 + 2\log m_{ii}) \tag{24}$$

where m_{ii} is the ith diagonal element of $\mathbf{M}$.

A summary of the steps in the computation of L is as follows:

(i) Compute $\mathbf{F}$, $\mathbf{G}$, $\mathbf{h}$, and $\boldsymbol{\Omega}$ from equations (15)–(18) and the formulae for the submatrices of $\boldsymbol{\Omega}$ given in Bergstrom 1986, p. 370.
(ii) Compute the elements of $\mathbf{M}$ recursively, row by row, from equation (22).
(iii) Compute $\boldsymbol{\eta}$ from equation (19).
(iv) Compute the elements of ε recursively from equation (23).
(v) Compute L from equation (24).

The Gaussian estimator $[\hat{\boldsymbol{\theta}}, \hat{\boldsymbol{\mu}}, \hat{\mathbf{y}}']$ of $[\boldsymbol{\theta}, \boldsymbol{\mu}, \mathbf{y}']$ can be obtained by a numerical optimization procedure involving successive evaluations of L. In this procedure we can take advantage of an explicit formula relating $\hat{\mathbf{y}}$ to $[\hat{\boldsymbol{\theta}}, \hat{\boldsymbol{\mu}}]$ and the data, i.e.

$$\begin{aligned} \hat{\mathbf{y}} = {} & [\mathbf{G}'(\hat{\boldsymbol{\theta}}, \hat{\boldsymbol{\mu}})\boldsymbol{\Omega}^{-1}(\hat{\boldsymbol{\theta}}, \hat{\boldsymbol{\mu}})G(\hat{\boldsymbol{\theta}}, \hat{\boldsymbol{\mu}})]^{-1}\mathbf{G}'(\hat{\boldsymbol{\theta}}, \hat{\boldsymbol{\mu}})\boldsymbol{\Omega}^{-1}(\hat{\boldsymbol{\theta}}, \hat{\boldsymbol{\mu}}) \\ & \times [\mathbf{F}(\hat{\boldsymbol{\theta}}, \hat{\boldsymbol{\mu}})\bar{\mathbf{x}} - (\hat{\boldsymbol{\theta}}, \hat{\boldsymbol{\mu}})]. \end{aligned} \tag{25}$$

[1] In the original version of this paper, published in *Computers and Mathematics with Applications* (1989), the vector ε was denoted by $\mathbf{z}$.

See Bergstrom (1985, p. 114 this vol.) for the derivation of equation (25) and a proposed iterative procedure in which we alternate between minimizing L with respect to $[\boldsymbol{\theta}, \boldsymbol{\mu}]$, for a given $\mathbf{y}$, using some numerical optimization procedure and minimizing L with respect to $\mathbf{y}$, for given $[\boldsymbol{\theta}, \boldsymbol{\mu}]$, using equation (25).

3. An Optimal Forecasting Algorithm

We shall assume, in this section, that the continuous time system represented by equation (1) and (2) holds over the interval $[0, T+k]$ where T and k are positive integers. Our aim is to obtain the optimal forecasts of $\bar{\mathbf{x}}_{T+1}, \bar{\mathbf{x}}_{T+2}, \ldots, \bar{\mathbf{x}}_{T+k}$ from the sample $\bar{\mathbf{x}}^s(0), \bar{\mathbf{x}}_1, \bar{\mathbf{x}}_2, \mathbf{x}_2, \ldots, \bar{\mathbf{x}}_T$ where $\bar{\mathbf{x}}_t$ $(t = 1, \ldots, T+k)$ is defined by equation (5). The estimation procedure described in the previous section is particularly convenient for this purpose. A simple extension of certain matrices and vectors used in the computation of the estimates of the parameters of the model provides the basis for a computationally efficient forecasting algorithm. This algorithm will now be described in detail, after which it will be shown (Theorem 2) that the forecasts which it generates are optimal in the sense described in the introduction.

The first step of the algorithm is based on an extension of the matrix $\hat{\boldsymbol{\Omega}} = \boldsymbol{\Omega}(\hat{\boldsymbol{\theta}}, \hat{\boldsymbol{\mu}})$ where $\boldsymbol{\Omega}$ is defined by equation (20) and the formulae for its submatrices are given in Bergstrom (1986, p. 139 this vol.) This extension can be made without any further computation since it involves only the submatrices $\hat{\boldsymbol{\Omega}}_0$, $\hat{\boldsymbol{\Omega}}_1$ and $\hat{\boldsymbol{\Omega}}_2$, which will have been computed as part of the parameter estimation procedure. Let the symmetric $n(T+k) \times n(T+k)$ matrix $\hat{\boldsymbol{\Omega}}^*$ be defined by

$$\hat{\boldsymbol{\Omega}}^* = \begin{bmatrix} \hat{\boldsymbol{\Omega}}_{11}^* & \hat{\boldsymbol{\Omega}}_{12}^* \\ \hat{\boldsymbol{\Omega}}_{21}^* & \hat{\boldsymbol{\Omega}}_{22}^* \end{bmatrix}, \tag{26}$$

where

$$\hat{\boldsymbol{\Omega}}_{11}^* = \hat{\boldsymbol{\Omega}} \tag{27}$$

and the $nT \times nk$ matrix $\hat{\boldsymbol{\Omega}}_{12}^*$ and the $nk \times nk$ matrix $\hat{\boldsymbol{\Omega}}_{22}^*$ are defined by

$$\hat{\boldsymbol{\Omega}}_{12}^* = \begin{bmatrix} \mathbf{0} & \mathbf{0} & \mathbf{0} & . & . & . & \mathbf{0} \\ . & . & . & . & . & . & . \\ . & . & . & . & . & . & . \\ . & . & . & . & . & . & . \\ \mathbf{0} & \mathbf{0} & \mathbf{0} & . & . & . & \mathbf{0} \\ \hat{\boldsymbol{\Omega}}_2' & \mathbf{0} & \mathbf{0} & . & . & . & \mathbf{0} \\ \hat{\boldsymbol{\Omega}}_1' & \hat{\boldsymbol{\Omega}}_2' & \mathbf{0} & . & . & . & \mathbf{0} \end{bmatrix}, \tag{28}$$

$$\hat{\boldsymbol{\Omega}}_{22}^{*} = \begin{bmatrix} \hat{\boldsymbol{\Omega}}_0 & \hat{\boldsymbol{\Omega}}_1' & \hat{\boldsymbol{\Omega}}_2' & \mathbf{0} & \cdot & \cdot & \cdot & \cdot & \mathbf{0} \\ \hat{\boldsymbol{\Omega}}_1 & \hat{\boldsymbol{\Omega}}_0 & \hat{\boldsymbol{\Omega}}_1' & \hat{\boldsymbol{\Omega}}_2' & \cdot & \cdot & \cdot & \cdot & \mathbf{0} \\ \hat{\boldsymbol{\Omega}}_2 & \hat{\boldsymbol{\Omega}}_1 & \hat{\boldsymbol{\Omega}}_0 & \hat{\boldsymbol{\Omega}}_1' & \cdot & \cdot & \cdot & \cdot & \mathbf{0} \\ \mathbf{0} & \hat{\boldsymbol{\Omega}}_2 & \hat{\boldsymbol{\Omega}}_1 & \hat{\boldsymbol{\Omega}}_0 & \cdot & \cdot & \cdot & \cdot & \mathbf{0} \\ \cdot & \cdot & \cdot & \cdot & \cdot & \cdot & \cdot & \cdot & \cdot \\ \cdot & \cdot & \cdot & \cdot & \cdot & \hat{\boldsymbol{\Omega}}_0 & \hat{\boldsymbol{\Omega}}_1' & \hat{\boldsymbol{\Omega}}_2' & \mathbf{0} \\ \cdot & \cdot & \cdot & \cdot & \cdot & \hat{\boldsymbol{\Omega}}_1 & \hat{\boldsymbol{\Omega}}_0 & \hat{\boldsymbol{\Omega}}_1' & \hat{\boldsymbol{\Omega}}_2' \\ \cdot & \cdot & \cdot & \cdot & \cdot & \hat{\boldsymbol{\Omega}}_2 & \hat{\boldsymbol{\Omega}}_1 & \hat{\boldsymbol{\Omega}}_0 & \hat{\boldsymbol{\Omega}}_1' \\ \mathbf{0} & \mathbf{0} & \mathbf{0} & \mathbf{0} & \mathbf{0} & \mathbf{0} & \hat{\boldsymbol{\Omega}}_2 & \hat{\boldsymbol{\Omega}}_1 & \hat{\boldsymbol{\Omega}}_0 \end{bmatrix}. \tag{29}$$

Then the first step is the Cholesky factorization of $\hat{\boldsymbol{\Omega}}^*$. Let $\hat{\mathbf{M}}^*$ be the $n(T+k) \times n(T+k)$ lower triangular matrix such that

$$\hat{\boldsymbol{\Omega}}^* = \hat{\mathbf{M}}^*[\hat{\mathbf{M}}^*]'. \tag{30}$$

Then $\hat{\mathbf{M}}^*$ can be written in the partitioned form

$$\hat{\mathbf{M}}^* = \begin{bmatrix} \hat{\mathbf{M}}_{11}^* & \mathbf{0} \\ \hat{\mathbf{M}}_{21}^* & \hat{\mathbf{M}}_{22}^* \end{bmatrix}, \tag{31}$$

where

$$\hat{\mathbf{M}}_{11}^* = \hat{\mathbf{M}}$$

and

$$\hat{\mathbf{M}}\hat{\mathbf{M}}' = \hat{\boldsymbol{\Omega}}.$$

Since $\hat{\mathbf{M}}$ will have been computed as part of the estimation procedure the remaining elements of $\hat{\mathbf{M}}^*$ can be computed recursively, row by row, starting at row $nT + 1$. In fact, for the algorithm, we need to compute only up to row $n(T+2)$.

The next step is the computation of the $nk \times 1$ vector $\hat{\boldsymbol{\eta}}_p$ defined by

$$\hat{\boldsymbol{\eta}}_p = \hat{\mathbf{M}}_{21}^* \hat{\boldsymbol{\varepsilon}} \tag{32}$$

where $\hat{\boldsymbol{\varepsilon}}$ is the vector satisfying

$$\hat{\mathbf{M}}\hat{\boldsymbol{\varepsilon}} = \hat{\boldsymbol{\eta}} = \mathbf{F}(\hat{\boldsymbol{\theta}})\bar{\mathbf{x}} - \mathbf{G}(\hat{\boldsymbol{\theta}})\hat{\mathbf{y}} - \mathbf{h}(\hat{\boldsymbol{\theta}}).$$

Both $\hat{\boldsymbol{\eta}}$ and $\hat{\boldsymbol{\varepsilon}}$ will have been computed as part of the estimation procedure. Moreover, since the only non-zero elements of $\hat{\mathbf{M}}_{21}^*$ are in the $2n \times 2n$ submatrix at the upper right hand corner of that matrix, the vector $\hat{\boldsymbol{\eta}}_p$ will have only $2n$ non-zero elements, and it can be written in the form

$$\hat{\boldsymbol{\eta}}_p' = [\hat{\boldsymbol{\eta}}_{p1}', \hat{\boldsymbol{\eta}}_{p2}', \mathbf{0}, \ldots, \mathbf{0}],$$

where $\hat{\boldsymbol{\eta}}_{p1}$ and $\hat{\boldsymbol{\eta}}_{p2}$ are $n \times 1$ vectors.

Finally, the forecasts $\hat{\mathbf{x}}_{p1}, \hat{\mathbf{x}}_{p2}, \ldots, \hat{\mathbf{x}}_{pk}$ of $\bar{\mathbf{x}}_{T+1}, \bar{\mathbf{x}}_{T+2}, \ldots, \bar{\mathbf{x}}_{T+k}$ respectively are computed recursively from the equations

$$\begin{aligned}
\hat{\mathbf{x}}_{p1} &= \mathbf{F}_1(\hat{\boldsymbol{\theta}})\bar{\mathbf{x}}_T + \mathbf{F}_2(\hat{\boldsymbol{\theta}})\bar{\mathbf{x}}_{T-1} + \hat{\boldsymbol{\eta}}_{p1},\\
\hat{\mathbf{x}}_{p2} &= \mathbf{F}_1(\hat{\boldsymbol{\theta}})\hat{\mathbf{x}}_{p1} + \mathbf{F}_2(\hat{\boldsymbol{\theta}})\bar{\mathbf{x}}_T + \hat{\boldsymbol{\eta}}_{p2},\\
\hat{\mathbf{x}}_{p3} &= \mathbf{F}_1(\hat{\boldsymbol{\theta}})\hat{\mathbf{x}}_{p2} + \mathbf{F}_2(\hat{\boldsymbol{\theta}})\bar{\mathbf{x}}_{p1},\\
&\vdots \qquad\qquad\qquad \vdots\\
\hat{\mathbf{x}}_{pk} &= \mathbf{F}_1(\hat{\boldsymbol{\theta}})\hat{\mathbf{x}}_{p,k-1} + \mathbf{F}_2(\hat{\boldsymbol{\theta}})\hat{\mathbf{x}}_{p,k-2}.
\end{aligned} \tag{33}$$

A summary of the steps in the forecasting algorithm, assuming that the parameter estimates have already been obtained by the procedure described in Section 2, is as follows:

(i) Compute rows $nT+1$ to $n(T+2)$ of the lower triangular matrix $\hat{\mathbf{M}}^*$ satisfying equation (30). The first nT rows of $\hat{\mathbf{M}}^*$ will have been computed as part of the parameter estimation procedure, and the next $2n$ rows can be computed recursively starting at row $nT+1$.
(ii) Compute $\hat{\boldsymbol{\eta}}_p$ from equation (32).
(iii) Compute $\hat{\mathbf{x}}_{p1}, \hat{\mathbf{x}}_{p2}, \ldots, \hat{\mathbf{x}}_{p+k}$ recursively from the system of equations (33).

For the optimality theorem we shall require the following additional assumption.

Assumption 4

The innovations defined by Assumption 1 are Gaussian. Assumptions 1 and 4 together imply that $\int_0^t \zeta(\mathrm{d}r)$ is Brownian motion.

Theorem 2

Let $\mathbf{x}(t)$ be the solution of the system (1) on the interval $[0, T+k]$ subject to the boundary condition given by equation (2). Then, under Assumptions 1–4, the vectors $\hat{\mathbf{x}}_{p1}, \hat{\mathbf{x}}_{p2}, \ldots, \hat{\mathbf{x}}_{pk}$ obtained from the system (33) are the maximum likelihood estimates of the conditional expectations of the vectors $\bar{\mathbf{x}}_{T+1}, \bar{\mathbf{x}}_{T+2}, \ldots, \bar{\mathbf{x}}_{T+k}$ respectively, conditional on $\bar{\mathbf{x}}_1, \bar{\mathbf{x}}_2, \ldots, \bar{\mathbf{x}}_T$, i.e. conditional on all the information in the sample.

Proof. Let the $n(T+k)\times 1$ vectors $\mathbf{x}^*$, $\boldsymbol{\eta}^*$, and $\mathbf{h}^*$ be defined by

$$\begin{aligned}
[\mathbf{x}^*]' &= [\bar{\mathbf{x}}_1', \bar{\mathbf{x}}_2', \ldots, \bar{\mathbf{x}}_{T+k}'],\\
[\boldsymbol{\eta}^*]' &= [\boldsymbol{\eta}_1', \boldsymbol{\eta}_2', \ldots, \boldsymbol{\eta}_{T+k}'],\\
[\mathbf{h}^*]' &= [\mathbf{h}_1', \mathbf{h}_2', \mathbf{0}, \ldots, \mathbf{0}],
\end{aligned}$$

where $\bar{\mathbf{x}}_1, \bar{\mathbf{x}}_2, \ldots, \bar{\mathbf{x}}_{T+k}$ are defined by equation (5) $(t = 1, \ldots, T+k)$, $\boldsymbol{\eta}_3$, $\boldsymbol{\eta}_4, \ldots, n_{T+k}$ are defined by equation (11) $(t = 3, \ldots, T+k)$, $\boldsymbol{\eta}_1$ and $\boldsymbol{\eta}_2$ are defined by equations (9) and (10), and $\mathbf{h}_1$ and $\mathbf{h}_2$ are defined by equations (15) and (16). Let $\mathbf{F}^*$ be the $n(T+k)\times n(T+k)$ matrix having the same form as the $nT \times nT$ matrix $\mathbf{F}$ defined by equation (17), and let $\mathbf{G}^*$ be the $n(T+k) \times (n^{\mathrm{f}} + n)$ matrix having the same form as the $nT \times n^f + n$ matrix $\mathbf{G}$ defined by equation (18). Then

$$\mathbf{F}^*\mathbf{x}^* - \mathbf{G}^*\mathbf{y} - \mathbf{h}^* = \boldsymbol{\eta}^*. \tag{34}$$

Moreover, the covariance matrix $\boldsymbol{\Omega}^*$ of the vector $\boldsymbol{\eta}^*$ has the same form as the matrix $\hat{\boldsymbol{\Omega}}^*$ defined by equations (26)–(29) the elements of $\boldsymbol{\Omega}^*$ and $\hat{\boldsymbol{\Omega}}^*$ being identical functions $[\boldsymbol{\theta}, \boldsymbol{\mu}]$ and $[\hat{\boldsymbol{\theta}}, \hat{\boldsymbol{\mu}}]$ respectively.

Now let $\mathbf{M}^*$ be the lower triangular matrix defined by the Cholesky factorization

$$\boldsymbol{\Omega}^* = \mathbf{M}^*[\mathbf{M}^*]', \tag{35}$$

and let $\boldsymbol{\varepsilon}^*$ be the $n(T+k) \times 1$ vector satisfying

$$\mathbf{M}^*\boldsymbol{\varepsilon}^* = \boldsymbol{\eta}^*. \tag{36}$$

Then it follows from equations (35) and (36), and the fact that $\boldsymbol{\Omega}^*$ is the covariance matrix of $\boldsymbol{\eta}^*$, that the covariance matrix of $\boldsymbol{\varepsilon}^*$ is the $n(T+k) \times n(T+k)$ identity matrix, and hence, because of Assumption 4, that $\boldsymbol{\varepsilon}^*$ is a vector of independent normally distributed random variables each with mean 0 and variance 1. Since $\mathbf{M}^*$ can be written in the partitioned form

$$\mathbf{M}^* = \begin{bmatrix} \mathbf{M} & \mathbf{0} \\ \mathbf{M}^*_{21} & \mathbf{M}^*_{22} \end{bmatrix}, \tag{37}$$

and the leading $nT \times 1$ subvector of $\boldsymbol{\eta}^*$ is $\boldsymbol{\eta}$, it follows from equations (23) and (36) that the leading $nT \times 1$ subvector of $\boldsymbol{\varepsilon}^*$ is $\boldsymbol{\varepsilon}$. The conditional expectation of $\boldsymbol{\varepsilon}^*$ conditional on information in the sample (the parameters being treated as known) is given, therefore, by

$$\begin{aligned} E(\boldsymbol{\varepsilon}^*|\bar{\mathbf{x}}) &= E(\boldsymbol{\varepsilon}^*|\boldsymbol{\varepsilon}) \\ &= \begin{bmatrix} \boldsymbol{\varepsilon} \\ \mathbf{0} \end{bmatrix}, \end{aligned} \tag{38}$$

where the first line follows from the fact that $\boldsymbol{\varepsilon}$ is obtained from $\bar{\mathbf{x}}$ by the non-singular transformation

$$\boldsymbol{\varepsilon} = \mathbf{M}^{-1}[\mathbf{F}\bar{\mathbf{x}} - \mathbf{G}\mathbf{y} - \mathbf{h}],$$

which is obtained from equations (19) and (23).

It then follows, from equations (23), (36), (37), and (38), that the conditional expectation of $\boldsymbol{\eta}^*$ conditional on the information in the sample is given by

$$E(\boldsymbol{\eta}^*|\bar{\mathbf{x}}) = \begin{bmatrix} \boldsymbol{\eta} \\ \boldsymbol{\eta}_p \end{bmatrix} \tag{39}$$

where $\boldsymbol{\eta}_p$ is the $nk \times 1$ vector defined by

$$\boldsymbol{\eta}_p = \mathbf{M}_{21}^* \boldsymbol{\varepsilon}. \tag{40}$$

A comparison of equations (32) and (40) shows that $\boldsymbol{\eta}_p$ has the same form as $\hat{\boldsymbol{\eta}}_p$. The first $2n$ elements of the vectors $\boldsymbol{\eta}_p$ and $\hat{\boldsymbol{\eta}}_p$ are identical functions of $[\boldsymbol{\theta}, \boldsymbol{\mu}]$ and $[\hat{\boldsymbol{\theta}}, \hat{\boldsymbol{\mu}}]$ respectively and the remaining elements of both vectors are zero.

Finally, let $\mathbf{x}_{p1}, \mathbf{x}_{p2}, \ldots, \mathbf{x}_{pk}$ by the $n \times 1$ vectors satisfying the system:

$$\begin{aligned}
\mathbf{x}_{p1} &= \mathbf{F}_1(\boldsymbol{\theta})\bar{\mathbf{x}}_T + \mathbf{F}_2(\boldsymbol{\theta})\bar{\mathbf{x}}_{T-1} + \boldsymbol{\eta}_{p1},\\
\mathbf{x}_{p2} &= \mathbf{F}_1(\boldsymbol{\theta})\mathbf{x}_{p1} + \mathbf{F}_2(\boldsymbol{\theta})\bar{\mathbf{x}}_T + \boldsymbol{\eta}_{p2},\\
\mathbf{x}_{p3} &= \mathbf{F}_1(\boldsymbol{\theta})\mathbf{x}_{p2} + \mathbf{F}_2(\boldsymbol{\theta})\mathbf{x}_{p1},\\
&\vdots \qquad\qquad \vdots\\
\mathbf{x}_{pk} &= \mathbf{F}_1(\boldsymbol{\theta})\mathbf{x}_{p,k-1} + \mathbf{F}_2(\boldsymbol{\theta})\mathbf{x}_{p,k-2},
\end{aligned} \tag{41}$$

which has the same form as the system (33). It follows from equations (34), (39), and (41) that the vectors $\mathbf{x}_{p1}, \mathbf{x}_{p2}, \ldots, x_{pk}$ are the conditional expectations of $\bar{\mathbf{x}}_{T+1}, \bar{\mathbf{x}}_{T+2}, \ldots, \bar{\mathbf{x}}_{T+k}$, respectively, conditional on the information in the sample, i.e.

$$E(\bar{\mathbf{x}}_{T+i}|\bar{\mathbf{x}}) = \mathbf{x}_{pi} \quad (i = 1, \ldots, k). \tag{42}$$

Moreover, since $[\hat{\boldsymbol{\theta}}, \hat{\boldsymbol{\mu}}]$ is the maximum likelihood estimator of $[\boldsymbol{\theta}, \boldsymbol{\mu}]$ and $\hat{\mathbf{x}}_{pi}$ and $\mathbf{x}_{pi}$ $(i = 1, \ldots, k)$ are identical functions of $[\hat{\boldsymbol{\theta}}, \hat{\boldsymbol{\mu}})$ and $[\boldsymbol{\theta}, \boldsymbol{\mu}]$ respectively, $\hat{\mathbf{x}}_{p1}, \hat{\mathbf{x}}_{p2}, \ldots, \hat{\mathbf{x}}_{pk}$ are the maximum likelihood estimates of $\mathbf{x}_{p1}, \mathbf{x}_{p2}, \ldots, \mathbf{x}_{pk}$ respectively. End of proof.

4. The Treatment of Exogenous Variables

The forecasting algorithm described in the preceding section can easily be modified to take account of exogenous variables. We shall deal, very briefly, with the open second order system

$$\begin{aligned}
\mathrm{d}[\mathrm{D}\mathbf{x}(t)] &= \{\mathbf{A}_1(\boldsymbol{\theta})\mathrm{D}\mathbf{x}(t) + \mathbf{A}_2(\boldsymbol{\theta})\mathbf{x}(t) + \mathbf{B}(\boldsymbol{\theta})\mathbf{z}(t)\}\mathrm{d}t + \zeta(\mathrm{d}t) \quad (t \geqslant 0),\\
\mathbf{x}(0) &= \mathbf{y}_1, \quad \mathrm{D}\mathbf{x}(0) = \mathbf{y}_2,
\end{aligned} \tag{43}$$

where $\mathbf{z}(t)$ is an $m \times 1$ vector of non-random functions (exogenous variables) and ζ is defined by Assumption 1. This system can be interpreted as meaning that

$$\mathrm{D}\mathbf{x}(t) - \mathrm{D}\mathbf{x}(0) = \int_0^t [\mathbf{A}_1\mathrm{D}\mathbf{x}(r) + \mathbf{A}_2\mathbf{x}(r) + \mathbf{B}\mathbf{z}(r)]\mathrm{d}r + \int_0^t \zeta(\mathrm{d}r) \tag{44}$$

for all $t > 0$.

We assume that the exogenous variables include m^s stock variables which are observed at points of time $0, 1, 2, \ldots, T$ and m^{f} flow variables which are observed as integrals over the intervals $[0, 1]$, $[1, 2]$, $\ldots, [T-1, T]$ where $m^{\mathrm{s}} + m^{\mathrm{f}} = m$. Let the elements of $\mathbf{z}(t)$ be ordered (without loss of generality) so that it can be written in the form

$$\mathbf{z}(t) = \begin{bmatrix} \mathbf{z}^{\mathrm{s}}(t) \\ \mathbf{z}^{\mathrm{f}}(t) \end{bmatrix}, \tag{45}$$

where $\mathbf{z}^{\mathrm{s}}(t)$ is an $m^{\mathrm{s}} \times 1$ vector of stock variables and $\mathbf{z}^{\mathrm{f}}(t)$ is an $m^{\mathrm{f}} \times 1$ vector of flow variables. Then we can define a sequence of observable vectors $\bar{\mathbf{z}}_1, \bar{\mathbf{z}}_2, \ldots, \bar{\mathbf{z}}_T$ by

$$\bar{\mathbf{z}}_t = \begin{bmatrix} \frac{1}{2}\{\mathbf{z}^{\mathrm{s}}(t) + \mathbf{z}^{\mathrm{s}}(t-1)\} \\ \displaystyle\int_{t-1}^{t} \mathbf{z}^{\mathrm{f}}(r)\mathrm{d}r \end{bmatrix}. \tag{46}$$

The estimation and forecasting procedure is based on the discrete model

$$\bar{\mathbf{x}}_1 = \mathbf{G}_{11}\mathbf{y}_1 + \mathbf{G}_{12}\mathbf{y}_2 + \mathbf{E}_{11}\bar{\mathbf{z}}_1 + \mathbf{E}_{12}\bar{\mathbf{z}}_2 + \mathbf{E}_{13}\bar{\mathbf{z}}_3 + \boldsymbol{\eta}_1, \tag{47}$$

$$\bar{\mathbf{x}}_2 = \mathbf{C}_{11}\bar{\mathbf{x}}_1 + \mathbf{G}_{21}\mathbf{y}_1 + \mathbf{G}_{22}\mathbf{y}_2 + \mathbf{E}_{21}\bar{\mathbf{z}}_1 + \mathbf{E}_{22}\bar{\mathbf{z}}_2 + \mathbf{E}_{23}\bar{\mathbf{z}}_3 + \boldsymbol{\eta}_2, \tag{48}$$

$$\bar{\mathbf{x}}_t = \mathbf{F}_1\bar{\mathbf{x}}_{t-1} + \mathbf{F}_2\bar{\mathbf{x}}_{t-2} + \mathbf{E}_0\bar{\mathbf{z}}_t + \mathbf{E}_1\bar{\mathbf{z}}_{t-1} + \mathbf{E}_2\bar{\mathbf{z}}_{t-2} + \boldsymbol{\eta}_t, \quad (t = 3, \ldots, T), \tag{49}$$

where the vectors $\bar{\mathbf{x}}_t$ and $\bar{\mathbf{z}}_t$ $(t = 1, \ldots, T)$ are defined by equations (5) and (46) respectively, $\boldsymbol{\eta}_1$ and $\boldsymbol{\eta}_2$ are defined by equations (9) and (10) respectively, and $\boldsymbol{\eta}_t$ $(t = 3, \ldots, T)$ is defined by equation (11). It is shown in Bergstrom (1986, Theorems 2.3 and 2.4) that equations (47) and (48) hold exactly if the elements of $\mathbf{z}(t)$ are polynomials in t of degree not exceeding two over the time interval $[0, 3]$ and equation (49) holds exactly over any subinterval $[t-3, t]$ of $[0, T]$ in which the elements of $\mathbf{z}(t)$ are polynomials in t of degree not exceeding two. It is, of course, impossible to obtain a discrete model which holds exactly without making some assumption about the continuous time path of $\mathbf{z}(t)$. It should be emphasized, however, that the coefficient matrices $\mathbf{E}_0$, $\mathbf{E}_1$, $\mathbf{E}_2$ and $\mathbf{E}_{ij}$ $(i, j = 1, 2, 3)$ in the above system depend only on the parameters of the continuous time system (43) and not on the coefficients of the quadratic functions defining the continuous time path of $\mathbf{z}(t)$. The discrete model represented by equations (47)–(49) will be a very good approximation, therefore, even when the behaviour of the exogenous variables varies widely as between different parts of the sample period. Indeed, it can be shown that if the elements of $\mathbf{z}(t)$ are thrice continuously differential functions of t then the errors in the discrete model are of ord δ^4 as the unit observation period δ tends to zero.

The estimation and forecasting procedure with an open model is the same as that described in Sections 2 and 3 except that we use equations (47)–(49) instead of (6)–(8). In order to obtain forecasts of $\bar{\mathbf{x}}_{T+1}, \bar{\mathbf{x}}_{t+2}, \ldots, \bar{\mathbf{x}}_{T+k}$ we must, of course, make some assumption about the post sample vectors $\bar{\mathbf{z}}_{t+1}, \bar{\mathbf{z}}_{t+2}, \ldots, \bar{\mathbf{z}}_{T+k}$. The final step in the forecasting algorithm will be to compute the forecasts recursively from the system

$$
\begin{aligned}
\hat{\mathbf{x}}_{p1} &= \mathbf{F}_1(\hat{\boldsymbol{\theta}})\bar{\mathbf{x}}_T + \mathbf{F}_2(\hat{\boldsymbol{\theta}})\bar{\mathbf{x}}_{T-1} + \mathbf{E}_0(\hat{\boldsymbol{\theta}})\bar{\mathbf{z}}_{T+1} + \mathbf{E}_1(\hat{\boldsymbol{\theta}})\bar{\mathbf{z}}_T + \mathbf{E}_2(\hat{\boldsymbol{\theta}})\bar{\mathbf{z}}_{T-1} + \hat{\boldsymbol{\eta}}_{p1}, \\
\hat{\mathbf{x}}_{p2} &= \mathbf{F}_1(\hat{\boldsymbol{\theta}})\hat{\mathbf{x}}_{p1} + \mathbf{F}_2(\hat{\boldsymbol{\theta}})\bar{\mathbf{x}}_T + \mathbf{E}_0(\hat{\boldsymbol{\theta}})\bar{\mathbf{z}}_{T+2} + \mathbf{E}_1(\hat{\boldsymbol{\theta}})\bar{\mathbf{z}}_{T-1} + \mathbf{E}_2(\hat{\boldsymbol{\theta}})\bar{\mathbf{z}}_T + \hat{\boldsymbol{\eta}}_{p2}, \\
\hat{\mathbf{x}}_{p3} &= \mathbf{F}_1(\hat{\boldsymbol{\theta}})\bar{\mathbf{x}}_{p2} + \mathbf{F}_2(\hat{\boldsymbol{\theta}})\bar{\mathbf{x}}_{p1} + \mathbf{E}_0(\hat{\boldsymbol{\theta}})\bar{\mathbf{z}}_{T+3} + \mathbf{E}_1(\hat{\boldsymbol{\theta}})\bar{\mathbf{z}}_{T+2} + \mathbf{E}_2(\hat{\boldsymbol{\theta}})\bar{\mathbf{z}}_{T-1}, \\
&\vdots \qquad\qquad\qquad\qquad\qquad \vdots \\
\hat{\mathbf{x}}_{pk} &= \mathbf{F}_1(\hat{\boldsymbol{\theta}})\hat{\mathbf{x}}_{p,k-1} + \mathbf{F}_2(\hat{\boldsymbol{\theta}})\hat{\mathbf{x}}_{p,k-2} + \mathbf{E}_0(\hat{\boldsymbol{\theta}})\bar{\mathbf{z}}_{T+k} \\
&\quad + \mathbf{E}_1(\hat{\boldsymbol{\theta}})\bar{\mathbf{z}}_{T+k-1} + \mathbf{E}_2(\hat{\boldsymbol{\theta}})\bar{\mathbf{z}}_{T+k-2},
\end{aligned}
\tag{50}
$$

where $\hat{\boldsymbol{\eta}}_{p1}$ and $\hat{\boldsymbol{\eta}}_{p2}$ are the leading $n+1$ subvectors of $\hat{\boldsymbol{\eta}}_p = \hat{\mathbf{M}}_{21}^{*}\hat{\boldsymbol{\varepsilon}}$. The vector $\hat{\boldsymbol{\varepsilon}}$, which will now take account of the sample values of the exogenous variables, will already have been computed as part of the parameter estimation procedure which is described in Bergstrom 1986.

5. Conclusion

We have developed an algorithm for forecasting discrete stock and flow data generated by a higher order continuous time system whose parameters have been estimated by the method developed in Bergstrom 1985; 1986 and described, briefly, in Section 2 of this paper. It has been shown that the algorithm yields exact maximum likelihood estimates of the conditional expectation of the post sample discrete observations, conditional on all the information in the sample, when the innovations are Gaussian and, either the model is closed, or the exogenous variables are polynomials in time of degree not exceeding two. It is also highly efficient computationally, since it depends on simple extensions of certain matrices and vectors used in the estimation procedure.

The method developed in this paper has obvious potential practical applications in economic forecasting. Since it yields forecasts of observable discrete data it, also, provides a basis for the comparison of the predictive power of a continuous time model of the type described in this paper with various types of model formulated in discrete time. This could be of considerable importance in helping to resolve some of the methodological debates of recent years on the relative merits of various types of econometric models.

References

AGBEYEGBE, T. D. (1987), 'An exact discrete analog to a closed linear first-order continuous-time system with mixed sample', *Econometric Theory,* **3**, 143–9.

BERGSTROM, A. R. (1983), 'Gaussian estimation of structural parameters in higher order continuous time dynamic models', *Econometrica,* **51**, 117–52.

—— (1984), 'Continuous time stochastic models and issues of aggregation over time', in *Handbook of Econometrics* (eds. Z. Griliches and M. D. Intriligator) (Amsterdam, North-Holland) pp. 1146–212.

—— (1984), 'Monetary fiscal and exchange rate policy in a continuous time model of the United Kingdom', in *Contemporary Macroeconomic Modelling* (eds. P. Malgrange and P. Muet) (Oxford, Blackwell), pp. 183–206.

—— (1985), 'The estimation of parameters in nonstationary higher-order continuous time dynamic models', *Econometric Theory,* **1**, 369–85.

—— (1986), 'The estimation of open higher order continuous time dynamic models with mixed stock and flow data', *Econometric Theory,* **2**, 350–73.

—— (1987), 'Optimal control in wide-sense stationary continuous-time stochastic models', *J. Econ. Dynam. Control,* **11**, 425–43.

—— and C. R. WYMER (1976), 'A model of disequilibrium neoclassical growth and its application to the United Kingdom', in *Statistical Inference in Continuous Time Econometric Models* (ed. A. R. Bergstrom) (Amsterdam, North-Holland), pp. 267–327.

GANDOLFO, G. and P. C. PADOAN (1984), *A Disequilibrium Model of Real and Financial Accumulation in an Open Economy* (Berlin, Springer).

—— —— (1987), *The Mark V Version of the Italian Continuous Time Model* (Instituto di Economia della Facolta di Scienze Economiche e Bancarie, Siena).

HARVEY, A. C. and J. H. STOCK (1985), 'The estimation of higher-order continuous time autoregressive models', *Econometric Theory,* **1**, 97–117.

JONSON, P. D., E. R. MOSES, and C. R. WYMER (1977), 'The RBA 76 model of the Australian economy', *Conf. Applied Economic Research* (Reserve Bank of Australia).

—— W. J. MCKIBBIN, and R. G. TREVOR (1982), 'Exchange rates and capital flows', *Can. J. Econ.* **15**, 669–92.

KIRKPATRICK, G. (1987), *Employment Growth and Economic Policy: An Econometric Model of Germany* (Tübingen, Mohr).

KNIGHT, M. D. and C. R. WYMER (1978), 'A macroeconomic model of the United Kingdom', *IMF Staff Papers,* **25**, 742–78.

SIMS, C. A. (1980), 'Macroeconomics and reality', *Econometrica,* **48**, 1–48.

TULLIO, G. (1981), 'Demand management and exchange rate policy: the Italian experiences', *IMF Staff Papers,* **28**, 80–117.

ROZANOV, Y. A. (1967), *Stationary Random Processes* (San Francisco, Holden-Day).

9

Optimal Control in Wide-Sense Stationary Continuous-Time Stochastic Models

1. Introduction

DURING the last few years there has been a growing number of applications of continuous-time macroeconometric models to policy analysis: see, for example, Bergstrom 1984*b*, [Ch. 11 of this volume], Gandolfo and Padoan 1982, 1984, Gandolfo and Petit 1986, Jonson and Trevor 1981, Jonson, McKibbin and Trevor 1982, Kirkpatrick 1986, Sassanpour and Sheen 1984, Stefansson 1981 and Tullio 1981. These studies include both simulations of the behaviour of the estimated models under various policy assumptions and the mathematical analysis of the effects of various types of policy feedback on their asymptotic stability.

The latter type of analysis has been greatly facilitated by the methodology used in the design of these models. They have been designed in such a way that the variables have steady-state exponential growth paths, whose levels and growth rates can be obtained as explicit functions of the structural parameters, and so that the deviations of the logarithms of the variables from their steady-state paths satisfy a differential equations system whose non-linear part meets the Poincaré–Liapounov–Perron condition (see Bellman 1953, p. 93 and Coddington and Levinson 1955, p. 314). See Gandolfo 1981, ch. 2 for a clear and detailed exposition of this methodology which was developed in Bergstrom 1967 and applied to the prototype continuous-time econometric model of Bergstrom and Wymer 1976 [Ch. 10 of this volume].

The construction of the models according to this methodology also provides a potential basis for optimal control by methods applicable to LQ problems (minimization of quadratic costs subject to linear constraints). Such methods could be applied by treating the linear (in logarithms) approximation about the steady state as if it were the true model. Indeed, this procedure has already been applied by Stefansson (1981) in the optimal control of a small continuous-time econometric model of the Icelandic economy.

Such a procedure will not, of course, give the exact optimal feedbacks for controlling the non-linear model. But some sort of approximation is essential in controlling a complicated nonlinear continuous-time stochastic model with many equations. Attempts to apply stochastic dynamic programming to such a model will normally yield complicated partial differential equations which cannot be solved analytically. An alternative approximate method would be to use an approximate discrete-time dynamic model and apply the method developed by Chow (1975, ch. 12).

For the remainder of this paper we shall treat the approximate linear continuous-time model as if it were the true model. The main purpose of the paper is to provide a rigorous treatment of the problem of controlling a continuous-time linear stochastic model, with an infinite-horizon quadratic cost function, under the most general assumptions about the white-noise innovations.[1] In particular, we shall avoid the assumption that the integrated innovations are Brownian motion, an assumption which has been made in nearly all of the recent literature on the optimal control of continuous-time stochastic models.

It will often be realistic to assume, that, at least, some of the innovations in a continuous-time econometric model are generated not by Brownian motion, but by other types of process. For example, in the case of some financial variables it will be realistic to assume that the innovations are generated by a Poisson process. Our treatment will be sufficiently general to allow for innovations that are generated by a mixture of Brownian motion and Poisson processes and more general processes whose increments are not independent but merely orthogonal.

The solution of the deterministic control problem whose stochastic equivalent will be considered in this paper is well known. In the simplest case we assume that

$$\frac{\mathrm{d}}{\mathrm{d}t}y(t) = Ay(t) + Bx(t), \tag{1}$$

where $y(t)$ is an $n \times 1$ vector of state variables, $x(t)$ is an $r \times 1$ vector of control variables, and A and B are constant matrices. We then seek the feedback of the type

$$x(t) = Ky(t), \tag{2}$$

where K is a constant matrix, that minimizes the integral

$$J_1 = \int_0^\infty \{y'(t)Qy(t) + x'(t)Rx(t)]\mathrm{d}t, \tag{3}$$

where Q and R are constant matrices, Q being non-negative definite and

[1] We use the term 'white noise' to describe any innovation process whose increments are orthogonal even if these increments are not independent. An innovation process whose increments are independent is sometimes described as 'pure white noise'. See Hannan 1970, p. 15.

R positive definite. It is well known that the optimal feedback of type (2) is given by putting $K = K^*$, where

$$K^* = -R^{-1}B'P^*, \tag{4}$$

and P^* is the non-negative definite solution of the algebraic Riccati equation

$$Q + P^*A + A'P^* - P^*BR^{-1}B'P^* = 0. \tag{5}$$

See Aoki (1983, p. 231) for a simple proof of this result.

The above control problem could be made more general by replacing eq. (1) by

$$\frac{\mathrm{d}}{\mathrm{d}t}y(t) = Ay(t) + Bx(t) + Cz(t), \tag{1'}$$

where $z(t)$ is an $m \times 1$ vector of exogenous non-control variables. If a non-linear continuous-time model is linearized about the steady state corresponding to assumed exponential growth paths of the exogenous variables, then the exogenous variables will disappear and the logarithms of the deviations of the remaining variables from their steady-state paths will satisfy the system (1). But if the model is linearized about some other path, the exogenous variables will not disappear, and the linearized system will be of the form (1'). (See for example Gandolfo and Petit 1986 who linearize their system about an assumed 'ideal path' rather than the steady state.) A discussion of the complications resulting from the presence of exogenous non-control variables is beyond the scope of this paper. We shall confine our attention to the problem of controlling a linear stochastic model in which the only exogenous variables are the control variables.

The standard approach to continuous time stochastic control problems in recent years has been to use stochastic dynamic programming and the Ito calculus. (See Friedman 1975 for a rigorous modern exposition of the Ito calculus and the theory of stochastic differential equations of the Ito type.) For this purpose we can replace eq. (1) by a stochastic differential equation:

$$\mathrm{d}y(t) = \{Ay(t) + Bx(t)\}\mathrm{d}t + \mathrm{d}w(t), \tag{6}$$

where $w(t)$ is an n-dimensional Wiener process, i.e., a separable version of n-dimensional Brownian motion (see Friedman 1975, ch. 2). We can then define the cost function J_2 by

$$J_2 = E_t\left[\int_t^T \{y'(t)Qy(t) + x'(t)Rx(t)\}\mathrm{d}t + y'(T)Sy(T)\right], \tag{7}$$

where E_t denotes the expectation conditional on information up to time t, and Q, R and S are constant matrices, Q and S being non-negative definite and R positive definite. It has been shown (see Chow 1979) that the feedback that minimizes J_2 subject to the constraint (6) is

$$x(t) = -R^{-1}B'P^*(t)y(t), \tag{8}$$

where $P^*(t)$ is the solution of the Riccati differential equation

$$-\frac{\mathrm{d}P}{\mathrm{d}t} = Q + PA + A'P - PBR^{-1}B'P, \tag{9}$$

subject to

$$P(T) = S.$$

There has also been some discussion in the engineering literature (for example, Kwakernaak and Sivan 1972) of the problem of minimizing the cost function J_2 subject to a stochastic constraint similar to (6) except that the white-noise innovations are not required to be generated by Brownian motion. It has been shown that, even when the innovations are not generated by Brownian motion, the feedback (8) (with $P^*(t)$ given by the solution of (9) subject to $P(T) = S$) is optimal in the class of feedbacks of the type

$$x(t) = K(t)y(t),$$

where $K(t)$ is a matrix whose elements are functions of t. But this part of the literature is heuristic in that the precise conditions on the white noise that are necessary to justify the type of mathematical argument used are not made explicit. The conditions are certainly stronger than we shall need to assume in this paper. For it is implicitly assumed (see, for example, Kwakernaak and Sivan 1972, p. 254 and theorems 1.54 and 3.9) that the sample paths (realizations) of the quadratic functions of the state variables are integrable, at least in the Lebesgue sense, and that the conditions of Fubini's Theorem (Kolmogorov and Fomin 1961, vol. 2, p. 73) are satisfied so that the order of integration with respect to the probability measure and time can be interchanged, i.e. expectations can be taken under the integral sign. In this paper we shall not even assume that the sample paths are integrable.

Now let us assume an infinite-time horizon and consider the $(n + r)$-dimensional stationary random process $\{[x'(t), y'(t)], -\infty < t < \infty\}$ generated by the system

$$\mathrm{d}y(t) = \{Ay(t) + Bx(t)\}\mathrm{d}t + \mathrm{d}w(t), \qquad -\infty < t < \infty,$$
$$x(t) = Ky(t), \qquad -\infty < t < \infty,$$

where $w(t)$ is an n-dimensional Wiener process. The assumption that $w(t)$ is a Wiener process ensures (because a Wiener process is Gaussian) that the process $\{[x(t), y(t)], -\infty < t < \infty\}$ is metrically transitive (Rozanov 1967, p. 163) and hence, by the ergodic theorem (Rozanov 1967, theorem 5.1, p. 156), that

$$\lim_{T\to\infty} \frac{1}{T}\int_0^T \{y'(t)Qy(t) + x'(t)Rx(t)\}\mathrm{d}t$$
$$= E\{y'(t)Qy(t) + x'(t)Rx(t)\}, \tag{10}$$

with probability 1. An infinite-horizon quadratic cost function can be defined, therefore, in two equivalent ways: it can be defined as the time average

$$J_3 = \lim_{T\to\infty} \frac{1}{T}\int_0^T \{y'(t)Qy(t) + x'(t)Rx(t)\}\mathrm{d}t, \tag{11}$$

or the unconditional expectation

$$J_4 = E\{(y'(t)Qy(t) + x'(t)Rx(t)\}. \tag{12}$$

From the results given by eqs. (8) and (9) for the finite-horizon case we may infer heuristically, by putting $\mathrm{d}P/\mathrm{d}t = 0$ in (9), that the cost function J_3 (or J_4) is minimized subject to the constraint (6) by the feedback

$$x(t) = R^{-1}B'P^*y(t), \tag{13}$$

where P^* is the solution of the algebraic Riccati equation (5).

In section 3 we shall prove a similar result under much weaker assumptions than are necessary for the application of the Ito calculus. In particular, we shall avoid the assumption that the innovations are generated by a Wiener process. This assumption implies not only that the distribution of $w(t)$ is Gaussian, but also that its sample paths are continuous with probability 1 [see Friedman (1975, theorem 1.2, corol. 1.3, p. 38)). Indeed, this continuity of the sample paths plays an essential role in the development of the Ito calculus.

Instead of assuming that the process $\{[x'(t), y'(t)], -<t<\infty\}$ is strictly stationary we shall assume that it is stationary only in the wide sense (Rozanov 1967, p. 1). We shall use the theory of stochastic differential equations based on random measures, as developed in Bergstrom (1983, sect. 3) [Ch. 4 of this volume] rather than the Ito calculus, which is not applicable under our weaker assumptions. In particular, we shall rely on the existence and uniqueness theorem (Bergstrom 1983, theorem 1). For the purpose of this paper the theorem will be modified so that it applies to the stationary solution of the system rather than the solution subject to a boundary condition. The modified version of the theorem is presented in the next section (Theorem 1 of this paper).

In section 3 it will be shown (Theorem 2) that the feedback (13), where P^* is the solution of (5), is optimal in the class of feedbacks of type (2) under very general assumptions. It will be shown also (Theorem 3) that it is optimal in a more general class of linear feedbacks. These results will be extended to higher-order systems in section 4, where we shall also take account of adjustment costs.

2. The Theory of Linear Stochastic Differential Equations

We shall be concerned, in this section, with a system of linear stochastic differential equations with constant coefficient matrices and white-noise innovations. In the special case in which the innovations are generated by a Wiener process, it is a system of differential equations of the Ito type, and we can rely on the theory of stochastic differential equations as developed by Ito (1946, 1951), Doob (1953) and later authors working with similar assumptions (see Friedman 1975, ch. 5). But the assumption that the innovations are generated by a Wiener process is unnecessarily restrictive and, for reasons discussed in the introduction, will often be an inappropriate assumption in econometric modelling.

In a recent article, Bergstrom (1983), I developed a new theory of linear stochastic differential equations based on the theory of random measures. (See Bergstrom 1984*a*, pp. 1157–61) for an introduction to the theory of random measures.) I shall now present a modified version of the main theorem of that paper (Bergstrom 1983, theorem 1) so that it is directly applicable to a wide-sense stationary system.

We shall consider the kth-order system

$$\begin{aligned} \mathrm{d}[\mathrm{D}^{k-1}x(t)] = [&A_1\mathrm{D}^{k-1}x(t) + \ldots + A_{k-1}\mathrm{D}x(t) \\ &+ A_k x(t) + b]\mathrm{d}t + \zeta(\mathrm{d}t), \qquad -\infty < t < \infty, \end{aligned} \tag{14}$$

where $x(t)$ is a real n-dimensional continuous-time random process, $A_1, \ldots, A_k$ are $n \times n$ constant matrices, b is an $n \times 1$ constant vector, and D is the mean-square differential operator defined by

$$\lim_{h \to 0} \mathrm{E}\left[\mathrm{D}x_i(t) - \frac{x_i(t+h) - x_i(t)}{h}\right]^2 = 0, \qquad i = 1, \ldots, n.$$

With respect to the innovation vector $\zeta(\mathrm{d}t)$ we shall make the following assumption:

Assumption 1. $\zeta = [\zeta_1, \ldots, \zeta_n]'$ is a vector of random measures, defined on all subsets of the line $-\infty < t < \infty$ with finite Lebesgue measure, such that $\mathrm{E}[\zeta(\mathrm{d}t)] = 0$, $\mathrm{E}[\zeta(\mathrm{d}t)\zeta'(\mathrm{d}t)] = (\mathrm{d}t)\Sigma$, where Σ is a non-negative definite matrix, and $\mathrm{E}[\zeta_i(\Delta_1)\zeta_j(\Delta_2)] = 0$ $(i, j = 1, \ldots, n)$ for any disjoint sets Δ_1 and Δ_2 on the line $-\infty < t < \infty$.

Eq. (14) will be interpreted as meaning that, for any $-\infty < t_1 < t_2 < \infty$, $x(t)$ satisfies the integral equation

$$\begin{aligned} \mathrm{D}^{k-1}x(t_2) - \mathrm{D}^{k-1}x(t_1) = \int_{t_1}^{t_2} &[A_1\mathrm{D}^{k-1}x(t) + \ldots + A_{k-1}\mathrm{D}x(t) \\ &+ A_k x(t) + b]\mathrm{d}t + \int_{t_1}^{t_2} \zeta(\mathrm{d}t), \end{aligned} \tag{15}$$

where

$$\int_{t_1}^{t_2} \zeta(\mathrm{d}t) = \zeta[t_1, t_2],$$

and the first integral on the right side of (15) is the integral of a random process as defined by Rozanov 1967, p. 11. This type of integral, which is essentially due to Bochner 1933, does not require the sample paths of the process (corresponding to the elementary events in the space on which the probability measure is defined) to be integrable as are the sample paths of a Wiener process. We shall refer to it as the *wide-sense* integral and say that a random process for which the wide-sense integral exists is *integrable in the wide sense*.

We may, as in Bergstrom 1983, assume that the elements of $A_1, \ldots, A_n$ and b are functions of a more basic vector of structural parameters. But the properties of these functions will not concern us in this paper. We shall, however, require the following assumption which was unnecessary in the 1983 paper because of the boundary condition at $t = 0$.

Assumption 2. The eigenvalues of the $kn \times kn$ matrix $\bar{A}$ defined by (16) have negative real parts:

$$\bar{A} = \begin{bmatrix} 0 & I & 0 & \cdots & 0 & 0 \\ 0 & 0 & I & \cdots & 0 & 0 \\ \vdots & \vdots & \vdots & & \vdots & \vdots \\ 0 & 0 & 0 & \cdots & 0 & I \\ A_k & A_{k-1} & A_{k-2} & \cdots & A_2 & A_1 \end{bmatrix}. \tag{16}$$

We can now define a stationary solution of (14).

Definition. We shall say that the n-dimensional continuous-time random process $x(t)$ is a *stationary solution* of (14) if, for any $-\infty < t_1 < t_2 < \infty$, $x(t)$ satisfies the integral equation (15) and $\mathrm{D}^{k-1}x(t)$ is mean-square continuous on $[t_1, t_2]$ (i.e., $\mathrm{E}|\mathrm{D}^{k-1}x(t) - \mathrm{D}^{k-1}x(t-h)|^2 \to 0$, uniformly in t on $[t_1, t_2]$, as $h \to 0$).

The following theorem shows that there exists a unique [in the sense of (19)] stationary solution of (14) and gives its precise form.

Theorem 1. Under Assumptions 1 and 2, (a) a stationary solution of the system (14) exists and is given by

$$x(t) = y_1(t), \qquad -\infty < t < \infty, \tag{17}$$

where $y_i(t)$ is obtained from the partition of the $nk \times 1$ vector $y(t)$ into k $n \times 1$ vectors $y(t_1), \ldots, y(t_k)$,

$$y(t) = \int_{\infty}^{t} e^{(t-r)\bar{A}}\zeta(\mathrm{d}r) - \bar{A}^{-1}b, \tag{18}$$

$\bar{A}$ is defined by (16),

$$\bar{b} = \begin{bmatrix} 0 \\ 0 \\ \vdots \\ b \end{bmatrix}, \qquad \bar{\zeta} = \begin{bmatrix} 0 \\ 0 \\ \vdots \\ \zeta \end{bmatrix},$$

for any matrix A,

$$e^A = \sum_{r=0}^{\infty} \frac{1}{r!} A^r,$$

and the first term on the right-hand side of (18) is the integral of a measurable function with respect to a random measure (Rozanov 1967, p. 7).

(b) The solution (17) is unique in the sense that if $\hat{x}(t)$ is any other stationary solution of (14) then

$$P\{(\hat{x}(t) - x(t)) = 0\} = 1, \qquad -\infty < t < \infty. \tag{19}$$

Proof. (a) Let $\bar{\Sigma}$ be the $nk \times nk$ matrix defined by

$$\begin{bmatrix} 0 & 0 & . & . & 0 \\ 0 & 0 & . & . & 0 \\ . & . & . & . & . \\ . & . & . & . & . \\ 0 & 0 & . & . & \Sigma \end{bmatrix}.$$

Then Assumption 2 implies the existence of the integral

$$\int_{-\infty}^{t} e^{(t-r)\bar{A}}\,\bar{\Sigma} e^{(t-r)\bar{A}'}\mathrm{d}r = \int_{0}^{\infty} e^{r\bar{A}}\,\bar{\Sigma} e^{r\bar{A}'}\mathrm{d}r,$$

and hence (by a multi-dimensional generalization of condition 2.12 of Rozanov 1967, p. 6) the existence of the integral

$$\int_{-\infty}^{t} e^{(t-r)\bar{A}}\bar{\zeta}(\mathrm{d}r).$$

An kn-dimensional wide-sense stationary random process $\{y(t), \infty < t < \infty\}$ is well-defined, therefore, by eq. (18). Moreover, from (18) we obtain, for any $-\infty < t_1 < \infty$,

$$y(t) = \int_{t_1}^{t} e^{(t-r)\bar{A}}\bar{\zeta}(\mathrm{d}r) + e^{(t-t_1)\bar{A}} \times \left[\int_{-\infty}^{t_1} e^{(t_1-r)\bar{A}}\bar{\zeta}(\mathrm{d}r) - \bar{A}^{-1}\bar{b} + \bar{A}^{-1}\bar{b}\right] - \bar{A}^{-1}\bar{b}, \qquad t \geq t_1. \tag{20}$$

It follows, therefore, by Theorem 1 of Bergstrom 1983 (see eq. (6) of that paper) that the random process $x(t)$ defined by eqs. (17) and (18) is the solution of (14) subject to the boundary condition

$$\begin{bmatrix} x(t_1) \\ \mathrm{D}x(t_1) \\ \vdots \\ \mathrm{D}^{k-1}x(t_1) \end{bmatrix} = \int_{-\infty}^{t_1} e^{(t_1-r)\bar{A}}\bar{\zeta}(\mathrm{d}r) - \bar{A}^{-1}\bar{b}.$$

It then follows from the definition of the solution of (14) subject to a boundary condition (Bergstrom 1983, p. 68 this volume) that $x(t)$ satisfies (15) and $\mathrm{D}^{k-1}x(t)$ is mean-square continuous on $[t_1, t_2]$, i.e. $x(t)$ satisfies the conditions defining a stationary solution.

(b) It also follows [using Bergstrom (1983, theorem 1, part (b)] that, if $\hat{x}(t)$ is any other stationary solution of (14), then, for any $-\infty < t_1 < t_2 < \infty$,

$$P\{(\hat{x}(t) - x(t)) = 0\} = 1, \qquad t_1 < t < t_2,$$

and hence that (19) is satisfied. ■

3. Optimal Control of a First-Order System

We shall now consider the first-order system

$$\mathrm{d}y(t) = \{Ay(t) + Bx(t)\}\mathrm{d}t + \zeta(\mathrm{d}t), \quad -\infty < t < \infty, \tag{21}$$

where $y(t)$ is an $n \times 1$ vector of state variables, $x(t)$ is an $r \times 1$ vector of control variables, A and B are constant matrices, and ζ is a vector of random measures satisfying Assumption 1. Provided that the process $\{x(t), -\infty < t < \infty\}$ is integrable in the wide sense, we can interpret (21) as meaning that, for any $-\infty < t < \infty$,

$$y(t_2) - y(t_1) = \int_{t_1}^{t_2} \{Ay(t) + Bx(t)\}\mathrm{d}t + \int_{t_1}^{t_2} \zeta(\mathrm{d}t). \tag{22}$$

In particular, if $x(t)$ is related to $y(t)$ by a linear feedback of the type given by eq. (2), then

$$\mathrm{d}y(t) = (A + BK)y(t)\mathrm{d}t + \zeta(\mathrm{d}t), \qquad -\infty < t < \infty, \tag{23}$$

which is a special case of (14) and is interpreted as meaning that, for any $-\infty < t_1 < t_2 < \infty$,

$$y(t_2) - y(t_1) = (A + BK)\int_{t_1}^{t_2} y(t)\mathrm{d}t + \int_{t_1}^{t_2} \zeta(\mathrm{d}t). \tag{24}$$

We shall not require the eigenvalues of A to have negative real parts, but shall make the following weaker assumption.

Assumption 3. The column vectors of the $n \times nr$ controllability matrix $[B, AB, \ldots, A^{n-1}B]$ span the n-dimensional space R^n.

Under Assumption 3 the matrix K can be chosen so that the eigenvalues of $A + BK$ have negative real parts (see Aoki 1976, pp. 80, 135). Then the n-dimensional random process $\{y(t), -\infty < t < \infty\}$ generated by (23) is stationary in the wide sense, and a quadratic cost function J_4 can be defined by the unconditional expectation (12). Moreover, if the elements of $y(t)$ and $x(t)$ have finite fourth moments (a sufficient condition for which is that the random variables defining ζ have finite fourth moments), then, by the ergodic theorem for a wide-sense stationary process (Rozanov 1967, theorem 6.2, p. 25), we have $J_4 = J_5$ where

$$J_5 = \underset{T\to\infty}{\text{l.i.m.}} \frac{1}{T}\int_0^T \{y'(t)Qy(t) + x'(t)Rx(t)\}\mathrm{d}t, \tag{25}$$

where l.i.m. denotes the limit in the mean square and the integral is defined in the wide sense.

We shall now show (Theorem 2) that the feedback of type (2) that minimizes J_4 (and J_5, provided that the state variables have finite fourth moments) is given by (4) and (5).

Theorem 2. Let $\{[x'(t), y'(t)], -\infty < t < \infty\}$ be the wide-sense stationary process generated by the system (21) and a feedback of type (2) under Assumptions 1 and 3. Then the cost function J_4 (and J_5, provided that the state variables have finite fourth moments) is minimized, with respect to K, when $K = -R^{-1}B'P^*$ and P^* is the non-negative definite matrix satisfying (5).

Proof.

$$\begin{aligned} J_4 &= E\{y'(t)Qy(t) + x'(t)Rx(t)\} \\ &= E\{y'(t)Qy(t) + y'(t)K'RKy(t)\} \\ &= \text{tr}(Q + K'RK)V, \end{aligned} \tag{26}$$

where

$$V = E[y(t)y'(t)].$$

But from (23), by Theorem 1,

$$y(t) = \int_{-\infty}^{t} e^{(t-r)(A-BK)}\zeta(\mathrm{d}r), \tag{27}$$

and hence (by a multi-dimensional generalization of Rozanov 1967, eq. 2.14, p. 7)

$$\begin{aligned} V &= \int_{-\infty}^{t} e^{(t-r)(A+BK)}\Sigma e^{(t-r)(A+BK)'}\mathrm{d}r \\ &= \int_0^{\infty} e^{r(A+BK)}\Sigma e^{r(A+BK)'}\mathrm{d}r. \end{aligned} \tag{28}$$

Therefore, substituting (28) into (26),

$$J_4 = \int_0^\infty \text{tr}[(Q + K'RK)e^{r(A+BK)}\Sigma e^{r(A+BK)'}]\text{d}r$$

$$= \int_0^\infty \text{tr}[e^{r(A+BK)'}(Q + K'RK)e^{r(A+BK)}\Sigma]\text{d}r. \tag{29}$$

Now using the spectral decomposition

$$\Sigma = \lambda_1 h_1 h_1' + \lambda_2 h_2 h_2' + \ldots + \lambda_n h_n h_n',$$

where $\lambda_1, \ldots, \lambda_n$ are the eigenvalues of Σ and $h_1, \ldots, h_n$ are their associated eigenvectors, we obtain from (29)

$$J_4 = \sum_{i=1}^{n} \lambda_i \int_0^\infty \text{tr}[e^{r(A+BK)'}(Q + K'RK)e^{r(A+BK)}h_i h_i']\text{d}r$$

$$= \sum_{i=1}^{n} \lambda_i \int_0^\infty h_i' e^{r(A+BK)'}(Q + K'RK)e^{r(A+BK)}h_i \text{d}r. \tag{30}$$

The formula (30) for J_4 enables us to complete the proof by relying on the well-known solution of the deterministic control problem discussed in the Introduction. It is easily verified that the solution of the deterministic system (1) with the feedback (2), subject to the initial condition $y(0) = h_i$, is

$$y(t) = e^{t(A+BK)}h_i. \tag{31}$$

Moreover, substituting from (31) and (2) into (3), we obtain

$$J_1 = \int_0^\infty h_i' e^{r(A+BK)'}(Q + K'RK)e^{r(A+BK)}h_i \text{d}r.$$

We know from the solution of the deterministic control problem that J_1 is minimized by putting $K = -R^{-1}B'P^*$, where P^* is the non-negative definite matrix satisfying (5). But, since Σ is a non-negative definite matrix, each of the eigenvalues $\lambda_1, \ldots, \lambda_n$ is non-negative. Therefore each of the n terms on the right-hand side of (30) is minimized when $K = -R^{-1}B'P^*$, where P^* is the non-negative definite matrix satisfying (5). ■

The use of the cost function J_2, defined by (7), for the finite-horizon stochastic control problem is usually motivated by saying that it is the average of the cost functions corresponding to infinitely many different realizations or sample paths (see, for example, Kwakernaak and Sivan 1972, p. 254). It is remarkable that, by using the spectral decomposition of the covariance matrix of the innovations, we have been able to show that the cost function for the infinite-horizon stochastic control problem is a weighted average of only n cost functions corresponding to n pseudo sample paths, which differ only with respect to their initial state vectors and, hence, are all minimized by the same feedback. Moreover, we have obtained this result under very weak assumptions which do not even require the true sample paths to be integrable.

It should be noticed that we have, essentially, proved a dynamic continuous-time version of the certainty equivalence theorem under very weak assumptions about the innovations. The theorem might be inferred heuristically, by analogy, from the dynamic discrete-time certainty equivalence theorem (see Simon 1956; Theil 1957; and Whittle 1983, p. 137). But this sort of analogy is not a substitute for a rigorous proof.

So far we have restricted the feedback to be of type (2). We shall now show that, for the system (21) with the cost function J_4 (and J_5, provided that the state variables have finite fourth moments), the feedback $x(t) = -R^{-1}B'P^*y(t)$, with P^* given by the solution of (5), is optimal in a much more general class of linear feedbacks. We shall consider the class of feedbacks of the type

$$x(t) = K_1 y(t) + K_2 \int_{-\infty}^{t} H(t-r) y(r) \mathrm{d}r, \tag{32}$$

where K_1 and K_2 are constant matrices, $H(r)$ is a matrix-valued function satisfying the linear differential equation

$$\frac{\mathrm{d}^k}{(\mathrm{d}r)^k} H(r) = F_1 \frac{\mathrm{d}^{k-1}}{(\mathrm{d}r)^{k-1}} H(r) + \ldots + F_k H(r), \tag{33}$$

and $F_1, \ldots, F_k$ are constant matrices such that the eigenvalues of $\bar{F}$ [defined as in (16) with F replacing A] have negative real parts. For example, in the simplest case, we have

$$\frac{\mathrm{d}}{\mathrm{d}r} H(r) = F H(r), \tag{34}$$

$$H(r) = e^{rF}. \tag{35}$$

In the general case $H(r)$ can be written explicitly as the Fourier transform

$$H(r) = \frac{1}{2\pi} \int_{-\infty}^{\infty} e^{irt} [(it)^k - (it)^{k-1} F_1 - \ldots - F_k]^{-1} \mathrm{d}t. \tag{36}$$

Theorem 3. Let $\{[x'(t), y'(t)], -\infty < t < \infty\}$ *be the wide-sense stationary process generated by the system (21) and a feedback of the type given by (32) and (33), under Assumptions 1 and 3. Then the cost function* J_4 *(and* J_5*, provided that the state variables have finite fourth moments) is minimized, with respect to* K_1, K_2, $F_1, \ldots, F_k$, *when* $K_1 = -R^{-1}B'P^*$, $K_2 = 0$ *and* P^* *is the non-negative definite matrix satisfying* (5).

Proof. For simplicity we shall prove the theorem for the simplest case where $H(r)$ is defined by (35). The proof for the general case is similar. Let $z(t)$ be defined by

$$z(t) = \int_{-\infty}^{t} H(t-r) y(r) \mathrm{d}r = \int_{-\infty}^{t} e^{(t-r)F} y(r) \mathrm{d}r.$$

Then

$$dz(t) = \{y(t) + Fz(t)\}dt, \tag{37}$$

which is interpreted as meaning that, for any $-\infty < t_1 < t_2 < \infty$,

$$z(t_2) - z(t_1) = \int_{t_1}^{t_2} \{y(t) + Fz(t)\}dt, \tag{38}$$

the integral on the right-hand side of (38) [and in the definition of $z(t)$] being defined in the wide sense. Combining (21) and (37) we obtain the system

$$d\begin{bmatrix} y(t) \\ z(t) \end{bmatrix} = \begin{bmatrix} A & 0 \\ I & F \end{bmatrix}\begin{bmatrix} y(t) \\ z(t) \end{bmatrix}dt + \begin{bmatrix} B \\ 0 \end{bmatrix}[x(t)]dt + \begin{bmatrix} \zeta(dt) \\ 0 \end{bmatrix}. \tag{39}$$

Moreover, the feedback of the type given by (32) and (35) can be written

$$x(t) = K_1 y(t) + K_2 z(t), \tag{40}$$

while the cost function J_4 can be written

$$J_4 = E\left\{[y'(t), z'(t)]\begin{bmatrix} Q & 0 \\ 0 & 0 \end{bmatrix}\begin{bmatrix} y(t) \\ z(t) \end{bmatrix}\right\} + E\{x'(t)Rx(t)\}. \tag{41}$$

We are now required to find the feedback of the type (40) that minimizes J_4, as defined by (41), subject to the constraint (39). By Theorem 2 the optimal feedback is

$$x(t) = K_1^* y(t) + K_2^* z(t),$$

where

$$[K_1^*, K_2^*] = -R^{-1}[B', 0]P^*,$$

and

$$P^* = \begin{bmatrix} P_{11}^* & P_{12}^* \\ P_{21}^* & P_{22}^* \end{bmatrix}$$

is the non-negative definite matrix satisfying the algebraic Riccati equation

$$\begin{bmatrix} Q & 0 \\ 0 & 0 \end{bmatrix} + \begin{bmatrix} P_{11}^* & P_{12}^* \\ P_{21}^* & P_{22}^* \end{bmatrix}\begin{bmatrix} A & 0 \\ I & F \end{bmatrix}$$

$$+ \begin{bmatrix} A' & I \\ 0 & F' \end{bmatrix}\begin{bmatrix} P_{11}^* & P_{12}^* \\ P_{21}^* & P_{22}^* \end{bmatrix}$$

$$- \begin{bmatrix} P_{11}^* & P_{12}^* \\ P_{21}^* & P_{22}^* \end{bmatrix}\begin{bmatrix} B \\ 0 \end{bmatrix}$$

$$R^{-1}\begin{bmatrix} B' & 0 \end{bmatrix}\begin{bmatrix} P_{11}^* & P_{12}^* \\ P_{21}^* & P_{22}^* \end{bmatrix} = 0. \tag{42}$$

It is easily verified that the solution of (42) is given by putting $P_{12}^* = P_{21}^* = P_{22}^* = 0$ and P_{11}^* equal to the non-negative definite matrix satisfying (5). ■

The intuitive explanation of Theorem 3 is that, since the cost function involves the levels only of the control variables and not their rates of change, it is optimal to adjust these variables instantaneously in response to new information. Indeed, this explanation suggests that, if costs do not depend on the rate of change of the control variables, then the feedback $x(t) = K_1 y(t)$ is optimal in an even more general class of feedbacks than that given by (32) and (33). The assumption that the cost function does not include adjustment costs (i.e. terms depending on the rates of change of the control variables) is, of course, rather unrealistic. But the basic results given by Theorems 2 and 3 can easily be applied to more general systems, with more realistic cost functions, as will be shown in the next section.

4. Higher-Order Systems and Adjustment Costs

We shall now show, in the context of an example, how the results of Theorems 2 and 3 can be applied to higher-order systems with cost functions that include adjustment costs. Suppose that the $n \times 1$ state vector $y(t)$ and $r \times 1$ control vector $x(t)$ are related by the second-order system

$$\mathrm{d}[\mathrm{D}y(t)] = \{A_1 \mathrm{D}y(t) + A_2 y(t) + Bx(t)\}\mathrm{d}t + \zeta(\mathrm{d}t), \qquad -\infty < t < \infty, \tag{43}$$

where A_1, A_2 and B are constant matrices, D is the mean square differential operator and ζ satisfies Assumption 1. Provided that the process $\{x(t), -\infty < t < \infty\}$ is integrable in the wide sense, we can interpret (43) as meaning that, for any $-\infty < t_1 < t_2 < \infty$,

$$\mathrm{D}y(t_2) - \mathrm{D}y(t_1) = \int_{t_1}^{t_2} [A_1 \mathrm{D}y(t) + A_2 y(t) + Bx(t)]\mathrm{d}t + \int_{t_1}^{t_2} \zeta(\mathrm{d}t). \tag{44}$$

Let the cost function J_6 be defined by the unconditional expectation

$$J_6 = \mathrm{E}\{y'(t)Q_1 y(t) + \mathrm{D}y'(t)Q_2 \mathrm{D}y(t) + x'(t)R_1 x(t) + \mathrm{D}x'(t)R_2 \mathrm{D}x(t)\}. \tag{45}$$

Then, provided that the elements of $y(t)$, $\mathrm{D}y(t)$, $x(t)$ and $\mathrm{D}x(t)$ have finite fourth moments, we have $J_6 = J_7$ where J_7 is the wide-sense time average

$$J_7 = \text{l.i.m.} \frac{1}{T} \int_0^T \{y'(t)Q_1 y(t) + \mathrm{D}y'(t)Q_2 \mathrm{D}y(t) + x'(t)R_1 x(t) + \mathrm{D}x'(t)R_2 \mathrm{D}x(t)\}\mathrm{d}t. \tag{46}$$

It should be noted, incidentally, that since we are now dealing with a second-order system, the elements of $\mathrm{D}y(t)$ (D being the mean square diferential operator) are proper random variables, whereas when $y(t)$ is generated by the first-order system (21) $\mathrm{D}y(t)$ does not exist.

We now seek the linear feedback that minimizes J_6 (or J_7) subject to the constraint (43). Let the vectors $w(t)$, $\bar{\zeta}(\mathrm{d}t)$ and $z(t)$ and the matrices A, B, Q and R be defined by

$$w(t) = \begin{bmatrix} y(t) \\ \mathrm{D}y(t) \\ x(t) \end{bmatrix}, \qquad \bar{\zeta}(\mathrm{d}t) = \begin{bmatrix} 0 \\ \zeta(\mathrm{d}t) \\ 0 \end{bmatrix}, \tag{47}$$

$$z(t) = \mathrm{D}x(t), \tag{48}$$

$$A = \begin{bmatrix} 0 & I & 0 \\ A_2 & A_1 & B \\ 0 & 0 & 0 \end{bmatrix}, \tag{49}$$

$$B = \begin{bmatrix} 0 \\ 0 \\ I \end{bmatrix}, \tag{50}$$

$$Q = \begin{bmatrix} Q_1 & 0 & 0 \\ 0 & Q_2 & 0 \\ 0 & 0 & R_1 \end{bmatrix}, \tag{51}$$

$$R = R_2. \tag{52}$$

Then, from (43), (47), (48), (49) and (50), we obtain the system

$$\mathrm{d}w(t) = \{Aw(t) + Bz(t)\}\mathrm{d}t + \bar{\zeta}(\mathrm{d}t), \tag{53}$$

and, from (45), (51) and (52), we obtain

$$J_6 = \mathrm{E}\{w'(t)Qw(t) + z'(t)Rz(t)\}. \tag{54}$$

We have now reduced the problem to one of finding the linear feedback that minimizes J_6, as defined by (54), subject to the constraint (53), where $w(t)$ is regarded as a state vector and $z(t)$ as a control vector. Provided that the matrices A and B defined by (49) and (50) satisfy Assumption 3, we can make a direct application of Theorems 2 and 3 to obtain

$$\begin{aligned} z(t) &= -R^{-1}B'P^*w(t) \\ &= -R^{-1}[0, 0, I]P^*w(t), \end{aligned}$$

where P^* is the non-negative definite $3n \times 3n$ matrix satisfying (5) with A, B, Q and R defined by (49), (50), (51) and (52), respectively. The optimal feedback is, therefore, a differential equation of the form

$$\mathrm{D}x(t) = K_1\mathrm{D}y(t) + K_2 y(t) + K_3 x(t). \tag{55}$$

From (55) we obtain

$$\begin{aligned} x(t) &= [D - K_3]^{-1}[K_1 D + K_2]y(t) \\ &= [D - K_3]^{-1}\{[D - K_3]K_1 + [K_3 K_1 + K_2]\}y(t) \\ &= K_1 y(t) + \int_{-\infty}^{t} e^{(t-r)K_3}[K_3 K_1 + K_2]y(r)\mathrm{d}r, \end{aligned} \tag{56}$$

which is a feedback belonging to the general class of linear feedbacks defined by (32) and (33).

5. Conclusions

We have solved the optimal control problem for a continuous-time stochastic model with an infinite-horizon quadratic cost function under weaker assumptions concerning the innovation process than have been made in the literature on this subject. Since the optimal linear feedback is identical with that for the corresponding deterministic control problem, we have, essentially, proved a continuous-time infinite-horizon version of the dynamic certainty equivalence theorem under very weak assumptions.

Although we have, formally, assumed wide-sense stationarity the results of the main theorems of the paper are applicable under more general assumptions. For the wide-sense time average J_5 exists as a mean-square limit (provided that the fourth moments of the state variables exist) when the initial state vector $y(0)$ is an arbitrary fixed vector and its value does not depend on the initial state vector. Theorem 2 implies, therefore, that, if the optimal feedback given by this theorem is applied from the time $t = 0$ onwards, then the cost function J_5 will be minimized. The cost function J_4 can then be regarded as the unconditional expectation associated with a pseudo wide-sense stationary process, which describes how the variables would have behaved if the feedback had been applied over the infinite past and is introduced only to facilitate the proof through the use of the ergodic theorem.

In section 4 the main theorems of the paper have been applied in the derivation of the optimal linear feedback for a second-order system when the cost function involves both the levels and rates of change of the state and control variables. The particular model considered in section 4 is likely to be of great practical importance in applied econometric work. Although the continuous-time econometric models that have been developed so far (see the references in the Introduction) have been mainly first-order systems, it seems likely that, even with the data series as short as those normally available for econometric modelling, it will be necessary to model an economy as a second-order system in order to achieve a sufficiently realistic specification of the dynamic adjustment processes to make the best use of the data. Moreover, with the recent developments in computing technology and econometric methodology (see Bergstrom 1983, 1985, 1986) [Chs. 4, 5, and 6 of this volume] and Harvey and Stock 1985, it is now feasible to estimate a second-order system of about ten equations by methods which take account of the exact restrictions on the discrete stock and flow data implied by the continuous-time model and yield consistent estimates under the same weak assumptions concerning the innovations as have been made in this paper. An optimal feedback of the form (55) could then be obtained from the estimated parameters of the continuous-time model by the method followed in section 4.

References

AOKI, M. (1976), *Optimal Control and System Theory in Dynamic Economic Analysis* (Amsterdam, North-Holland).

—— (1983), *Notes on Economic Time Series Analysis: System Theoretic Perspectives* (Berlin, Springer-Verlag).

BELLMAN, R. (1953), *Stability Theory of Differential Equations* (New York, McGraw-Hill).

BERGSTROM, A. R. (1967), *The Construction and use of Economic Models* (London, English Universities Press).

—— (1983), 'Gaussian estimation of structural parameters in higher order continuous time dynamic models', *Econometrica*, **51**, 1117–52.

—— (1984*a*), 'Continuous time stochastic models and issues of aggregation over time', in Zvi Griliches and Michael D. Intriligator (eds.), *Handbook of Econometrics* (Amsterdam, North-Holland), 1146–212.

—— (1984*b*), 'Monetary, fiscal and exchange rate policy in a continuous time model of the United Kingdom, in Pierre Malgrange and Pierre-Alain Muet (eds.), *Contemporary Macroeconomic Modelling* (Oxford, Blackwell), 183–206.

—— (1985), 'The estimation of parameters in nonstationary higher-order continuous-time dynamic models', *Econometric Theory*, **1**, 369–85.

—— (1986), 'The estimation of open higher order continuous time dynamic models with mixed stock and flow data', *Econometric Theory*, **2**, 350–73.

—— and C. R. WYMER (1976), 'A model of disequilibrium neoclassical growth and its application to the United Kingdom', in A. R. Bergstrom (ed.) *Statistical Inference in Continuous Time Economic Models* (Amsterdam, North-Holland).

BOCHNER, S. (1933), 'Integration von Funktionen, deren Werte die Elemente eines Vektorraumes sind', *Fundamenta Mathematicae*, **20**, 262–76.

CHOW, G. C. (1975), *Analysis and Control of Dynamic Economic Systems* (New York, Wiley).

—— (1979), 'Optimal control of stochastic differential equation systems', *Journal of Economic Dynamics and Control*, **1**, 143–75.

CODDINGTON, E. A. and N. LEVINSON (1955), *Ordinary Differential Equations* (New York, McGraw-Hill).

DOOB, J. L. (1953), *Stochastic Processes* (New York, Wiley).
FRIEDMAN, A. (1975), *Stochastic Differential Equations and Applications* (New York, Academic Press).
GANDOLFO, G. (1981), *Qualitative Analysis and Econometric Estimation of Continuous Time Dynamic Models* (Amsterdam, North-Holland).
—— and P. C. PADOAN (1982), 'Policy simulations with a continuous time macrodynamic model of the Italian economy: A preliminary analysis', *Journal of Economic Dynamics and Control,* **4**, 205–24.
—— —— (1984), *A Disequilibrium Model of Real and Financial Accumulation in an Open Economy* (Berlin, Springer-Verlag).
—— and M. L. PETIT (1986), 'Optimal control in a continuous time macroeconometric model of the Italian economy, C.N.R. progetto finalizzato struttura ed evoluzione dell'economia italiana', Working paper no. 6, Nov.
HANNAN, E. J. (1970), *Multiple Time Series* (New York, Wiley).
HARVEY, A. C. and J. H. STOCK (1985), 'The estimation of higher order continuous time autoregressive models', *Econometric Theory,* **1**, 97–117.
ITO, K. (1946), 'On a stochastic integral equation', *Proceedings of the Japanese Academy,* **1**, 32–5.
—— (1951), 'On stochastic differential equations', *Memoir of the American Mathematical Society,* **4**.
JONSON, P. D. and R. G. TREVOR (1981), 'Monetary rules: A preliminary analysis', *Economic Record,* **57**, 150–67.
—— W. J. MCKIBBIN and R. G. TREVOR (1982), 'Exchange rates and capital flows: A sensitivity analysis', *Canadian Journal of Economics,* **15**, 669–92.
KIRKPATRICK, G. (1986), 'Structural modelling of dynamic macro-economic systems: General considerations, techniques and an application to the Federal Republic of Germany', Paper presented at the International Conference on Modelling Dynamic Systems, Paris.
KOLMOGOROV, A. N. and S. V. FOMIN (1981), *Elements of the Theory of Functions and Functional Analysis,* vol. 2 (New York, Graylock Press).
KWAKERNAAK, H. and R. SIVAN (1972), *Linear Optimal Control Systems* (New York, Wiley).
ROZANOV, Y. A. (1967), *Stationary Random Processes* (San Francisco, Holden-Day).
SASSANPOUR, C. and J. SHEEN (1984), 'An empirical analysis of the effect of monetary disequilibrium in open economies', *Journal of Monetary Economics,* **13**, 127–63.
SIMON, H. A. (1956), 'Dynamic programming under uncertainty with a quadratic criterion function', *Econometrica,* **24**, 74–81.
STEFANSSON, S. B. (1981), 'Inflation and economic policy in a small open economy: Iceland in the post-war period', Ph.D. thesis (University of Essex, Colchester).
THEIL, H. (1957), 'A note on certainty equivalence in dynamic planning', *Econometrica,* **25**, 346–9.
TULLIO, G. (1981), 'Demand management and exchange rate policy: The Italian experience, *IMF Staff Papers,* **28**, 80–117.
WHITTLE, P. (1983), *Prediction and Regulation* (Oxford, Blackwell).

PART III

Applications

10

A Model of Disequilibrium Neoclassical Growth and its Application to the United Kingdom[1]

A. R. Bergstrom and C. R. Wymer

1. Introduction

THERE is a growing literature on methods of estimating the structural parameters of stochastic economic models in continuous time from discrete observations of the variables (see Bergstrom 1966*a*; Phillips 1972, 1974*a*, and 1974*b*; Sargan 1974; Wymer 1972. In this chapter we describe the application of these methods in fitting a disequilibrium neoclassical growth model to United Kingdom data.

The model is an extension and generalization of a prototype cyclical growth model discussed in Bergstrom 1967 and 1966*b* [Ch. 2 of this volume]. That model was derived from the simple neoclassical growth model by discarding the assumption of full employment and introducing a price feed-back mechanism which tends to limit fluctuations in employment but cannot ensure a continuous state of full employment. The feed-back mechanism is essentially Keynesian, its components being a liquidity preference function, an investment function and wage and price adjustment relations. The prototype model is a closely interlocked system of differential equations which synthesizes neoclassical and Keynesian theory and is capable of generating stable cyclical growth paths for plausible values of the parameters. But it is too simple for direct econometric application to the United Kingdom, particulary since it assumes a closed economy. It is extended and generalized in this study by the introduction of foreign trade relations, the replacement of the Cobb–Douglas production function by a more general relation and various consequential changes.

[1] A. R. Bergstrom and C. R. Wymer would like to thank the Science Research Council Atlas Computing Laboratory and, in particular, Mr M. E. Claringbold, for providing the computing facilities necessary for this work and extensive computing assistance.

The study has three broad aims. The first is to discover to what extent a model of this type is capable of explaining the behaviour of the United Kingdom economy. This is not just a matter of considering its goodness of fit over the sample period. It involves, in addition, a consideration of the extent to which, for parameter values that are consistent with the sample, the model is capable of explaining certain characteristics of the economy which have been observed over a much longer period. Because of the smallness of the samples that are normally available for estimation of the parameters of econometric models, it is very important that they should be tested in this way. And, for this reason, it is desirable that they should be designed in such a way that they are amenable to mathematical analysis. One of the distinguishing features of the model discussed in this chapter as compared with most econometric models of complete economies is the extent to which its properties can be studied mathematically. A substantial part of the chapter will be devoted to a rigorous derivation of the steady state and asymptotic stability properties of the model.

The second aim is to test, on a model of a real economy, the methods of estimation referred to in the opening paragraph. An important part of the study has been the development of a computer programme for obtaining full information maximum likelihood estimates of the parameters of a stochastic differential equation system which is non-linear in the parameters and involves cross equation restrictions (see Wymer[2]). This is a considerably more complex computational problem than that of obtaining full information maximum likelihood estimates of the parameters of a standard simultaneous equations model, even when the latter is subject to non-linear cross equation restrictions.

A final aim is to produce a prototype medium-term forecasting model of the United Kingdom economy and one which could, with some modification, be used as a basis for macroeconomic regulation by monetary and fiscal policy. Such a model should take account of both the trade cycle and growth mechanisms in the economy. Moreover, even though it is not intended primarily for long-term prediction, it should be capable of producing long-run behaviour which seems plausible in the light of past experience. Failure to produce plausible long-run behaviour could indicate a structural defect such as the omission of an important feed-back. Such a defect could seriously affect the predictive power of the model and its usefulness for either medium-term forecasting or monetary and fiscal regulation.

Before the introduction of any assumptions about the stochastic elements of the model, it will be presented and discussed in a purely deterministic form. In section 2 the complete model will be presented and the individual equations discussed. The steady state solution, the way in

[2] C. R. Wymer, Computer programs: Discon manual and: Continuous systems manual, mimeo.

which this is affected by changes in the various structural parameters, the dynamic mechanism generating the transitory solution and the role of monetary phenomena in this mechanism will all be discussed in section 3. The asymptotic stability analysis, including the derivation of the asymptotic stability conditions in terms of the structural parameters will be carried out in section 4. The method of estimation and various problems which arise in its application to this model will be briefly discussed in section 5 while the parameter estimates and the behaviour implied by these estimates will be discussed in section 6. Section 7 will be devoted to a study of the predictive performance of the model including simulations over the sample period and a post sample forecasting period.

2. The Model

The model is highly aggregative being, formally, a model of a single sector economy whose product can be used for consumption (private and public) or capital formation (private and public) or exported. It comprises a system of 13 first-order differential equations including 10 behaviour relations and 3 identities. There are no exogenous variables except for 3 trends terms which allow for technical progress, the growth of the labour supply and the gradual increase in income and prices in the rest of the world. Hence the model completely determines the future paths of the variables for any set of initial values. Economic theory is used intensively in order to restrict the number of parameters to be estimated. There are, altogether, only 35 parameters including a vector β of 16 long-run elasticities and propensities, a vector γ of 16 speed of adjustment parameters and a vector λ of 3 trend parameters.

The equations of the model are

$$\frac{DC}{C} = \gamma_1 \log\left(\frac{\beta_1 Y}{C}\right), \tag{1}$$

$$\frac{DL}{L} = \gamma_2 \log\left[\frac{\beta_2 e^{-\lambda_1 t}\{Y^{-\beta_4} - \beta_3 K^{-\beta_4}\}^{-1/\beta_4}}{L}\right], \tag{2}$$

$$Dk = \gamma_3\left[\gamma_4\left\{\beta_3\left(\frac{Y}{K}\right)^{1+\beta_4} - r + \frac{Dp}{p}\right\} + \beta_5 - k\right], \tag{3}$$

$$\begin{aligned} DY = {} & \gamma_5\{(1-\beta_6)(C + DK + E) - Y\} \\ & + \gamma_6\{\beta_7(C + DK + E) - S\}, \end{aligned} \tag{4}$$

$$\dot{D}I = \gamma_7\{\beta_6(C + DK + E) - I\} + \gamma_8\{\beta_7(C + DK + E) - S\}, \tag{5}$$

$$\frac{DE}{E} = \gamma_9 \log\left\{\frac{\beta_8 p^{-\beta_9} e^{\lambda_2 t}}{E}\right\}, \tag{6}$$

$$\frac{Dp}{p} = \gamma_{10} \log\left[\frac{\beta_{10}\beta_2 w e^{-\lambda_1 t}\left\{1 - \beta_3\left(\frac{Y}{K}\right)^{\beta_4}\right\}^{-(1+\beta_4)/\beta_4}}{p}\right], \tag{7}$$

$$\frac{Dw}{w} = \gamma_{11} \log\left[\frac{\beta_2 e^{-\lambda_1 t}\{Y^{-\beta_4} - \beta_3 K^{-\beta_4}\}^{-1/\beta_4}}{\beta_{11} e^{\lambda_3 t}}\right], \tag{8}$$

$$\frac{Dr}{r} = \gamma_{12} \log\left[\frac{\beta_{12} p Y^{\beta_{13}} r^{-\beta_{14}}}{M}\right], \tag{9}$$

$$Dm = \gamma_{13} \log\left(\frac{E}{\beta_{15} I}\right) + \gamma_{14} \log\left(\frac{\beta_{16} e^{\lambda_3 t}}{L}\right) + \gamma_{15} D \log\left(\frac{E}{\beta_{15} I}\right) + \gamma_{16} D \log\left\{\frac{\beta_{16} e^{\lambda_3 t}}{L}\right\}, \tag{10}$$

$$DS = Y + I - C - DK - E, \tag{11}$$

$$\frac{DK}{K} = k, \tag{12}$$

$$\frac{DM}{M} = m; \tag{13}$$

where

C = real consumption,
Y = real net income or output,
K = amount of fixed capital,
E = real exports,
I = real expenditure on imports,
S = stocks,
L = employment,
M = volume of money,
r = interest rate,
p = price level,
w = wage rate,
k = proportional rate of increase of fixed capital,
m = proportional rate of increase in volume of money,
t = time
$D = \mathrm{d}/\mathrm{d}t$,

$0 < \beta_1 < 1$, $\beta_4 > -1$, $0 < \beta_6 < 1$, and all other parameters positive (except posssibly λ_3).

Equation (1) is a form of consumption function in which consumption is assumed to adjust gradually rather than instantaneously to a change in income. It is assumed that corresponding to any given level of income Y there is a partial equilibrium level of consumption $\beta_1 Y$ and that, at each point of time, the proportional rate of increase in consumption is an increasing function of the proportional excess of partial equilibrium consumption over actual consumption. The parameter β_1 is the long-run propensity to consume or, in other words, the limit to which the ratio of consumption to income would converge if income remained stationary for a sufficiently long period. The speed of adjustment parameter γ_1 can be assumed to depend partly on the psychological cost of adjusting consumption and partly on the speed with which expectations with regard to future income adjust in response to changes in current income.

It might appear, at first sight, that equation (1) assumes consumers to be myopic so that when income is growing steadily the ratio of consumption to income will always be less than the desired ratio. But this is not necessarily so. Suppose that the desired ratio of consumption to income is β_1' and that the expected rate of growth of income is g. Then we might postulate a consumption relation

$$\frac{DC}{C} = \gamma_1 \log\left(\frac{\beta_1' Y}{C}\right) + g, \tag{1'}$$

which implies that if consumption is at its desired level in relation to income it will be increasing at a proportional rate equal to the expected rate of growth of income. But equation (1′) can be written in the form

$$\frac{DC}{C} = \gamma_1 \log\left(\frac{\beta_1' e^{g/\gamma_1} Y}{C}\right), \tag{1''}$$

which is identical with (1) when $\beta_1 = \beta_1' e^{g/\gamma_1}$. A similar comment applies to most of the other equations in the model.

The employment equation (2) has the same general form as (1). The partial equilibrium level of employment is assumed to be related to output and the stock of fixed capital by a constant elasticity of substitution production function which includes a trend term to allow for technical progress. The parameter β_3 is a measure of the capital intensity of the production function while $1/(1+\beta_4)$ is the elasticity of substitution between labour and capital. Technical progress is assumed to be neutral in the Harrod sense and λ_1 is the rate of decrease per unit of time (through technical progress) in the amount of labour required to produce a given output with a given stock of capital. The speed of adjustment parameter γ_2 can be assumed to depend on costs associated with varying the number of persons employed (for example costs of hiring and training). These must be balanced by the firm against costs associated with either excessive overtime work or excess labour capacity within the firm.

It should be emphasized that the term partial equilibrium level of employment is used in the preceding paragraph with reference to a given stock of capital and level of output. The adjustment of the ratio of labour to capital in response to a change in their relative costs is brought about through variations in the capital–output ratio via the investment function. We shall return to this point at a later stage, after discussing other relations reflecting firms' behaviour.

Since k is defined by equation (12) as the proportional rate of increase in K, equation (3) is a second-order non-linear differential equation in K. It implies, therefore, that the stock of fixed capital depends with a rather complicated distributed time lag on output Y and the real interest rate $r - Dp/p$. More directly, it is assumed that, corresponding to any real interest rate and capital–output ratio, there is a partial equilibrium proportional rate of increase in the stock of fixed capital and that, at each point of time, the actual proportional rate of increase in capital is increasing at a rate proportional to the excess of the partial equilibrium rate over the actual rate. The partial equilibrium rate of capital formation is assumed to be a linear function (with constants γ_4 and β_5) of the excess of the marginal product of capital $\beta_3(Y/K)^{1+\beta_4}$ over the real interest rate. The parameter β_5 is the partial equilibrium rate of capital formation associated with equality between the marginal product of capital and the real interest rate. If there were perfect competition and no risk it would equal the rate at which entrepreneurs expected output to grow. The speed of adjustment parameter γ_3 can be assumed to depend on costs associated with variations in the rate of capital formation.

Equations (4) and (5) are closely related and will be discussed together. The term $C + DK + E$ equals total sales for consumption, capital formation and exports. The parameters β_6 and β_7 are the partial equilibrium ratios of imports and stocks respectively to sales. Equation (4) assumes, therefore, that the rate of increase in output depends on the excess demand for home produced goods to meet current sales and on excess stocks, while equation (5) assumes that the rate of increase in imports depends on the excess demand for imported goods to meet current sales and on excess stocks.

It might appear, at first sight, that these equations make no allowance for price substitution between imports and home-produced goods. But this is not so. As pointed out in the introduction, the model is essentially a model of a single-sector economy. There is a single internal price variable p which is identified, in the econometric application of the model, with the implicit price deflator of the gross domestic product. The real variables C, Y, DK, E and I are all measured by deflating the corresponding items in the national accounts by this single deflator. Thus, if Q_i is the volume of imports and p_i is the price of imports, then $I = Q_i(p_i/p)$. Equation (5) implies, therefore, that the price elasticity of demand for imports (which measures the response of Q_i to a change in p_i when p, $C + DK + E$ and S are held constant) is unity rather than zero.

Equation (6) is essentially a demand equation for exports. The partial equilibrium level of exports is assumed to depend on the price level, with elasticity β_9, and a trend term which allow for the gradual change in income and prices in the rest of the world. The parameter λ_2 can be regarded as a weighted sum of the growth rates of real income and the price level in the rest of the world the weights being the income and price elasticity of demand respectively. The introduction of a trend term to allow for changes outside the economy is a simple device for avoiding the introduction of exogenous variables. But there is a good econometric justification for its use. Income and prices in the rest of the world are not genuine exogenous variables, and to treat them as such would introduce an asymptotic bias into our estimates. In a model of the world economy they would be jointly determined together with the variables in the domestic economy. By solving the model of the world economy for its reduced form the income and price level in each country could be expressed as the sum of two components, the first being a function of genuine exogenous variables which may be assumed to change gradually over time, and the second a function of the random disturbances. In equation (6) the former component is allowed for by the term $\lambda_2 t$, while the second is included in the disturbance term.

When using the model for forecasting beyond the sample period, the parameter λ_2 can be changed if, for example, there is good reason to believe that the rate of growth in the rest of the world will be different from what it was during the sample period. Moreover, a variation in the exchange rate can be allowed for by changing the parameter β_8. We do, in fact, make such an adjustment to allow for the devaluation of sterling in 1967 when testing the predictive performance of the model on the United Kingdom economy.

In the price adjustment equation (7) the partial equilibrium price level is assumed to equal marginal cost $w(\partial L/\partial Y)$ multiplied by a constant β_{10}. The case in which $\beta_{10} = 1$ corresponds to perfect competition. Although we have formally assumed a single-sector economy, the possibility of imperfect competition can be allowed for by assuming that the products of the various firms are differentiated but that they have identical production functions and demand functions and, hence, a common selling price p. We have then $\beta_{10} = e/(e + 1) > 1$, where e is the own-price elasticity of demand for each firm's product. As e tends to minus infinity β_{10} tends to 1 and we approach a state of perfect competition.

The partial derivative $\partial L/\partial Y$ used in the definition of marginal cost is obtained by differentiating the production function in equation (2). Marginal cost has been identified, therefore, with the cost of producing a small additional unit of output when capital is fixed but employment is allowed to adjust optimally. This is a somewhat arbitrary choice which is made for simplicity. How much weight should be attached to this concept of marginal cost, or to very short-run marginal cost which takes account of the costs (such as overtime payments) of using a given number of employees more intensively, or to long-run marginal cost which allows for the optimal adjustment of capital, would depend in a more sophisticated model on such factors as the speed of response of demand to variations in price and the administrative costs of changing prices.

It is convenient, at this stage, to return to the point raised in the discussion of equation (2) about the adjustment of factor proportions to their relative costs. In a perfectly competitive, riskless economy in which capital is growing steadily at a rate at which entrepreneurs expect output to grow and the price level is constant, we have $\beta_{10} = 1$, $k = \beta_5$ and $Dp = Dk = 0$. Then, by equation (3), the marginal product of capital equals the interest rate, and by equation (7), the marginal product of labour equals the wage rate. It follows that the marginal rate of substitution between labour and capital equals the ratio of the wage rate to the interest rate.

In the wage rate adjustment equation (8) the proportional rate of increase in the money wage rate is assumed to be an increasing function of the proportional excess demand for labour. But the measurement of the excess demand for labour differs, in two respects, from that used in the usual formulation of the Phillips curve type of relation. First, the demand for labour is identified with the partial equilibrium level of employment defined by equation (2) rather than with the actual level of employment. It is assumed, therefore, that the willingness of an employer to bid up the wage rate in order to attract labour (or retain labour or avoid strikes) depends not on the number of persons currently employed by him, but on the number of persons that it would be most profitable to employ at the current wage rate, given his output and the stock of capital. Secondly, the supply of labour is represented by a simple trend function with growth rate λ_2. We recognize that it would be preferable to represent the supply of labour by demographic variables. But a trend is preferable to the commonly used estimate of the labour force obtained by adding the official estimates of employment and unemployment. For, since people tend to enter or leave the labour force (so measured) as the demand for labour varies, it would be unrealistic to treat this variable as exogenous.

The interest rate adjustment equation (9) is a dynamic representation of the Keynesian liquidity preference theory of interest rate determination. It is assumed that, at each point of time, the proportional rate of increase in the interest rate is an increasing function of the proportional excess demand for money holdings and that the demand for real money holdings depends on real income and the interest rate with elasticities β_{13} and β_{14}, respectively. This type of relation might be formally justified by assuming that all investment in new capital goods is financed by the issue of bonds and that the only assets that can be held by individuals are money and bonds. In a more sophisticated model equities could be introduced as an additional asset in the individual's portfolio. But this would involve the introduction into the model of an additional variable to represent the price of equities and an additional price adjustment equation. It must be emphasized that the variable p cannot be taken to represent the price of equities. Since the model formally describes a single-sector economy, p can be taken to represent the price of newly produced capital goods. But the price of equities must be regarded as the price of second-hand capital goods which have become embodied in a corporate enterprise. The price of equities and the price of new capital goods are, in fact, capable of very divergent movements.

It was shown in Bergstrom 1967 that the prototype model on which this model is based is capable of producing plausible long-run behaviour even if the money supply is assumed to grow at an arbitrary constant rate. But this is no longer true when the model is extended for application to an open economy. For, unless the rate of increase in the money supply is correctly related to the rate of increase in prices and income in the rest of the world, the ratio of exports to imports will tend to zero or infinity. The fact that United Kingdom historical data exhibit no such tendency can be attributed to the influence of the balance of payments on the money supply. This influence has occurred to some extent automatically through variations in bank holdings of gold and foreign exchange and to some extent through monetary policy. It is essential that this feature of reality should be incorporated in the model if it is to produce plausible long-run behaviour.

The influence of the balance of payments on the money supply is represented by equation (10) which is the only policy relation in the model. The parameter β_{15} can be interpreted as the ratio of exports to imports aimed at by the policy-maker, so that $\beta_{15} = 1$ unless it is the policy to become a long-term borrower or lender. The rate of increase in the proportional rate of increase in the money supply is assumed to be influenced by both the ratio of exports to imports and the proportional rate of change in this ratio. The remaining terms in the equation allow for the possible influence of the employment level and its rate of change, β_{16}/β_{11} being the proportional level of employment aimed at. But the inclusion of these terms is not essential in order for the model to produce plausible long-run behaviour. Indeed, it will be shown in section 3 that their introduction has no effect on the steady state level of employment and that, unless the level of employment aimed at is that which necessarily occurs in the steady state (whether these terms are included or not), their only effect is to prevent the long-run attainment of the desired balance of payments. This result is a consequence of the implicit assumption of a fixed exchange rate. The inclusion of the employment terms in equation (10) may, however, have a stabilizing influence by reducing the amplitude of the trade cycle.

The remaining equations in the model are definitional. Although equations (3) and (10) are essentially second-order differential equations in K and M, respectively, it is mathematically convenient to define k and m by (12) and (13) in order to express the complete model as a system of 13 first-order differential equations.

3. The Steady State and the Working of the Model

The system (1) to (13) has a particular solution:

$$C = C^* e^{(\lambda_1+\lambda_3)t}, \tag{14}$$

$$L = L^* e^{\lambda_3 t}, \tag{15}$$

$$k = \lambda_1 + \lambda_3 = k^*, \tag{16}$$

$$Y = Y^* e^{(\lambda_1+\lambda_3)t}, \tag{17}$$

$$I = I^* e^{(\lambda_1+\lambda_3)t}, \tag{18}$$

$$E = E^* e^{(\lambda_1+\lambda_3)t}, \tag{19}$$

$$p = p^* e^{\{(\lambda_2-\lambda_1-\lambda_3)/\beta_9\}t}, \tag{20}$$

$$w = w^* e^{\{(\lambda_2-\lambda_1-\lambda_3)/\beta_9+\lambda_1\}t}, \tag{21}$$

$$r = r^*, \tag{22}$$

$$m = (\lambda_2 - \lambda_1 - \lambda_3)/\beta_9 + \beta_{13}(\lambda_1 + \lambda_3) = m^*, \tag{23}$$

$$S = S^* e^{(\lambda_1+\lambda_3)t} \tag{24}$$

$$K = K^* e^{(\lambda_1+\lambda_3)t} \tag{25}$$

$$M = M^* e^{\{(\lambda_2-\lambda_1-\lambda_3)/\beta_9+\beta_{13}(\lambda_1+\lambda_3)\}t}, \tag{26}$$

where

$$\log C^* = \log\left[\frac{\beta_1\beta_{11}}{\beta_2}\{1-\beta_3 q^{-\beta_4}\}^{1/\beta_4}\right] - \frac{\lambda_1+\lambda_3}{\gamma_1} + \frac{\lambda_2-\lambda_3-\lambda_1(1-\beta_9)}{\beta_9\gamma_{11}},$$

$$\log L^* = \log\beta_{11} + \frac{\lambda_2-\lambda_3-\lambda_1(1-\beta_9)}{\beta_9\gamma_{11}} - \frac{\lambda_3}{\gamma_2},$$

$$\log Y^* = \log\left[\frac{\beta_{11}}{\beta_2}\{1-\beta_3 q^{-\beta_4}\}^{1/\beta_4}\right] + \frac{\lambda_2-\lambda_3-\lambda_1(1-\beta_9)}{\beta_9\gamma_{11}}$$

$$\log I^* = \log\left\{\frac{\theta\beta_1 e^{-(\lambda_1+\lambda_3)/\gamma_1} + (\lambda_1+\lambda_3)(\gamma_8+\theta q) + \gamma_5\gamma_8}{\gamma_6(\lambda_1+\lambda_3+\gamma_7) - \theta b}\right\}$$

$$+ \log Y^*,$$

$$\log E^* = \log b + \log I^*,$$

$$\log p^* = \frac{1}{\beta_9}\left\{\log\left(\frac{\beta_8}{E^*}\right) - \frac{\lambda_1+\lambda_3}{\gamma_9}\right\},$$

$$\log w^* = \log\left\{\frac{p^*}{\beta_2\beta_{10}}\right\} + \frac{\lambda_2-\lambda_1-\lambda_3}{\beta_9\gamma_{10}} + \frac{1+\beta_4}{\beta_4}$$

$$\times \log\{1-\beta_3 q^{-\beta_4}\},$$

$$r^* = \beta_3 q^{-(1+\beta_4)} + \frac{\lambda_2-\lambda_1-\lambda_3}{\beta_9} - \frac{\lambda_1+\lambda_3-\beta_5}{\gamma_4},$$

$$\log S^* = \log\left[\left\{\frac{\gamma_5(1-\beta_6)}{\gamma_6} + \beta_7\right\}\left\{\beta_1 e^{-(\lambda_1+\lambda_3)/\gamma_1} + q(\lambda_1+\lambda_3)\right.\right.$$

$$\left. + \frac{b\{\theta\beta_1 e^{-(\lambda_1+\lambda_3)/\gamma_1} + (\lambda_1+\lambda_3)(\gamma_8+\theta q) + \gamma_5\gamma_8}{\gamma_6(\lambda_1+\lambda_3+\gamma_7) - \theta b}\right\}$$

$$\left. - \frac{\lambda_1+\lambda_3+\gamma_5}{\gamma_6}\right] + \log Y^*,$$

$$\log K^* = \log q + \log Y^*,$$

$$\log M^* = \log\beta_{12} + \log p^* + \beta_{13}\log Y^* - \beta_{14}\log r^*,$$

$$q = -\frac{\{(\lambda_1+\lambda_3)(\lambda_1+\lambda_3+\gamma_5)+\gamma_6\} \times \{\theta b - \gamma_6(\lambda_1+\lambda_3+\gamma_7)\} - \gamma_8(\lambda_1+\lambda_3+\gamma_5)(\gamma_6-\pi b)}{\gamma_6(\lambda_1+\lambda_3)\{\theta - \pi(\lambda_1+\lambda_3+\gamma_7)\}}$$

$$- \frac{\beta_1 e^{-(\lambda_1+\lambda_3)/\gamma_1}}{\lambda_1+\lambda_3},$$

$$\log b = \log\beta_{15}$$

$$+ \frac{\gamma_{14}}{\gamma_{13}}\left\{\log\left(\frac{\beta_{11}}{\beta_{16}}\right) + \frac{\lambda_2-\lambda_3-\lambda_1(1-\beta_9)}{\beta_9\gamma_{11}} - \frac{\lambda_3}{\gamma_2}\right\},$$

$$\theta = \gamma_6\gamma_7\beta_6 - \gamma_5\gamma_8(1-\beta_6),$$

$$\pi = (\lambda_1+\lambda_3)\{\gamma_5(1-\beta_6) + \gamma_6\beta_7\} + \gamma_6.$$

In this solution, which we describe as the steady state, each variable grows at a constant proportional rate (possibly zero). The formulae for these growth rates are summarized in the following table:

Variable	Steady state growth rate
L	λ_3
C, Y, K, S, I, E	$\lambda_1 + \lambda_3$
p	$(\lambda_2 - \lambda_1 - \lambda_3)/\beta_9$
w	$(\lambda_2 - \lambda_1 - \lambda_3)/\beta_9 + \lambda_1$
M	$(\lambda_2 - \lambda_1 - \lambda_3)/\beta_9 + \beta_{13}(\lambda_1 + \lambda_3)$
r	0

In the steady state employment L grows at the same rate λ_3 as the labour supply, so that the proportional level of employment is constant. The real variables C, Y, K, S, I and E all grow at the rate $\lambda_1 + \lambda_3$ which can be interpreted as the rate of growth of the labour supply measured in efficiency units. This is identical with the steady state growth rate that would be produced by a pure neoclassical growth model in which there is exogenous growth of employment at the rate λ_3 and a production function incorporating Harrod neutral technical progress at the rate λ_1. We note that the steady state growth rate of these variables is unaffected by the propensity to consume β_1, entrepreneurial expectations or willingness to take risks as reflected in β_5, or exogenous growth in the demand for exports as reflected in γ_2. Other things will change in order to accommodate any change in these or other parameters to the growth rate $\lambda_1 + \lambda_3$.

The formula for the steady state rate of increase in the price level can be most easily interpreted by considering first the case in which prices in the rest of the world are constant and the income elasticity of demand for exports is unity, so that λ_2 equals the rate of growth of real income in the rest of the world. The formula then implies that the domestic price level will be increasing or decreasing, in the steady state, according as the rate of real growth in the rest of the world is greater than or less than it is in the domestic economy. The explanation of this result is that if foreign income were growing faster than domestic income and the domestic price level were constant, exports would tend to grow faster than imports; and a mechanism, involving the feed-back relation (10) and other relations in the model, would produce an increasing price level which would tend to correct the imbalance. The mechanism will be discussed more fully at a later stage in this section. If foreign prices are increasing rather than constant, as assumed in the above argument, this will be reflected in a greater value of λ_2 which, by the formula, implies a higher steady state rate of increase in p. This too can be explained by the adjustments required in order to prevent a divergence between the growth rates of imports and exports. The dependence of the steady state growth rate of p on the parameter β_9 can be explained by the fact that the change in the domestic price level that is required in order to correct a given proportional disequilibrium in the balance of payments is smaller the greater is the price elasticity of demand for exports.

In the steady state the rate of increase in the wage rate equals the rate of increase in the price level plus the rate of technical progress. This implies that the real wage per efficiency unit of labour is constant. The rate of increase in the money supply equals the rate of increase in money income, so that the ratio of money holdings to money income is constant. We note also that the rate of increase in the interest rate is zero or, in other words, that the steady state interest rate is constant. Since the rate of increase in the price level is constant the steady state real interest rate is also constant.

We turn now to a consideration of the levels of the steady state growth paths. These levels, at time zero, are represented by the constant terms denoted with an asterisk in the steady state solution. Since they have been expressed as explicit functions of the structural parameters of the model, it is easy to compute the steady state behaviour of the complete system for any given numerical values of the structural parameters. Moreover, since the signs of many of the partial derivatives of these functions are independent of the numerical values of the parameters, it is possible to determine definitely the effect of changes in certain parameters on the steady state paths of some variables. It will be illuminating to consider the effect of changes in selected parameters on the steady state levels of employment and output, the ratio of exports to imports and the interest rate.

The partial derivatives of L^* satisfy the conditions

$$\frac{\partial \log L^*}{\partial \log \beta_{11}} = 1,$$

$$\frac{\partial L^*}{\partial \lambda_2} > 0, \quad \frac{\partial L^*}{\partial \lambda_3} < 0,$$

$$\frac{\partial L^*}{\partial \gamma_{14}} = \frac{\partial L^*}{\partial \gamma_{16}} = \frac{\partial L^*}{\partial \beta_{16}} = 0.$$

The condition $\partial \log L^*/\partial \log \beta_{11} = 1$ implies that any increase in labour supply will cause an equal proportional increase in the steady state level of employment, so that it will have no effect on the steady state proportional level of unemployment. The condition $\partial L^*/\partial \lambda_2 > 0$ implies that the steady state level of employment is greater the greater is the rate of inflation or real growth in the rest of the world. This is explained by the fact that the greater the rate of increase in foreign prices or real income the greater will be the rate of increase in demand for exports and, hence, the greater the rate of domestic inflation that can occur without causing a divergence between the growth rates of imports and exports. The level of employment is related to the rate of domestic inflation via the price adjustment equation (7) and the wage rate adjustment equation (8). The condition $\partial L^*/\partial \lambda_3 < 0$ which means that the steady state level of employment is a decreasing function of the rate of increase in the labour supply can be explained in a similar way. The greater the rate of increase in the labour supply the greater will be the rate of growth of real domestic income or imports and the lower, therefore, the rate of domestic inflation that can occur without causing a divergence between the rates of growth of imports and exports.

The conditions $\partial L^*/\partial \gamma_{14} = \partial L^*/\partial \gamma_{16} = \partial L^*/\partial \beta_{16} = 0$ mean that the introduction of the employment terms into the monetary policy feed-back relation (10) has no effect on the steady state level of employment. As we have already mentioned this is a consequence of the assumption of a fixed exchange rate. In an economy with a fixed exchange rate and free trade the need to relate the rate of domestic inflation to the rate of inflation and real growth in the rest of the world precludes the possibility of choosing the balance between unemployment and inflation that would be considered optimal by other criteria.

The partial derivatives of Y^* satisfy the conditions

$$\frac{\partial Y^*}{\partial \beta_1} < 0, \quad \frac{\partial Y^*}{\partial \lambda_2} > 0.$$

The interpretation of the condition $\partial Y^*/\partial \beta_1 < 0$ is that the steady state level of output is an increasing function of the long-run propensity to save. This is a familiar neoclassical result and depends on the influence of the saving propensity on the capital-output ratio. The condition $\partial Y^*/\partial \lambda_2 > 0$ has no counterpart in the pure neoclassical growth model. It reflects the influence of λ_2 on the steady state employment level and the influence of the employment level on output.

A necessary condition for

$$E^* = \beta_{15} I^*$$

is that either $\gamma_{14} = 0$ or $\beta_{16} = L^*$. This means that the desired export–import ratio will not be attained, in the steady state, unless either the level of employment has no influence on the acceleration of the money supply or the level of employment aimed at is that which necessarily occurs in the steady state. We have already seen that the steady state level of employment is not affected by the presence of the employment term in the feed-back relation (10). Moreover, the experience with the prototype model (see Bergstrom 1967) suggests that allowing the level of employment to influence the acceleration of the money supply is likely to reduce the stability of the system. We may conclude, therefore, that the inclusion of the term with coefficient γ_{14} in the feed-back relation (10) has no effect on the steady employment level, prevents the attainment of the desired export-import ratio and is likely to be destabilizing. The effect of allowing the rate of change in the employment level to affect the acceleration of the money supply is rather different. The parameter γ_{16} has no effect on the steady state levels of either employment or the export–import ratio. And the experience with the prototype model suggests that a positive value of this parameter could have a stabilizing influence.

The partial derivatives of r^* satisfy the conditions

$$\frac{\partial r^*}{\partial \beta_1} > 0, \quad \frac{\partial r^*}{\partial \beta_3} > 0,$$

$$\frac{\partial r^*}{\partial \beta_{12}} = \frac{\partial r^*}{\partial \beta_{13}} = \frac{\partial r^*}{\partial \beta_{14}} = 0.$$

These conditions imply that the steady state interest rate is higher the lower is the propensity to save and the more capital-intensive the production function, and that it is unaffected by the parameters of the liquidity preference function. Thus, although the path of the interest rate in the transient solution of the system is determined by a Keynesian mechanism, the steady state interest rate is determined, as in classical theory, by the degree of thrift and the productivity of capital. The explanation of this result is that, in the long run, the money wage rate will become so adjusted to other monetary phenomena in the system that their effect is neutralized and the interest rate will be determined by real phenomena. The Keynesian interest rate mechanism, nevertheless, plays a very important role in the working of the model. This will now be discussed.

The importance of studying the properties of econometric models and, in particular, of ensuring that they are capable of producing long-run behaviour that seems plausible in the light of past experience was stressed in the introduction. One historical fact which a realistic model of the United Kingdom should be capable of explaining is that, although the productive capacity of the economy has increased by about 4,000 per cent since the beginning of the 19th century (see Deane and Cole 1969) demand has approximately followed supply, so that unemployment has, for most of the time, been between 1 per cent and 10 per cent of the labour force. To some extent this could be explained by the fact that an increase in demand generates an increase in capacity through capital accumulation. But this can be no more than a partial explanation since most of the increase in capacity that has occurred is attributable to technical progress and the growth of the labour force rather than capital accumulation. One possible explanation is that there is an exogenous component of demand like Hicks' (1950) 'autonomous investment' which, by some fluke of nature, has grown at a rate equal to the sum of the rate of technical progress and the rate of growth of the labour force. A more plausible explanation is that there is some sort of feed-back mechanism which tends to limit unemployment. Such a mechanism is incorporated in our model and, in this mechanism the interest rate and other monetary phenomena play an important role.

It is appropriate to draw attention, at this stage, to an important difference between our model and other econometric models of complete economies. In most such models certain components of expenditure are treated as exogenous variables and play a similar role in the generation of growth to that of 'autonomous investment' in Hicks' model of the trade cycle. Moreover, considerable experimentation with such variables is necessary in order to produce plausible long-run post sample growth paths. (See, for example, the contributions on the Brookings, OBE and Wharton models in Hickman 1972.) In the model described in section 2 neither the consumption equation (1) nor the investment equation (3) contains any exogenous variables or trend terms. Moreover, although there is a trend term in the export demand equation (6), this has no influence on the steady state growth rate. The steady state growth rate depends, as in the pure neoclassical growth model, only on the rate of technical progress λ_1 and the rate of increase in the labour supply λ_3. If either of these parameters changes then, provided that certain stability conditions are satisfied, a transitory dynamic mechanism will bring about a gradual change in the rate of growth of demand so that demand continues to approximately follow the path of full employment output.

A rigorous derivation of the stability conditions, which are essential for the working of this mechanism, will be carried out in the next section. We shall conclude this section with a non-mathematical description of the way in which the mechanism works.

The effect of an increase in the rate of technical progress or growth in the labour supply is illustrated in figure 1. It is assumed that the system is initially in a steady state and that there is an increase at time t_1 in either λ_1 or λ_3. This will cause a shift in the steady growth path of output. The old and new steady state growth paths of log Y will, normally, be related as shown by the unbroken lines in figure 1(a), with the new path passing below the old path but having a steeper gradient. This is because an increase in either λ_1 or λ_3 will, normally, decrease the steady state ratio of capital to output so that the new steady state ratio will be less than the actual ratio at time t_1. The steady state interest rate will, normally, increase as shown in figure 1(b). The smoothed (i.e. ignoring oscillations) paths of log Y and r in the transition from the old to new steady states are represented by the broken curves. A simplified description of the economic changes reflected in these transitory paths is as follows.

The initial effect of the increase in λ_1 or λ_3 will be to cause a gradual fall in the ratio of the demand for labour to the supply of labour and, hence, a fall in the rate of increase in the money wage rate and the price level. Since

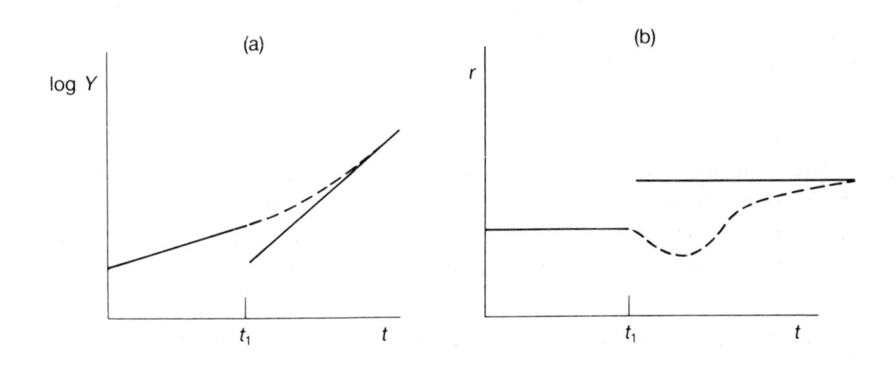

FIG. 1

there will be no instantaneous change in the rate of increase in the nominal money supply, the real money supply will start to increase more rapidly, and this will cause a gradual fall in the interest rate as shown in figure 1(b). Eventually this will be sufficient to offset the fall in the rate of increase in the price level, so that there will be a fall in the real rate of interest also. This will cause an increase in the rate of capital formation and an increase in the rate of growth of output as shown in figure 1(a). The increased rate of growth of output will eventually bring about an increase in the interest rate and the inducement to maintain the increased rate of capital formation will be provided by the higher marginal product of capital associated with the increased ratio of output to capital. The whole process will be reinforced by an acceleration in the nominal money supply resulting from a temporary increase in the ratio of exports to imports. Moreover, it will be complicated by oscillations which have been ignored in the above description.

4. Stability Analysis

In order to show that the model is capable of producing realistic long-run behaviour for values of the parameters that are consistent with the sample, it is necessary to show that, for such parameters, the mechanism just described does work, at least in some neighbourhood of the steady state. More precisely, it is necessary to show that the steady state is asymptotically stable so that if the vector of variables is initially in some neighbourhood of the steady state growth path it will converge to this path if undisturbed. Ideally we should go further than this and investigate the global stability of the system. But such an investigation is beyond the scope of this study.

The asymptotic stability analysis is based on the following theorem of Perron (see Coddington and Levinson 1955).

Theorem

The solution $x(t) = 0$ of the differential equation system $Dx(t) = Ax(t) + f(x, t)$, in which A is a matrix of constants and f a vector of functions, is asymptotically stable if $f(x, t)/|x|$ tends to zero uniformly in t as $|x|$ tends to zero and the eigenvalues of A have negative real parts.

It is important to notice that the convergence of $f(x, t)/|x|$ to zero must be uniform in t. The mere fact that $f(x, t)/|x|$ tends to zero for each value of t as $|x|$ tends to zero [as would be the case if $f(x, t)$ were the residual in a Taylor series expansion about the sample means] does not ensure that we can deduce anything about the asymptotic stability of the solution from the properties of A. A convenient property of our model, which greatly facilitates the asymptotic stability analysis, is that the logarithms of the ratios of the variables to their steady state paths satisfy a non-linear system of differential equations which does not involve t. This system is:

$$Dy_1 = \gamma_1(y_4 - y_1), \tag{27}$$

$$Dy_2 = \gamma_2\left[\frac{1}{\beta_4}\log\left\{\frac{1-\beta_3 q^{-\beta_4}}{e^{-\beta_4 y_4} - \beta_3 q^{-\beta_4} e^{-\beta_4 y_{12}}}\right\} - y_2\right], \tag{28}$$

$$Dy_3 = \gamma_3\left[\gamma_4\left\{\beta_3 q^{-(1+\beta_4)} e^{(1+\beta_4)(y_4 - y_{12})} - r^* e^{y_9} + Dy_7 + \frac{\lambda_2 - \lambda_1 - \lambda_3}{\beta_9}\right\} + \beta_5 - y_3 - k^*\right], \tag{29}$$

$$Dy_4 = \{\gamma_5(1-\beta_6) + \gamma_6\beta_7\}\left\{\frac{C^*}{Y^*}(e^{y_1 - y_4} - 1) + \frac{K^*}{Y^*}e^{y_{12} - y_4}y_3 + \frac{k^* K^*}{Y^*}(e^{y_{12} - y_4} - 1) + \frac{E^*}{Y^*}(e^{y_6 - y_4} - 1)\right\} - y_6\frac{S^*}{Y^*}(e^{y_{11} - y_4} - 1), \tag{30}$$

$$Dy_5 = (\gamma_7\beta_6 + \gamma_8\beta_7)\left\{\frac{C^*}{I^*}(e^{y_1-y_5} - 1) + \frac{K^*}{I^*}e^{y_{12}-y_5}y_3 + \frac{k^*K^*}{I^*}(e^{y_{12}-y_5} - 1) + \frac{E^*}{I^*}(e^{y_6-y_5} - 1)\right\} - \gamma_8\frac{S^*}{I^*}(e^{y_{11}-y_5} - 1), \tag{31}$$

$$Dy_6 = \gamma_9\{-\beta_9\gamma_7 - y_6\}, \tag{32}$$

$$Dy_7 = \gamma_{10}\left[y_8 - y_7 - \frac{1+\beta_4}{\beta_4}\log\left\{\frac{1 - \beta_3 q^{-\beta_4}e^{\beta_4(y_4-y_{12})}}{1 - \beta_3 q^{-\beta_4}}\right\}\right], \tag{33}$$

$$Dy_8 = \frac{y_{11}}{\beta_4}\log\left[\frac{1 - \beta_3 q^{-\beta_4}}{e^{-\beta_4 y_4} - \beta_3 q^{-\beta_4}e^{-\beta_4 y_{12}}}\right], \tag{34}$$

$$Dy_9 = \gamma_{12}(y_7 + \beta_{13}y_4 - \beta_{14}y_9 - y_{13}), \tag{35}$$

$$Dy_{10} = \gamma_{13}(y_6 - y_5) - \gamma_{14}y_2 + \gamma_{15}(Dy_6 - Dy_5) - \gamma_{16}Dy_2, \tag{36}$$

$$Dy_{11} = \frac{Y^*}{S^*}(e^{y_4-y_{11}} - 1) + \frac{I^*}{S^*}(e^{y_5-y_{11}} - 1) - \frac{C^*}{S^*}(e^{y_1-y_{11}} - 1) - \frac{K^*}{S^*}e^{y_{12}-y_{11}}y_3 - \frac{k^*K^*}{S^*}(e^{y_{12}-y_{11}} - 1) - \frac{E^*}{S^*}(e^{y_6-y_{11}} - 1), \tag{37}$$

$$Dy_{12} = y_3, \tag{38}$$

$$Dy_{13} = y_{10}, \tag{39}$$

where

$$y_1 = \log\{C/C^*e^{(\lambda_1+\lambda_3)t}\},$$
$$y_2 = \log\{L/L^*e^{\lambda_3 t}\},$$
$$y_3 = k - k^*,$$
$$y_4 = \log\{Y/Y^*e^{(\lambda_1+\lambda_3)t}\},$$
$$y_5 = \log\{I/I^*e^{(\lambda_1+\lambda_3)t}\},$$
$$y_6 = \log\{E/E^*e^{(\lambda_1+\lambda_3)t}\},$$
$$y_7 = \log\{p/p^*e^{\{(\lambda_2-\lambda_1-\lambda_3)/\beta_9\}t}\},$$
$$y_8 = \log\{w/w^*e^{\{(\lambda_2-\lambda_1-\lambda_3)/\beta_9+\lambda_1\}t}\},$$
$$y_9 = \log(r/r^*),$$

$$y_{10} = m - m^*,$$
$$y_{11} = \log\{S/S^* e^{(\lambda_1+\lambda_3)t}\},$$
$$y_{12} = \log\{K/K^* e^{(\lambda_1+\lambda_3)t}\},$$
$$y_{13} = \log\{M/M^* e^{\{(\lambda_2-\lambda_1-\lambda_3)/\beta_9+\beta_{13}(\lambda_1+\lambda_3)\}t}\}.$$

Taking a Taylor series expansion of the above system we obtain the approximate linear system:

$$Dy_1 = \gamma_1(y_4 - y_1), \tag{40}$$

$$Dy_2 = \gamma_2\left[\frac{y_4 - \beta_3 q^{-\beta_4} y_{12}}{1 - \beta_3 q^{-\beta_4}} - y_2\right], \tag{41}$$

$$Dy_3 = \gamma_3[\gamma_4\{\beta_3 q^{-(1+\beta_4)}(y_4 - y_{12}) - r^* y_9 + Dy_7\} - y_3], \tag{42}$$

$$\begin{aligned} Dy_4 = {} & \{\gamma_5(1-\beta_6) + \gamma_6\beta_7\}\left\{\frac{C^*}{Y^*}(y_1 - y_4) + \frac{K^*}{Y^*}y_3 \right. \\ & \left. + \frac{k^*K^*}{Y^*}(y_{12} - y_4) + \frac{E^*}{Y^*}(y_6 - y_4)\right\} \\ & - \gamma_6\frac{S^*}{Y^*}(y_{11} - y_4), \end{aligned} \tag{43}$$

$$\begin{aligned} Dy_5 = {} & (\gamma_7\beta_6 + \gamma_8\beta_7)\left\{\frac{C^*}{I^*}(y_1 - y_5) + \frac{K^*}{I^*}y_3 \right. \\ & \left. + \frac{k^*K^*}{I^*}(y_{12} - y_5) + \frac{E^*}{I^*}(y_6 - y_5)\right\} - \gamma_8\frac{S^*}{I^*}(y_{11} - y_5), \end{aligned} \tag{44}$$

$$Dy_6 = \gamma_9(-\beta_9 y_7 - y_6), \tag{45}$$

$$Dy_7 = \gamma_{10}\left[y_8 - y_7 + \left\{\frac{(1+\beta_4)\beta_3 q^{-\beta_4}}{1 - \beta_3 q^{-\beta_4}}\right\}(y_4 - y_{12})\right], \tag{46}$$

$$Dy_8 = \gamma_{11}\left\{\frac{y_4 - \beta_3 q^{-\beta_4} y_{12}}{1 - \beta_3 q^{-\beta_4}}\right\}, \tag{47}$$

$$Dy_9 = \gamma_{12}(y_7 + \beta_{13}y_4 - \beta_{14}y_9 - y_{13}), \tag{48}$$

$$Dy_{10} = \gamma_{13}(y_6 + y_5) - \gamma_{14}y_2 + \gamma_{15}(Dy_6 - Dy_5) - \gamma_{16}Dy_2, \tag{49}$$

$$Dy_{11} = \frac{Y^*}{S^*}(y_4 - y_{11}) + \frac{I^*}{S^*}(y_5 - y_{11}) - \frac{C^*}{S^*}(y_1 - y_{11})$$
$$- \frac{K^*}{S^*}y_3 - \frac{k^*K^*}{S^*}(y_{12} - y_{11}) - \frac{E^*}{S^*}(y_6 - y_{11}), \tag{50}$$

$$Dy_{12} = y_3, \tag{51}$$

$$Dy_{13} = y_{10}. \tag{52}$$

The remainder terms in the Taylor series expansion of the system (27)–(39) are of a lower order of smallness than $|y|$ as $|y|$ tends to zero. Moreover, since they do not involve t the question of uniform convergence does not arise. The asymptotic stability depends, therefore on the matrix of coefficients of the system (40)–(52). More precisely, if this system is written

$$Dy(t) = Ay(t), \tag{53}$$

then provided that the eigenvalues of A have negative real parts the steady state of the non-linear model will be asymptotically stable in the sense that $y_1(t), \ldots, y_{13}(t)$ will tend to zero as t tends to infinity provided that their initial values are sufficiently small. Since the elements of A have been expressed as explicit functions of the structural parameters of the system (1)–(13), the asymptotic stability conditions can be expressed explicitly in terms of these structural parameters by using the Routh–Hurwicz conditions.

5. The Estimation Procedure

The complete system, including stochastic terms, can be written

$$Dx(t) = f[x(t), t, \beta, \gamma, \lambda] + \zeta(t), \tag{54}$$

where

$$x(t) = \begin{bmatrix} x_1(t) \\ \vdots \\ x_{13}(t) \end{bmatrix}, \quad \beta = \begin{bmatrix} \beta_1 \\ \vdots \\ \beta_{16} \end{bmatrix}, \quad \gamma = \begin{bmatrix} \gamma_1 \\ \vdots \\ \gamma_{16} \end{bmatrix}, \quad \lambda = \begin{bmatrix} \lambda_1 \\ q\lambda_2 \\ \lambda_3 \end{bmatrix}q,$$

$$\zeta(t) = \begin{bmatrix} \zeta_1(t) \\ \vdots \\ \zeta_{10}(t) \\ 0 \\ 0 \\ 0 \end{bmatrix};$$

where

$$x_1 = \log C, \quad x_2 = \log L, \quad x_3 = k, \quad x_4 = \log Y,$$
$$x_5 = \log I, \quad x_6 = \log E, \quad x_7 = \log p, \quad x_8 = \log w,$$
$$x_9 = \log r, \quad x_{10} = m, \quad x_{11} = \log S, \quad x_{12} = \log K,$$
$$x_{13} = \log M,$$

and the $\zeta_i(t)$ are random disturbance functions and f denotes the vector of functions defined by the system (1) to (13).

In the present state of our knowledge it is not possible to obtain consistent estimates of the parameters of a non-linear stochastic differential equation system of this degree of complexity from a sample of discrete observations. Our estimates were obtained, therefore, by replacing (54) by the linear approximation,

$$Dx(t) = B(\bar{x}, \beta, \gamma, \lambda)x(t) + c(\bar{x}, \beta, \gamma, \lambda) + d(\bar{x}, \beta, \gamma, \lambda)t + \zeta(t), \quad (55)$$

where $\bar{x}$ is the sample mean of $x(t)$ and the elements of the matrix B and vectors c and d are known functions of $\bar{x}$, β, γ and λ obtained by a Taylor series expansion of f about $\bar{x}$. In this approximation $\zeta(t)$ is assumed to be a pure noise vector so that if $t_1 < t_2 < t_3 < t_4$, then

$$E\left\{\int_{t_1}^{t_2} \zeta(t)\mathrm{d}t\right\} = 0, \quad (56)$$

$$E\left\{\int_{t_1}^{t_2} \zeta(t)\mathrm{d}t \int_{t_3}^{t_4} \zeta(t)'\mathrm{d}t\right\} = 0, \quad (57)$$

$$E\left\{\int_{t_1}^{t_2} \zeta(t)\mathrm{d}t \int_{t_1}^{t_2} \zeta(t)'\mathrm{d}t\right\} = (t_2 - t_1)\Sigma, \quad (58)$$

where Σ is a positive semidefinite matrix. The estimates were obtained by treating (55) as if it were the true model.

It should be emphasized that the system (55) can tell us nothing about the dynamic properties of the system (54) and its capacity to produce plausible long-run behaviour. For this purpose we must make use of the results obtained in sections 3 and 4. We first obtain estimates and asymptotic confidence intervals for the elements of β, γ and λ by methods which would be appropriate if (55) were the true model. Next we substitute these estimates into the formulae given in section 3 in order to obtain estimates and confidence intervals for the steady state paths generated by the system (54). And, finally, we investigate the asymptotic stability of the estimated steady state by using the results of section 4 and, in particular, obtaining confidence intervals for the eigenvalues of the matrix A in (53). The use of the linear approximation (55) for estimation purposes will, of course, produce a bias in the various estimates and confidence intervals involved in the above procedure, and this should be remembered when interpreting the results.

An alternative to the above procedure is to add a vector of random disturbance functions to the system (53) and treat this as if it were the true model. The elements of the matrix A are functions only of the structural parameters and do not involve the sample means. But, because of the complexity of these functions, which involve the complicated formulae for the steady state growth levels, the use of this system for estimation would provide even more arduous computational problems than the use of (55). Moreover, it could be expected to provide better estimates of the parameters only if the system were close to the steady state.

We turn now to the problem of estimating (55). For simplicity of exposition we shall consider first the problem of estimating the system

$$Dx(t) = B(\alpha)x(t) + \zeta(t), \tag{59}$$

where the elements of B are known functions of an unknown parameter vector α, and $\zeta(t)$ is a pure noise vector. It can be shown that a sample $x(1), \ldots, x(T)$ generated by (59) satisfies the autogressive system

$$x(t) = e^{B(\alpha)}x(t-1) + \xi(t), \tag{60}$$

where

$$e^B = I + B + \tfrac{1}{2}B^2 + \ldots + \frac{1}{r!}B^r + \ldots.$$

and

$$\xi(t) = \int_{t-1}^{t} e^{B(t-r)}\zeta(r)\mathrm{d}r,$$

so that

$$E\{\xi(t)\xi(t)'\} = \int_0^1 e^{Br}\Sigma e^{B'r}\mathrm{d}r = \Omega.$$

It follows that quasi-maximum likelihood estimates could be obtained from such a sample by minimising the determinant of

$$\sum_{t=1}^{T} \{x(t) - e^{B(\alpha)}x(t-1)\}\{x(t) - e^{B(\alpha)}x(t-1)\}',$$

with respect to α.

But, since most of the variables in our model are flow variables whose rates of flow at a point of time cannot be measured, a sample of observations $x(1), \ldots, x(T)$ is not available to us. We cannot, for example, measure the rate of consumption at a point of time. Our consumption data are measurements of the integral of this rate over quarterly intervals. Our problem is, therefore, to estimate α from a sample of observations of the integrals

$$\int_0^1 x(t)\mathrm{d}t, \ldots, \int_{T-1}^T x(t)\mathrm{d}t.$$

From (60) we obtain

$$\begin{aligned}\int_{t-1}^t x(\tau)\mathrm{d}\tau &= \mathrm{e}^{B(\alpha)}\int_{t-2}^{t-1} x(\tau)\mathrm{d}\tau + \int_{t-1}^t \xi(\tau)\mathrm{d}\tau \\ &= \mathrm{e}^{B(\alpha)}\int_{t-2}^{t-1} x(\tau)\mathrm{d}\tau + \int_{t-1}^t\int_{\tau-1}^{\tau} \mathrm{e}^{B(\alpha)(\tau-\theta)}\zeta(\theta)\mathrm{d}\theta\mathrm{d}\tau. \qquad (61)\end{aligned}$$

Since the disturbance vector in (61) is a double integral which is affected by the behaviour of $\zeta(t)$ over the interval $t-2$ to t, its first-order autocovariance matrix is clearly non-zero and depends on the unknown parameter α. Consequently the problem of obtaining a consistent estimator of α from flow data is very complicated. One possibility might be to use an iterative procedure in which we first estimate α using an assumed autocovariance matrix, next compute the autocovariance matrix implied by this estimate of α, then obtain a new estimate of α using this autocovariance matrix, and so on. But nothing is known about the convergence properties of this procedure.

If the time interval between observations is small then the elements of B will be small and the disturbance in (61) can be approximated by the double integral $\int_{t-1}^t\int_{\tau-1}^{\tau}\zeta(\theta)\mathrm{d}\theta\mathrm{d}\tau$. We did, in fact, make use of this approximation. When the elements of B have the values implied by our estimates, the bias resulting from its use is very small. (See Phillips 1974 for a rigorous analysis of the consequences of this approximation in obtaining estimates from flow data.)

The autocorrelation properties of $\int_{t-1}^t\int_{\tau-1}^{\tau}\zeta(\theta)\mathrm{d}\theta\mathrm{d}\tau$ can easily be obtained from the assumptions (56) to (58) relating to the integral of $\zeta(t)$. The autocovariance matrix is given by

$$\begin{aligned}&E\left[\int_{t-1}^t\int_{\tau-1}^{\tau}\zeta(\theta)\mathrm{d}\theta\mathrm{d}\tau\int_t^{t+1}\int_{\tau-1}^{\tau}\zeta(\theta)'\mathrm{d}\theta\mathrm{d}\tau\right] \\ &= \lim_{n\to\infty} E\left\{\frac{1}{n}\sum_{i=1}^n\int_{t-2+i/n}^{t-1+i/n}\zeta(\theta)\mathrm{d}\theta\frac{1}{n}\sum_{i=1}^n\int_{t-1+i/n}^{t+i/n}\zeta(\theta)'\mathrm{d}\theta\right\} \\ &= \lim_{n\to\infty}\frac{1}{n^2}E\left\{\sum_{i=1}^n\sum_{j=1}^n\int_{t-2+(i+j-1)/n}^{t-2+(i+j)/n}\zeta(\theta)\mathrm{d}\theta\right. \\ &\qquad\left.\times\sum_{i=1}^n\sum_{j=1}^n\int_{t-1+(i+j-1)/n}^{t-1+(i+j)/n}\zeta(\theta)'\mathrm{d}\theta\right\} \\ &= \lim_{n\to\infty}\frac{1}{n}\sum_{m=1}^n\frac{m}{n}\left(1-\frac{m}{n}\right)\Sigma \\ &= \int_0^1 x(1-x)\mathrm{d}x\Sigma \\ &= \tfrac{1}{6}\Sigma. \qquad (62)\end{aligned}$$

Similarly, the covariance matrix is given by

$$E\left[\int_{t-1}^{t}\int_{\tau-1}^{\tau}\zeta(\theta)\mathrm{d}\theta\mathrm{d}\tau\int_{t-1}^{t}\int_{\tau-1}^{\tau}\zeta(\theta)'\mathrm{d}\theta\mathrm{d}\tau\right]$$
$$= 2\int_{0}^{1}(1-x)^2\mathrm{d}x\Sigma$$
$$= \tfrac{2}{3}\Sigma. \tag{63}$$

The first-order autocorrelation coefficient is therefore $\frac{1}{4}$. And, since the higher-order autocorrelation coefficients are clearly zero, $\int_{t-1}^{t}\int_{\tau-1}^{\tau}\zeta(\theta)\mathrm{d}\theta\mathrm{d}\tau$ has the same autocorrelation properties as the moving average vector $u_t = \varepsilon_t + 0.268\varepsilon_{t-1}$, where ε_t is serially uncorrelated (Wymer 1972).

This vector approximately satisfies the autoregressive relation

$$u_t - 0.268u_{t-1} + (0.268)^2u_{t-2} - (0.268)^3u_{t-3} = \varepsilon_t, \tag{64}$$

so that by replacing $\int_{t-1}^{t}x(\tau)\mathrm{d}\tau$ in (61) by the moving average

$$\int_{t-1}^{t}x(\tau)\mathrm{d}\tau - 0.268\int_{t-2}^{t-1}x(\tau)\mathrm{d}\tau + (0.268)^2\int_{t-3}^{t-2}x(\tau)\mathrm{d}\tau$$
$$- (0.268)^3\int_{t-4}^{t-3}x(\tau)\mathrm{d}\tau$$

we obtain a system in which the disturbance can, for practical purposes, be treated as serially uncorrelated. The parameter vector can then be estimated by minimising the determinant of the sample covariance matrix of the residuals of the transformed system.

So far we have been discussing the estimation of (59). The estimation of (55) is a little more complicated because of the presence of the constant vector c and the trend vector $\mathrm{d}t$. A sample $x(1), \ldots, x(T)$ generated by this system satisfies the system

$$\{x(t) + B^{-1}(c + B^{-1}d + \mathrm{d}t)\}$$
$$= \mathrm{e}^{B}\{x(t-1) + B^{-1}(c + B^{-1}d + d(t-1))\} + \xi(t), \tag{65}$$

in which the elements of B, c and d are functions of $\bar{x}$, β, γ and λ. From (65) we obtain

$$\left\{\int_{t-1}^{t}x(\tau)\mathrm{d}\tau + B^{-1}(c + B^{-1}d + \mathrm{d}(t-\tfrac{1}{2}))\right\}$$
$$= \mathrm{e}^{B}\left\{\int_{t-2}^{t-1}x(\tau)\mathrm{d}\tau + B^{-1}(c + B^{-1}d + \mathrm{d}(t-\tfrac{3}{2}))\right\}$$
$$+ \int_{t-1}^{t}\int_{\tau-1}^{\tau}\mathrm{e}^{B(\tau-\theta)}\zeta(\theta)\mathrm{d}\theta\mathrm{d}\tau. \tag{66}$$

By replacing $\int_{t-1}^{t}x(\tau)\mathrm{d}\tau$ by the moving average used in the previous paragraph the serial correlation of the disturbance in (66) can be approximately removed and the estimates of the parameters obtained as before.

The estimates presented in the next section were obtained by the procedure just outlined. The data $\int_{t-1}^{t} x_1(\tau)\mathrm{d}\tau, \ldots, \int_{t-1}^{t} x_{13}(\tau)\mathrm{d}\tau$ were first corrected for seasonal variation using the procedure proposed in Durbin 1963. Next they were transformed by taking the moving average that approximately eliminates the serial correlation in the disturbance. The moving averages were then substituted into (66) and quasi-maximum likelihood estimates obtained by minimising the determinant of the covariance matrix of the residuals with respect to the parameter vectors β, γ and λ.

Even if the model is correctly specified it is very unlikely that an estimate, from a small sample, of the vector of 35 parameters will satisfy all the a priori restrictions on sign and magnitude. It was, in fact, necessary to restrict 7 of the parameters to values which satisfy the a priori restrictions. The selection of these parameters was based on a series of preliminary estimates. Even these estimates were obtained by a complete system full information method, and there was no experimentation with single-equation methods. But, in obtaining the preliminary estimates, we used the somewhat simpler full information method which makes use of a discrete approximation to the differential equation system to obtain a simultaneous equations system. The properties of this method have been discussed in the various articles referred to in the introduction. But the fact that we have flow data makes the application of the method in this study somewhat more complicated than when point observations are available, as assumed in these articles.

Integrating (55) twice we obtain

$$\int_{t-1}^{t} x(\tau)\mathrm{d}\tau - \int_{t-2}^{t-1} x(\tau)\mathrm{d}\tau = B\int_{t-1}^{t}\int_{\tau-1}^{\tau} x(\theta)\mathrm{d}\theta\mathrm{d}\tau + c + \mathrm{d}(t-1) + \int_{t-1}^{t}\int_{\tau-1}^{\tau} \zeta(\theta)\mathrm{d}\theta\mathrm{d}\tau, \tag{67}$$

from which we obtain, by using Simpson's rule, the approximation relation

$$\int_{t-1}^{t} x(\tau)\mathrm{d}\tau - \int_{t-2}^{t-1} x(\tau)\mathrm{d}\tau = \tfrac{1}{2}B\left\{\int_{t-1}^{t} x(\tau)\mathrm{d}\tau + \int_{t-2}^{t-1} x(\tau)\mathrm{d}\tau\right\} + c + \mathrm{d}(t-1) + \int_{t-1}^{t}\int_{\tau-1}^{\tau} \zeta(\theta)\mathrm{d}\theta\mathrm{d}\tau. \tag{68}$$

This is a simultaneous equations system in which $\int_{t-1}^{t} x(\tau)\mathrm{d}\tau$ is the vector of jointly dependent variables. But it is, in two respects, more complicated than the simultaneous equations systems most commonly considered in the econometric literature. First, the coefficients are non-linear functions of more basic structural parameters (i.e. the elements of β, γ and λ). And, secondly, the disturbances are serially correlated. It is, nevertheless, simpler to estimate than (66) since it does not involve e^B. Again the serial correlation in the disturbance was approximately removed by replacing $\int_{t-1}^{t} x(\tau)d\tau$ by the third-order moving average given in (64). Estimates were then obtained by full information maximum likelihood using a program described in Wymer.[3]

6. The Estimates

Estimates of the structural parameters were obtained from United Kingdom quarterly data for the period 1955–66 and are shown, together with *t*-values derived from the estimates and their asymptotic variances, in table 1. The parameters whose values were restricted a priori to assumed values are denoted with an asterisk. They imply that the elasticity of substitution between labour and capital is 0.5 ($\beta_4 = 1.0$), the price elasticity of demand for exports is unity ($\beta_9 = 1.0$), there is perfect competition ($\beta_{10} = 1.0$), the income elasticity of demand for money is unity ($\beta_{13} = 1.0$), the ratio of exports to imports aimed at by the policy-makers is unity ($\beta_{15} = 1.0$), a stock deficiency is removed entirely by increased imports and has no direct influence on domestic production ($\gamma_6 = 0$) and the level of employment has no independent influence on monetary policy, which is influenced only by the balance of payments position ($\gamma_{14} = \gamma_{16} = 0$). With these restrictions the estimates of all 26 of the remaining parameters have the correct signs, and 21 of these are significant at the 1 per cent level.

The restriction of the elasticity of substitution between labour and capital to 0.5 (which, for the economy as a whole, seems plausible in the light of various cross section studies) has enabled us to obtain a highly significant estimate of the capital intensity parameter β_3. The long-run elasticity of output with respect to capital is $\beta_3(Y/K)^{\beta_4} = \beta_3(Y/K)$. The estimate of β_3 implies that the average value of this elasticity over the sample period was 0.13. It should be noted that the parameter β_3 occurs in equations (2), (3), (7) and (8), so that our estimate has made use of information relating, not only to production, but also to investment and price and wage adjustments.

The interpretation of the remaining parameters β_i is straightforward. The estimates imply that the long-run marginal propensity to consume is 0.92, the partial equilibrium ratio of imports to total sales 0.16, partial equilibrium stocks are equal to about $4\frac{1}{2}$ months sales ($\beta_7 = 1.52$) and the interest elasticity of demand for money is 0.83.

Two key parameters in the dynamic mechanism discussed in section 4

[3] C. R. Wymer, Full-information maximum likelihood estimates with non-linear restrictions; and: Computer programs: Resimul manual, mimeo.

TABLE 1. Estimates of structural parameters from United Kingdom; quarterly data, 1955–66. The parameters restricted a priori to assumed values are denoted with an asterisk

Parameter	Estimate	t-value
β_1	0.9206	215.18
β_2	0.0037	108.04
β_3	1.2180	24.07
β_4	1.0000*	
β_5	0.0042	1.52
β_6	0.1632	75.46
β_7	1.5263	66.15
β_8	1365.1909	62.11
β_9	1.0000*	
β_{10}	1.0000*	
β_{11}	22.1379	34.57
β_{12}	0.0537	0.62
β_{13}	1.0000*	
β_{14}	0.8289	2.08
β_{15}	1.0000*	
γ_1	0.2454	5.71
γ_2	0.0990	4.08
γ_3	0.0886	3.36
γ_4	1.1492	3.87
γ_5	0.6437	5.25
γ_6	0.0000*	
γ_7	0.3486	3.51
γ_8	0.0448	7.72
γ_9	0.2926	4.18
γ_{10}	0.1312	3.40
γ_{11}	0.1047	2.85
γ_{12}	0.1990	1.33
γ_{13}	0.0422	2.69
γ_{14}	0.0000*	
γ_{15}	0.1562	4.68
γ_{16}	0.0000*	
λ_1	0.0058	17.29
λ_2	0.0113	13.49
λ_3	0.0008	0.85

are γ_4, which measures the effect of variations in the real interest rate on investment, and γ_{11}, which measures the effect of a change in the demand for labour on the rate of increase in the wage rate. The estimates of both of these parameters are significant having t-values of about 4 and 3, respectively. The interpretation of $\gamma_4 = 1.15$ is that an increase in the real interest rate from say 4 per cent per annum to 5 per cent per annum will reduce the partial equilibrium rate of capital accumulation from, say, 4.5 per cent per annum to 3.35 per cent per annum, or by about 25 per cent. The interpretation of $\gamma_{11} = 0.10$ is that a 1 per cent increase in the demand for labour will increase the rate of increase in the wage rate by 0.10 per cent per quarter, so that the annual rate of increase in the wage rate rises from say, 4 per cent per annum to 4.42 per cent per annum.

The estimates of the policy parameters γ_{13} and γ_{15} are also highly significant. The interpretation of $\gamma_{13} = 0.04$ is that a 1 per cent increase in the ratio of exports to imports will cause the acceleration in the money supply to increase by 0.04 per cent per quarter so that if, for example, the money supply had been increasing steadily by 1 per cent per quarter it will increase by approximately 1.04 per cent next quarter, 1.08 per cent the following quarter and so on. The interpretation of $\gamma_{15} = 0.16$ is that a 1 per cent increase in the quarterly rate of increase in the ratio of exports to imports will increase the acceleration in the money supply by 0.16 per cent per quarter.

Each of the remaining parameters γ_i can be interpreted as the reciprocal of the mean of an exponentially distributed time lag with which a variable, or its logarithm, adjusts to a change in the variables that influence it. The estimated mean lags, in this sense, are: 12 months for the adjustment of consumption to a change in income, $2\frac{1}{2}$ years for the adjustment of employment to a change in output or capital, 3 years for the adjustment of investment to a change in the real interest rate or the marginal product of capital, $4\frac{1}{2}$ months for the adjustment of output to a change in sales, 9 months for the adjustment of imports to a change in sales, $5\frac{1}{2}$ years for the adjustment of stocks (through imports) to a change in sales, 10 months for the adjustment of exports to a change in the price level, 2 years for the adjustment of the price level to a change in marginal cost and $1\frac{1}{2}$ years for the adjustment of the interest rate to a change in the money supply or income ($1/\gamma_{12}\beta_{14} = 6$). The t-values indicate that all of these lags, except the last, are measured with a high degree of precision. Comparison with the preliminary estimates and t-values obtained by using the approximate discrete model, discussed in section 5, suggests that one of the main advantages of the method used for obtaining the final estimates is the greater precision with which it measures speeds of response.

Finally, we have the estimates of the trend parameters λ_1, λ_2, and λ_3. These imply that the rate of technical progress is about 2.3 per cent per annum, the exogenous rate of growth of demand for exports is about 4.5 per cent per annum and the rate of growth in the labour supply about 0.3 per cent per annum. Using the formulae in of the table in section 3 we obtain the steady state growth rates shown in table 2.

The estimated levels of the steady state growth paths have been

TABLE 2. Estimated steady state growth rates

Variable	Growth rate (per cent per annum)	t-value
L	0.32	0.85
C, Y, K, S, I, E	2.62	8.67
p	1.91	4.95
w	4.23	8.98
M	4.52	13.49
r	0.00	

computed from the formulae given in section 3 and are expressed in table 3 as ratios of the actual 1966 levels of the variables. The estimates indicate that, in 1966 the ratio of capital to output was only about 60 per cent of its steady state level but that a high level of employment relative to the steady state level offset the low relative level of capital and resulted in the steady state level of output being produced.

The estimated value of the parameter β_8 implies that the partial equilibrium level of exports corresponding to the domestic price level during the sample period was lower than the actual value so that, if the domestic price level had been constant, exports would have grown at a rate much less than the steady state growth rate. But any such tendency would result in a slow growth in the volume of money, high interest rates and a low rate of fixed capital formation. This trend would continue until the domestic price level reaches a lower level relative to the price level of the rest of the world and the difference between the partial equilibrium and actual levels of exports is greatly reduced. In other words, the estimates imply that, in 1966, the potentiality of United Kingdom exports was such that, from a long-term point of view, sterling was over-valued but that this over-valuation would eventually be eliminated by reduced domestic inflation.

The eigenvalues of the estimate of matrix A in equation (53) are shown in table 4. There are 5 real eigenvalues all of which are negative and 4 pairs of complex eigenvalues including 3 pairs with negative real parts and 1 pair with positive real parts. At the 1 per cent level of significance we must reject the hypothesis that all the eigenvalues have negative real parts and so the model as estimated does not have an asymptotically stable steady state. This is likely to be the result of a mis-specification of the capital or financial behaviour of the economy. But if the parameters are as estimated, then the model generates an explosive 21-year cycle.

In order to provide a realistic description of the behaviour of the economy, it might be necessary to introduce some additional non-linearity.

TABLE 3. Estimated levels of steady state growth paths

Variable	Steady state level	Asymptotic standard error	Ratio of estimated steady state level to actual level in 1966	Asymptotic standard error
C	5427.8	120.34	1.01	0.014
L	24.3	0.49	0.96	0.015
Y	6055.8	128.17	1.00	0.016
I	1320.2	27.10	1.06	0.025
E	1320.2	27.10	1.09	0.025
p	1.0112	0.0252	0.89	0.026
w	237.0	7.54	1.34	0.059
r	0.0088	0.0032	0.52	0.187
S	9976.3	303.87	1.00	0.028
K	85829.6	11163.65	1.46	0.170
M	16571.4	6367.74	1.57	0.601

TABLE 4. Eigenvalues for linear approximation about steady state

Real part	t-value	Imaginary part	t-value
−0.8605	6.65		
−0.2946	4.13		
−0.1707	2.51		
−0.0990	4.08		
−0.0061	7.79		
−0.1791	5.30	±0.1425	3.92
−0.1826	4.52	±0.0599	2.10
0.0782	6.16	±0.0742	3.59
−0.0179	5.23	±0.0514	4.55

We might, for example, make the partial equilibrium rate of capital accumulation, in equation (3), a non-linear function of the excess of the marginal product of capital over the real rate of interest, the form of the function being chosen so that the rate of increase in the stock of capital has a lower bound equal to the rate of depreciation. But, even in this case the mechanism described in section 4 would have an important role to play in ensuring that, over a long period, the path about which the constrained oscillations occur converges to the steady state. The fact that all of the real roots are negative suggests that, for the estimated values of the parameters, the mechanism does work to this extent. The real root whose value is close to zero reflects the slow convergence of the capital–output ratio to its steady state value and does not indicate that the system is bordering on the sort of instability which might cause it to converge to the lower bound (see the discussion of the prototype model in Bergstrom 1967).

7. Predictive Performance of the Model

The model which has been specified and estimated has several advantages for forecasting purposes. Firstly, the full information maximum likelihood estimates use all a priori information which will increase the efficiency of the forecasts providing that the model has been specified correctly. Secondly, the use of a differential equation model allows forecasts to be made at shorter intervals than the observation interval of the sample. This might be useful for policy-makers and particularly for the monetary authorities. In practice it might often be difficult to obtain appropriate values of the exogenous variables for such forecasting but this difficulty does not arise here; the only exogenous variable is time. Thus it would be possible to provide consistent and efficient monthly forecasts, for example, even though the observation interval in the sample was one quarter. The model is designed, however, more for medium-term forecasting than short-term forecasting and it should be capable of providing plausible long-run behaviour.

The estimated exact discrete model (66) was used to produce forecasts of all variables for a post-sample period of two years. Both single-period forecasts (that is, forecasts for only one period ahead for each of the eight quarters) and multi-period or dynamic forecasts (that is, forecasts for several periods ahead) were calculated with their standard errors.[4] In order to obtain the forecasts an adjustment was made to the parameter β_8 to allow for the United Kingdom devaluation of 16.7 per cent in November 1967. The estimated value of β_8 was increased by this proportion. The post-sample data were prepared in a way similar to the data used for estimation. The moving average transformation shown in (64) was applied to the seasonally adjusted post-sample observations to eliminate the serial correlation in the disturbances. This transformation meant that the first three observations after devaluation could not be used for forecasting so that the post-sample forecasting period was from the first quarter of 1969 to the fourth quarter of 1970.

The single-period forecasts used the actual lagged transformed variables

[4] C. R. Wymer, The forecast error variance matrix of linear econometric models—A note, mimeo.

for the preceding period for producing the forecast for each period. The multi-period forecasts used the actual values of the transformed variables in the fourth quarter of 1968 as initial values and forecasts based on these values were produced for each of the following eight periods. As the only exogenous variable is a time trend the latter set of forecasts depend heavily on the dynamic structure of the estimated model. However, the forecasts derived from the linearized model might be poor if the initial values of the variables used for multi-period prediction were substantially different from the steady state values.

The root mean square errors of the single-period and multi-period forecasts are shown in table 5. The multi-period forecasts and their asymptotic standard errors are given in table 6. The single-period forecasts of some variables are shown in figure 2 (see pp. 239–43) which also includes the actual and estimated values of the transformed variables in the sample period.

The root mean square errors of the single-period forecasts were less than 5 per cent for all variables and less than 1 per cent for consumption, employment, prices, stocks, fixed capital and volume of money. The root mean square errors of the multi-period forecasts were less than 4 per cent for all variables except exports and the interest rate.

An indication of the standard errors of forecast of the single-period ahead forecasts is given in table 6 for the period 1969/i–70/iv; the single-period forecast errors for the other quarters of the post-sample period are of the same order. Thus the asymptotic standard errors of forecast for single-period prediction were less than 0.3 per cent for all

TABLE 5. Root mean square errors of post-sample forecasts: 1969–70

Variable	Root mean square error of single-period forecasts	Root mean square error of multi-period forecasts
C	0.0079	0.0285
L	0.0035	0.0194
Y	0.0165	0.0248
I	0.0492	0.0359
E	0.0494	0.0631
p	0.0075	0.0190
w	0.0269	0.0329
r	0.0463	0.1252
S	0.0051	0.0239
K	0.0003	0.0018
M	0.0051	0.0167

TABLE 6. Dynamic forecasts and their standard errors (initial value of variables was 1968/iv observation of transformed series)

Period	Variable	Forecast	Asymptotic standard error of forecast	Actual value
1969/i	*C*	6.902	0.0013	6.893
	L	2.540	0.0011	2.534
	Y	7.006	0.0016	6.975
	I	5.844	0.0061	5.843
	E	5.823	0.0101	5.791
	p	0.253	0.0017	0.263
	w	4.382	0.0029	4.358
	r	−2.004	0.0110	−1.924
	S	7.363	0.0006	7.357
	K	8.817	0.0001	8.817
	M	7.554	0.0013	7.552
1969/ii	*C*	6.911	0.0017	6.888
	L	2.543	0.0014	2.535
	Y	7.012	0.0023	6.989
	I	5.850	0.0078	5.849
	E	5.829	0.0130	5.884
	p	0.263	0.0023	0.268
	w	4.395	0.0042	4.384
	r	−2.008	0.0144	−1.870
	S	7.372	0.0014	7.362
	K	8.825	0.0003	8.825
	M	7.561	0.0042	7.552
1969/iii	*C*	6.920	0.0020	6.896
	L	2.546	0.0017	2.535
	Y	7.018	0.0029	7.003
	I	5.855	0.0088	5.824
	E	5.833	0.0148	5.890
	p	0.272	0.0026	0.280
	w	4.408	0.0052	4.389
	r	−2.010	0.0167	−1.867
	S	7.380	0.0021	7.366
	K	8.834	0.0004	8.832
	M	7.568	0.0079	7.545

variables in all periods except imports, exports, the wage rate and the interest rate. The forecast errors for imports and the wage rate were less than 1 per cent in all periods and the forecast error for exports and the interest rate were less than 1.6 per cent in all periods. These results show

TABLE 6. Cont.

Period	Variable	Forecast	Asymptotic standard error of forecast	Actual value
1969/iv	C	6.928	0.0024	6.907
	L	2.548	0.0020	2.533
	Y	7.024	0.0035	7.010
	I	5.860	0.0095	5.866
	E	5.837	0.0160	5.875
	p	0.282	0.0030	0.284
	w	4.421	0.0061	4.434
	r	−2.010	0.0186	−1.891
	S	7.388	0.0027	7.371
	K	8.842	0.0006	8.840
	M	7.573	0.0122	7.549
1970/i	C	6.935	0.0027	6.903
	L	2.551	0.0023	2.531
	Y	7.029	0.0041	6.984
	I	5.864	0.0102	5.832
	E	5.839	0.0168	5.877
	p	0.291	0.0033	0.302
	w	4.433	0.0069	4.429
	r	−2.008	0.0207	−1.915
	S	7.396	0.0032	7.371
	K	8.850	0.0009	8.848
	M	7.577	0.0171	7.564
1970/ii	C	6.942	0.0031	6.905
	L	2.553	0.0025	2.528
	Y	7.034	0.0046	7.012
	I	5.866	0.0108	5.927
	E	5.840	0.0174	5.916
	p	0.300	0.0037	0.323
	w	4.446	0.0077	4.485
	r	−2.004	0.0230	−1.885
	S	7.405	0.0036	7.373
	K	8.858	0.0011	8.856
	M	7.580	0.0226	7.579

the advantage of using the exact discrete model to obtain forecasts with high precision. However, the frequency with which the errors exceeded the standard error of forecast indicates that there may be some misspecification in the model.

TABLE 6. Cont.

Period	Variable	Forecast	Asymptotic standard error of forecast	Actual value
1970/iii	C	6.948	0.0035	6.918
	L	2.555	0.0028	2.528
	Y	7.038	0.0052	7.017
	I	5.868	0.0115	5.853
	E	5.842	0.0179	5.884
	p	0.309	0.0041	0.341
	w	4.459	0.0086	4.495
	r	−1.999	0.0259	−1.854
	S	7.414	0.0039	7.381
	K	8.865	0.0014	8.863
	M	7.582	0.0286	7.596
1970/iv	C	6.954	0.0039	6.915
	L	2.557	0.0031	2.527
	Y	7.042	0.0058	7.033
	I	5.868	0.0123	5.935
	E	5.843	0.0183	5.961
	p	0.318	0.0046	0.350
	w	4.471	0.0095	4.539
	r	−1.993	0.0293	−1.847
	S	7.424	0.0042	7.390
	K	8.873	0.0018	8.871
	M	7.582	0.0352	7.608

8. Conclusion

The estimates of the model justify the use which has been made of economic theory in specifying the model and the attention which has been paid to developing an appropriate estimator for such a differential equation system. We ought to emphasize that all estimates have the correct sign, plausible values and are highly significant, and the forecasting properties of the model suggest that the model and the estimation techniques used are capable of producing satisfactory forecasts with high precision.

Research is continuing on several aspects of the model. The model is being re-estimated to find under what conditions satisfactory estimates can be obtained with $\gamma_6 \neq 0$, and more work is being carried out to investigate the medium and long-term forecasting properties of the model. Moreover, the model is being used as a prototype for a larger model of the United Kingdom which includes a well-developed financial sector as well as behaviour functions for Government expenditure, taxation, Government securities and Bank Rate to allow an investigation of both fiscal and monetary policy in an open economy.

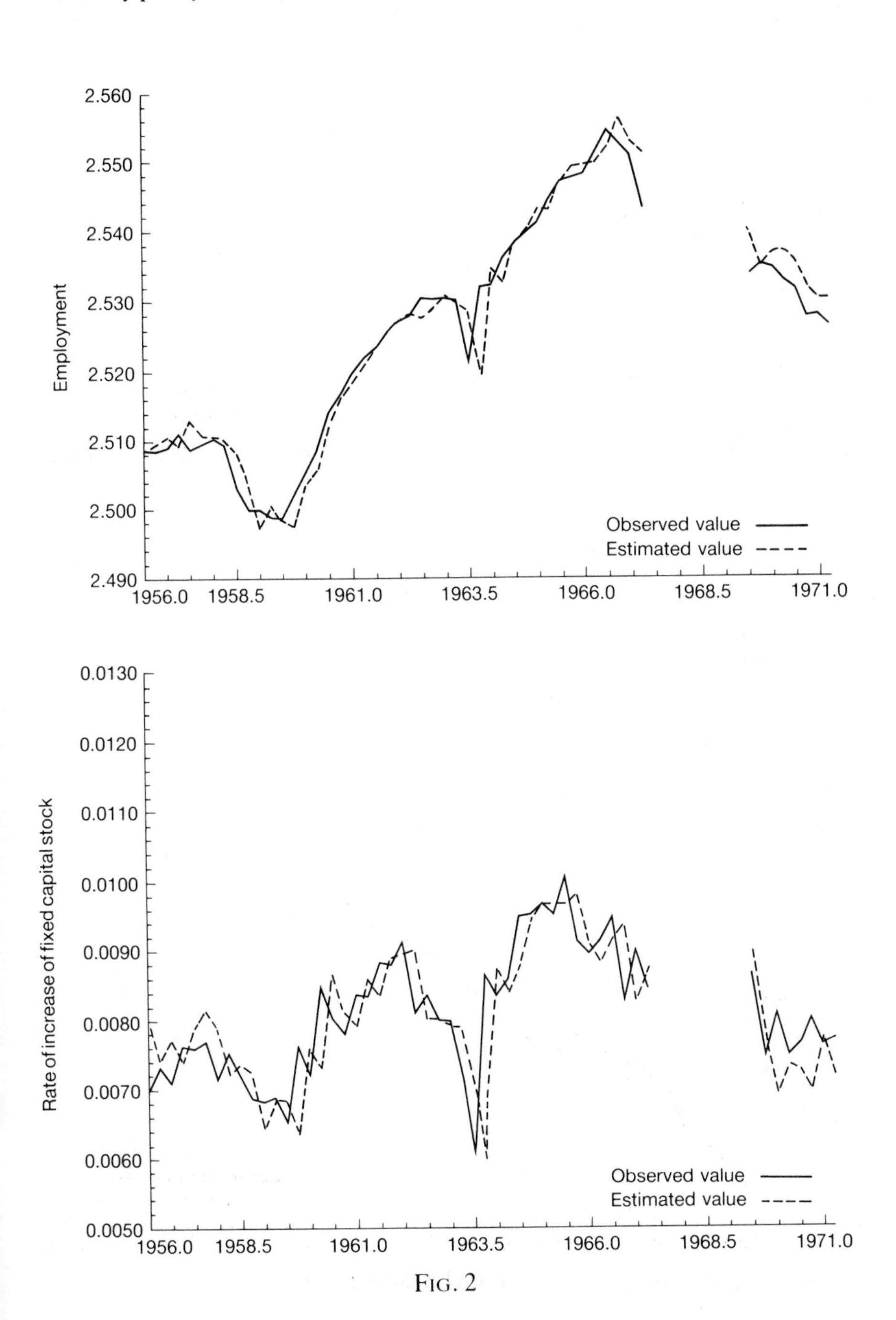

FIG. 2

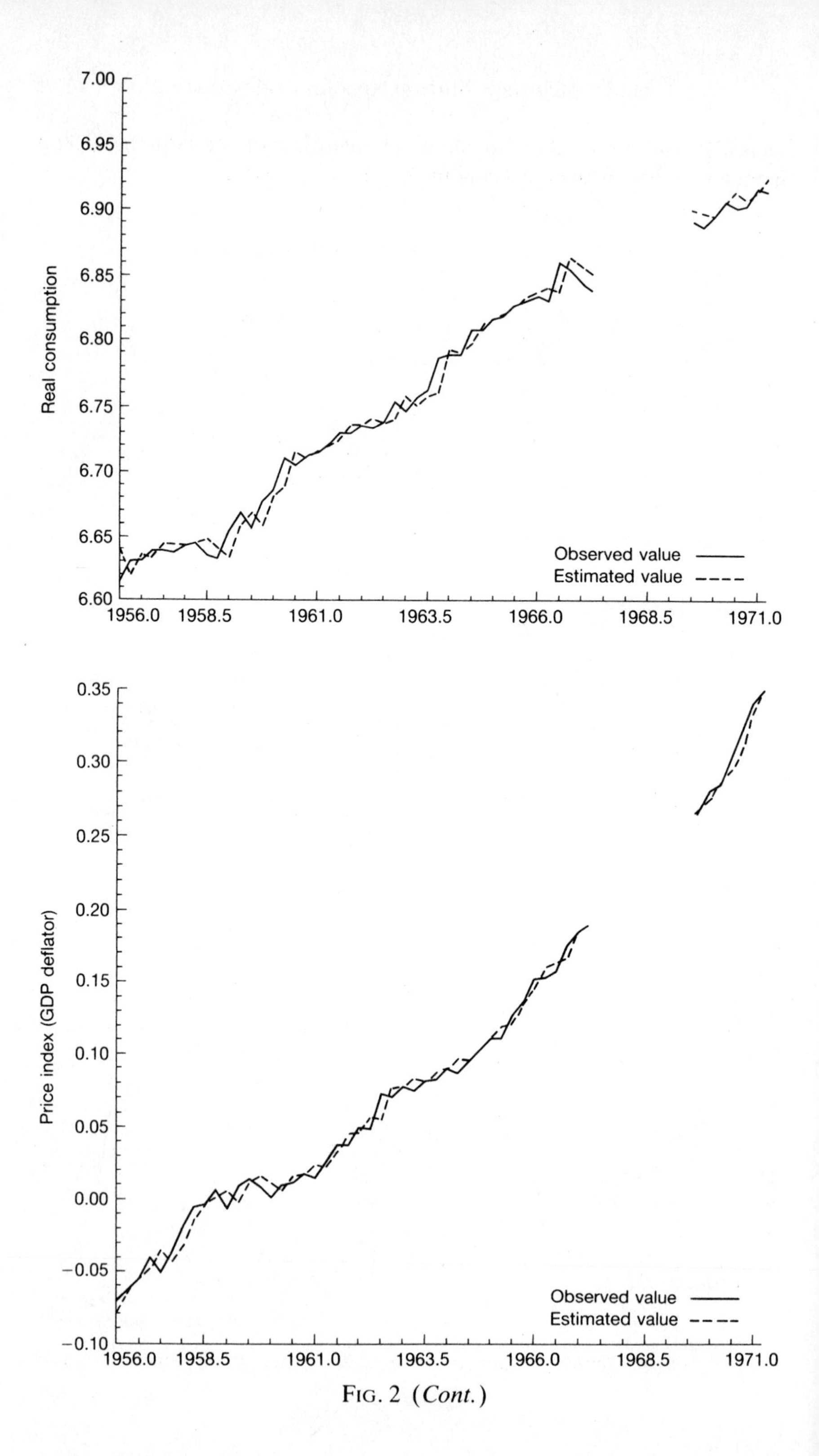

7.00
6.95
6.90
6.85
6.80
6.75
6.70
6.65
6.60
Real consumption
Observed value
Estimated value
1956.0
1958.5
1961.0
1963.5
1966.0
1968.5
1971.0
0.35
0.30
0.25
0.20
0.15
0.10
0.05
0.00
−0.05
−0.10
Price index (GDP deflator)
Observed value
Estimated value
1956.0
1958.5
1961.0
1963.5
1966.0
1968.5
1971.0

FIG. 2 (*Cont.*)

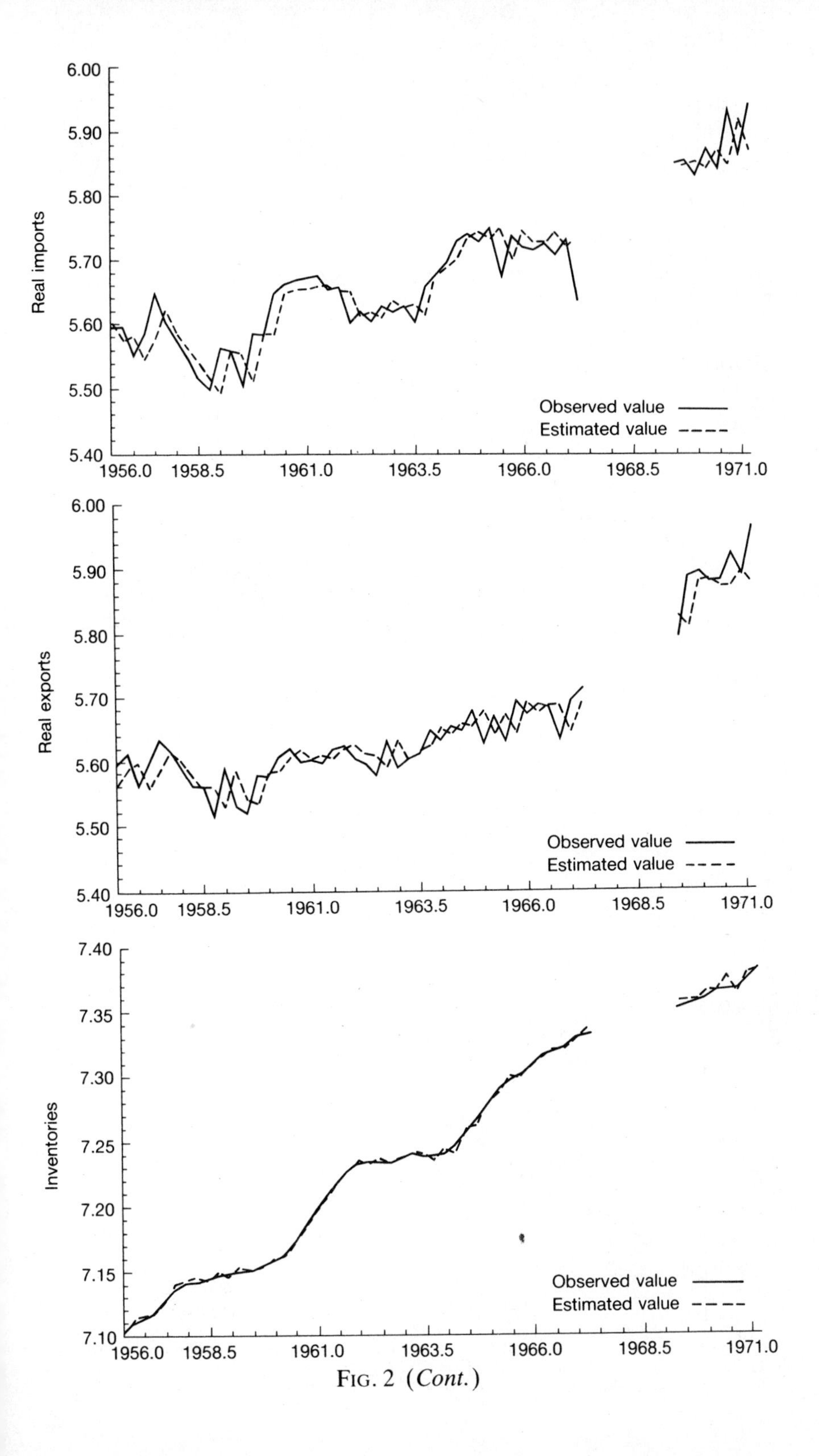
Real imports
6.00
5.90
5.80
5.70
5.60
5.50
5.40
1956.0
1958.5
1961.0
1963.5
1966.0
1968.5
1971.0
Observed value
Estimated value
Real exports
6.00
5.90
5.80
5.70
5.60
5.50
5.40
1956.0
1958.5
1961.0
1963.5
1966.0
1968.5
1971.0
Observed value
Estimated value
Inventories
7.40
7.35
7.30
7.25
7.20
7.15
7.10
1956.0
1958.5
1961.0
1963.5
1966.0
1968.5
1971.0
Observed value
Estimated value
FIG. 2 (*Cont.*)

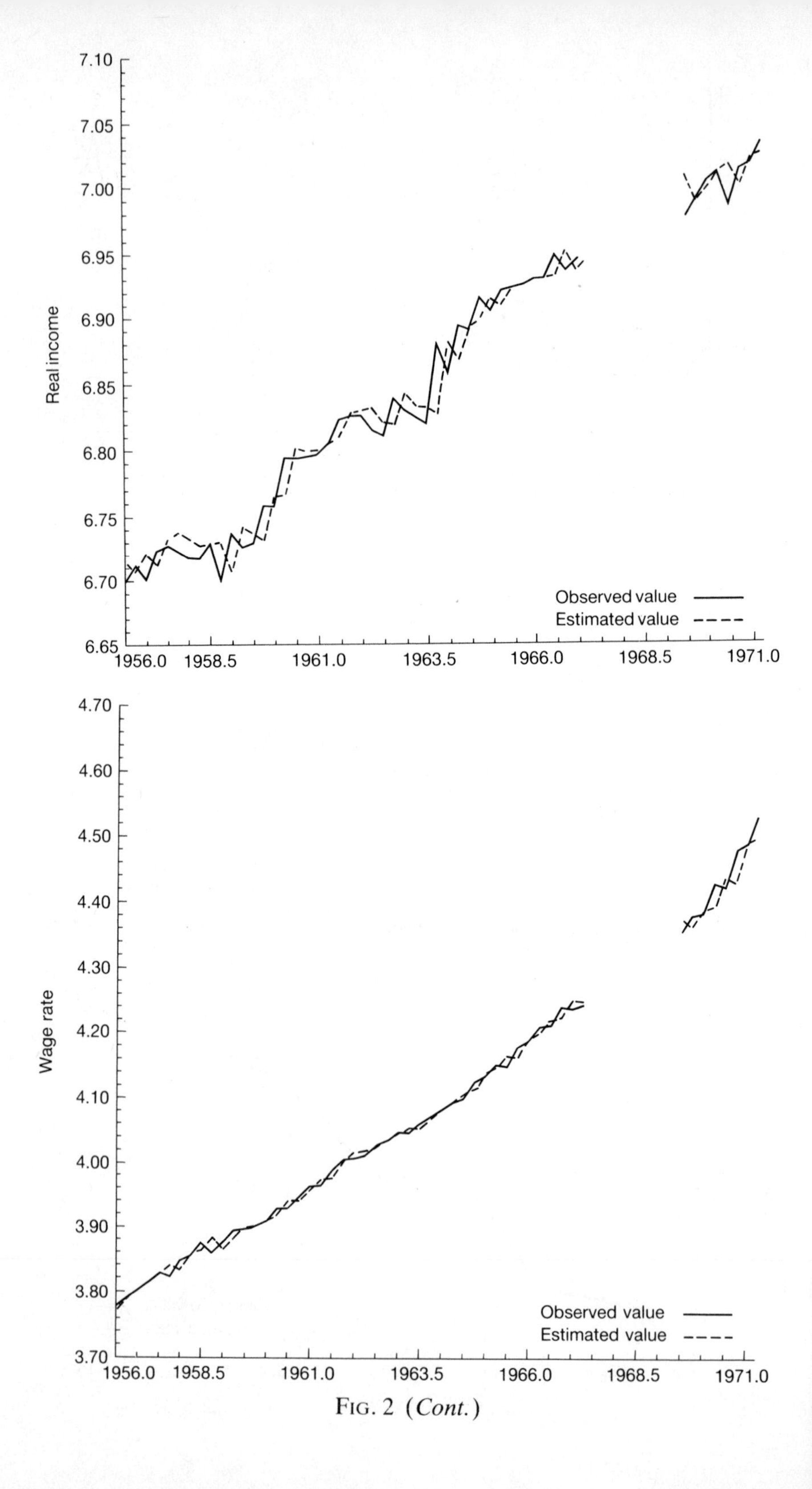

7.10
7.05
7.00
6.95
6.90
6.85
6.80
6.75
6.70
6.65
Real income
Observed value
Estimated value
1956.0
1958.5
1961.0
1963.5
1966.0
1968.5
1971.0
4.70
4.60
4.50
4.40
4.30
4.20
4.10
4.00
3.90
3.80
3.70
Wage rate
Observed value
Estimated value
1956.0
1958.5
1961.0
1963.5
1966.0
1968.5
1971.0

FIG. 2 (*Cont.*)

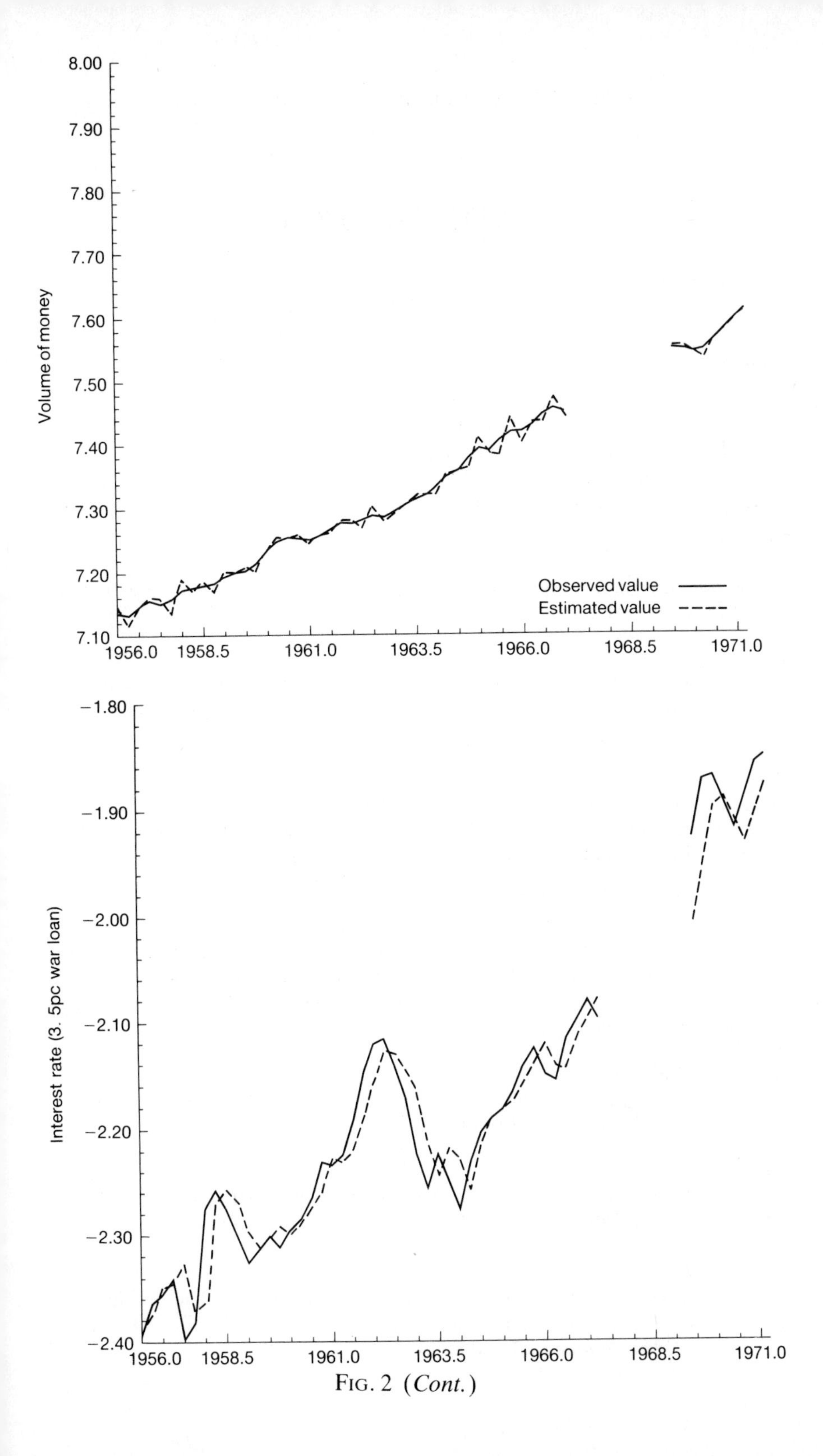
8.00
7.90
7.80
7.70
7.60
7.50
7.40
7.30
7.20
7.10
Volume of money
Observed value
Estimated value
1956.0
1958.5
1961.0
1963.5
1966.0
1968.5
1971.0
−1.80
−1.90
−2.00
−2.10
−2.20
−2.30
−2.40
Interest rate (3. 5pc war loan)
1956.0
1958.5
1961.0
1963.5
1966.0
1968.5
1971.0

FIG. 2 (*Cont.*)

Appendix 1

Data

Sources of data

AA	Annual Abstract of Statistics, HMSO
ET	Economic Trends, HMSO
MDS	Monthly Digest of Statistics, HMSO
FS	Financial Statistics, HMSO
NIE	National Income and Expenditure, HMSO

Definition of series

The series used in this study consisted of quarterly observations for the period 1955 to 1970 and were defined as follows:

C *Real consumption*
Consumers expenditure at current prices plus public authorities current expenditure at current prices deflated by the gross domestic product implicit price deflator, p, as defined below. *Source:* ET.

Y *Real income or output*
Gross domestic product at current factor cost plus taxes on expenditure less subsidies less interpolated quarterly capital consumption as defined below deflated by the implicit price deflator, p. *Source:* ET.

K *Real fixed capital formation*
Gross domestic fixed capital formation at current market prices less interpolated quarterly capital consumption as defined below deflated by the implicit price deflator, p, and cumulated on a base stock of £48400 m in 1958/iv. *Sources:* Gross capital formation data–ET; Capital consumption–AA; Fixed capital stock–NIE.

R *Real exports*
Exports of goods and services at current prices deflated by the implicit price deflator, p. *Source:* ET.

I *Real imports*
Imports of goods and services at current prices deflated by the implicit price deflator, p. *Source:* ET.

S *Inventories*
Value of physical increase in stocks at current prices deflated by the implicit price deflator, p, and cumulated on a base stock of £8990 m in 1958/iv. *Sources:* ET and MDS.

L *Employment*
Interpolated quarterly working population less average total registered unemployed in Great Britain and Northern Ireland. *Sources:* 1955 to 1967–AA; 1967 to 1970–MDS.

M *Volume of money*
End of quarter deposits of London Clearing Bank, Scottish and Northern Ireland banks plus end of quarter currency in circulation. *Source:* MDS and FS.

r *Interest rate*
Yield on $3\frac{1}{2}$% war loan. *Source:* FS.

p *Price level*
Gross domestic product at current market prices divided by gross domestic product at 1958 market prices. *Source:* ET.

w *Wage rate*
Income from employment divided by employment, L, as defined above. *Source:* ET.

The quadratic interpolation formulae used to interpolate quarterly capital consumption data from annual observations of capital consumption were obtained as follows:

If x_{t-1}, x_t, x_{t+1} are three successive annual observations of a continuous flow variable $x(t)$ the quadratic function passing through the three points is such that

$$\int_0^1 (as^2 + bs + c)\mathrm{d}s = x_{t-1},$$

$$\int_1^2 (as^2 + bs + c)\mathrm{d}s = x_t,$$

$$\int_2^3 (as^2 + bs + c)\mathrm{d}s = x_{t+1}.$$

Integrating and solving for a, b, c gives

$$a = 0.5x_{t-1} - 1.0x_t + 0.5x_{t+1},$$

$$b = -2.0x_{t-1} + 3.0x_t - 1.0x_{t+1},$$

$$c = 1.8333x_{t-1} - 1.1666x_t + 0.3333x_{t+1}.$$

The first two quarterly figures within any year can be interpolated by

$$\int_{1.0}^{1.25} (as^2 + bs + c)\mathrm{d}s = 0.0548x_{t-1} + 0.2343x_t - 0.0390x_{t+1},$$

$$\int_{1.25}^{1.50} (as^2 + bs + c)\mathrm{d}s = 0.0077x_{t-1} + 0.2657x_t - 0.0236x_{t+1},$$

and corresponding formulae give the third and fourth quarter interpolated figures.

Quarterly data for the years 1955 to 1970

Real Net Income	Real Consumption	Fixed Capital	Real Exports	Real Imports	Inventories
4887.0	4469.0	42131.0	1243.0	1349.0	8211.0
4887.0	4643.0	42493.0	1129.0	1217.0	8181.0
5118.0	4678.0	42879.0	1200.0	1311.0	8346.0
5299.0	4876.0	43332.0	1219.0	1264.0	8361.0
4941.0	4423.0	43723.0	1251.0	1263.0	8500.0
5173.0	4664.0	44101.0	1284.0	1246.0	8593.0
5058.0	4654.0	44492.0	1180.0	1216.0	8642.0
5385.0	4863.0	44961.0	1244.0	1191.0	8642.0
5118.0	4530.0	45396.0	1308.0	1317.0	8804.0
5275.0	4728.0	45814.0	1302.0	1273.0	8904.0
5159.0	4717.0	46228.0	1214.0	1246.0	8964.0
5372.0	4909.0	46712.0	1208.0	1154.0	8889.0
5117.0	4525.0	47143.0	1204.0	1140.0	8986.0
5149.0	4701.0	47532.0	1149.0	1105.0	9001.0
5226.0	4773.0	47935.0	1183.0	1195.0	9063.0
5450.0	5047.0	48400.0	1171.0	1160.0	8990.0
5128.0	4652.0	48806.0	1144.0	1132.0	9048.0
5462.0	4949.0	49238.0	1213.0	1201.0	9117.0
5421.0	4988.0	49685.0	1186.0	1246.0	9163.0
5863.0	5313.0	50255.0	1258.0	1273.0	9158.0

Quarterly data for the years 1955 to 1970 (*Cont.*)

Real Net Income	Real Consumption	Fixed Capital	Real Exports	Real Imports	Inventories
5567.0	4942.0	50775.0	1290.0	1354.0	9327.0
5763.0	5189.0	51254.0	1273.0	1361.0	9510.0
5691.0	5176.0	51782.0	1225.0	1393.0	9665.0
5990.0	5403.0	52383.0	1255.0	1343.0	9739.0
5753.0	5079.0	52965.0	1287.0	1352.0	9896.0
5995.0	5320.0	53529.0	1304.0	1341.0	10044.0
5909.0	5313.0	54137.0	1236.0	1293.0	10089.0
6102.0	5508.0	54764.0	1255.0	1243.0	10044.0
5701.0	5140.0	55344.0	1234.0	1266.0	10057.0
6056.0	5461.0	55882.0	1298.0	1284.0	10100.0
5955.0	5409.0	56440.0	1218.0	1304.0	10174.0
6167.0	5651.0	57022.0	1258.0	1258.0	10108.0
5763.0	5298.0	57474.0	1278.0	1266.0	10109.0
6324.0	5680.0	58035.0	1330.0	1324.0	10186.0
6189.0	5690.0	58646.0	1277.0	1389.0	10186.0
6657.0	5905.0	59340.0	1337.0	1365.0	10272.0
6303.0	5598.0	60043.0	1342.0	1462.0	10394.0
6672.0	5878.0	60738.0	1384.0	1484.0	10593.0
6550.0	5886.0	61474.0	1280.0	1498.0	10739.0
6925.0	6132.0	62283.0	1355.0	1463.0	10831.0
6563.0	5754.0	63075.0	1324.0	1403.0	10927.0
6804.0	6043.0	63790.0	1400.0	1461.0	11034.0
6737.0	6026.0	64508.0	1347.0	1484.0	11164.0
7042.0	6227.0	65320.0	1398.0	1410.0	11179.0

Quarterly data for the years 1955 to 1970 (*Cont.*)

Real Net Income	Real Consumption	Fixed Capital	Real Exports	Real Imports	Inventories
6746.0	5956.0	66105.0	1402.0	1462.0	11244.0
6924.0	6238.0	66789.0	1337.0	1432.0	11341.0
6855.0	6128.0	67532.0	1352.0	1486.0	11459.0
7163.0	6295.0	68336.0	1437.0	1302.0	11388.0
6841.0	6001.0	69159.0	1418.0	1444.0	11432.0
7077.0	6260.0	69928.0	1432.0	1485.0	11533.0
7026.0	6311.0	70699.0	1368.0	1454.0	11563.0
7264.0	6632.0	71522.0	1299.0	1472.0	11545.0
7059.0	6385.0	72411.0	1592.0	1706.0	11443.0
7149.0	6369.0	73162.0	1582.0	1664.0	11554.0
7246.0	6427.0	73955.0	1626.0	1706.0	11660.0
7698.0	6729.0	74878.0	1664.0	1655.0	11697.0
7076.0	6263.0	75708.0	1623.0	1715.0	11772.0
7367.0	6512.0	76407.0	1775.0	1718.0	11870.0
7383.0	6509.0	77167.0	1767.0	1709.0	11927.0
7791.0	6838.0	77989.0	1794.0	1699.0	11962.0
7156.0	6350.0	78745.0	1798.0	1710.0	11924.0
7557.0	6639.0	79497.0	1876.0	1852.0	12065.0
7534.0	6684.0	80261.0	1772.0	1796.0	12175.0
7999.0	6934.0	81125.0	1953.0	1834.0	12257.0

Quarterly data for the years 1955 to 1970

Employment	Volume of Money	Interest Rate	Price Level	Wage Rate
24.072	8913.0	0.0421	0.8486	113.742
24.190	9019.0	0.0434	0.8617	114.221
24.292	8945.0	0.0469	0.8810	116.335
24.332	9047.0	0.0456	0.8956	119.842
24.358	8682.0	0.0483	0.9059	122.834
24.449	8896.0	0.0491	0.9209	124.054
24.480	9027.0	0.0505	0.9337	125.735
24.486	9192.0	0.0507	0.9491	128.767
24.394	8867.0	0.0448	0.9347	131.180
24.477	9268.0	0.0510	0.9489	129.877
24.522	9274.0	0.0568	0.9734	132.697
24.451	9468.0	0.0559	0.9910	135.904
24.250	9212.0	0.0546	0.9890	139.134
24.208	9502.0	0.0527	1.0034	136.608
24.207	9544.0	0.0518	0.9979	137.935
24.129	9734.0	0.0532	1.0092	142.484
24.070	9525.0	0.0529	1.0109	143.706
24.230	9909.0	0.0525	1.0110	143.293
24.350	10124.0	0.0550	1.0058	144.435
24.404	10340.0	0.0539	1.0117	148.951

Quarterly data for the years 1955 to 1970 (*Cont.*)

Employment	Volume of Money	Interest Rate	Price Level	Wage Rate
24.507	10187.0	0.0569	1.0084	150.039
24.685	10289.0	0.0586	1.0196	151.509
24.801	10273.0	0.0582	1.0217	154.349
24.822	10444.0	0.0594	1.0326	156.756
24.824	10400.0	0.0617	1.0402	160.893
24.975	10675.0	0.0662	1.0477	163.363
25.048	10563.0	0.0676	1.0647	164.205
25.027	10870.0	0.0676	1.0660	166.780
25.044	10599.0	0.0638	1.0833	169.422
25.138	10799.0	0.0631	1.0926	170.141
25.164	10887.0	0.0571	1.1039	172.389
25.106	11155.0	0.0573	1.1029	174.540
24.831	11032.0	0.0594	1.1007	177.117
25.121	11273.0	0.0554	1.1084	178.337
25.220	11488.0	0.0560	1.1210	180.611
25.261	11788.0	0.0596	1.1195	184.870
25.290	11649.0	0.0601	1.1188	186.833
25.427	12264.0	0.0623	1.1350	190.506
25.497	12263.0	0.0620	1.1514	193.082
25.539	12333.0	0.0646	1.1539	198.716

Quarterly data for the years 1955 to 1970 (*Cont.*)

Employment	Volume of Money	Interest Rate	Price Level	Wage Rate
25.576	12537.0	0.0652	1.1637	199.249
25.688	12698.0	0.0679	1.1851	203.286
25.728	12796.0	0.0633	1.2115	206.623
25.753	13056.0	0.0660	1.2171	213.567
25.806	13172.0	0.0675	1.2138	215.299
25.866	13340.0	0.0701	1.2420	220.367
25.826	13252.0	0.0714	1.2646	221.134
25.573	13412.0	0.0680	1.2741	224.377
25.362	13430.0	0.0650	1.2773	223.326
25.482	13851.0	0.0687	1.2879	230.319
25.561	14082.0	0.0695	1.3057	232.111
25.386	14447.0	0.0714	1.3067	242.456
25.207	14546.0	0.0723	1.3113	243.900
25.283	14729.0	0.0778	1.3403	247.003
25.356	14813.0	0.0758	1.3619	249.526
25.306	15136.0	0.0808	1.3654	260.214
25.184	14897.0	0.0877	1.3824	259.688
25.284	15101.0	0.0939	1.4031	263.684
25.303	14882.0	0.0919	1.4258	266.094
25.202	15234.0	0.0894	1.4385	281.406
25.107	15192.0	0.0855	1.4557	283.586
25.082	15750.0	0.0942	1.4991	297.344
25.084	16014.0	0.0933	1.5408	304.217
24.998	16365.0	0.0968	1.5608	321.745

Appendix 2

Computer Programs

The computer programs used in this study form part of an inter-linked suite of programs developed by C. R. Wymer, for handling general econometric models but incorporating facilities for their application to continuous systems. The programs used were as follows:

Transf
Data manipulation program which includes options for quadratic interpolation, removal of seasonal variation and the transformation of data to eliminate the moving average inherent in the differential equation system.

Resimul
Calculates full information maximum likelihood estimates of general linear econometric models with any restrictions on the coefficients. The program contains special features to allow the approximate discrete model equivalent to a linear differential equation system to be estimated.

Discon
Calculates full information maximum likelihood estimates of the exact discrete model equivalent to the linearized differential equation system.

Continest
Calculates eigenvalues, eigenvectors and their standard errors, sensitivity matrices and the complementary function of a linear system.

Predic
Forecasting and simulation program for any linear model.

References

BERGSTROM, A. R. (1966*a*), 'Non-recursive models as discrete approximations to systems of stochastic differential equations', *Econometrica,* **34**, 173–82.

—— (1966*b*), 'Monetary phenomena and economic growth: A synthesis of neoclassical and Keynesian theories', *Economic Studies Quarterly,* **17**, 1–8.

—— (1967), *The Construction and use of Economic Models* (London, English Universities Press).

CODDINGTON, E. A. and N. LEVINSON (1955), *Theory of Ordinary Differential Equations* (New York, McGraw-Hill).

DEANE, P. and W. A. COLE (1969), *British Economic Growth 1688–1959*, 2nd edn. (Cambridge, Cambridge University Press).

DURBIN, J. (1963), 'Trend elimination for the purpose of estimating seasonal and periodic components in time series', in M. Rosenblatt (ed.), *Time Series Analysis* (New York, Wiley).

HICKMAN, B. G. (ed.) (1972), 'Econometric models of cyclical behaviour', in *NBER Studies in Income and Wealth*, no. 36 (Columbia University Press).

HICKS, J. R. (1950) *A Contribution to the Theory of the Trade Cycle* (Oxford, Oxford University Press).

PHILLIPS, P. C. B. (1972), 'The structural estimation of a stochastic differential equation system', *Econometrica*, **40**, 1021–42.

—— (1974), 'The estimation of some continuous time models', *Econometrica*, **42**, 803–24.

—— (1978), 'The treatment of flow data in the estimation of continuous time systems, in A. R. Bergstrom, A. J. L. Catt and M. Peston (eds.), *Stability and Inflation* (New York, Wiley).

SARGAN, J. D. (1974), 'Some discrete approximations to continuous time models', *Journal of the Royal Statistical Society*, Series B, **36**, 74–90.

WYMER, C. R. (1972), 'Econometric estimation of stochastic differential equation systems', *Econometrica*, **40** 565–78.

11

Monetary, Fiscal and Exchange Rate Policy in a Continuous-Time Econometric Model of the United Kingdom

It is only if we have complete confidence in our ability to control economic pressures, whatever their strength (and to control them at the precise time when they need to be controlled), that the sort of understanding given by Keynes is enough. Otherwise we must know something about the longer-run consequences of policies; we must have an assurance that policies adopted, to deal with a momentary emergency, will not set up, in the longer run, pressures which are greater than we can hope to withstand. (Hicks, *The Trade Cycle*)

Introduction

ALTHOUGH the above warning statement was written by Hicks (1950) 40 years ago, and may seem obvious today, I believe that it has been insufficiently heeded by economists and that some of the econometric models being used as a guide to policy can be very misleading with regard to the longer-run effects of these policies. Perhaps the main reason for this is that those structural parameters whose values influence mainly the long-run behaviour of the system are much more difficult to estimate than the parameters whose values have an important effect on its short-run behaviour; and, consequently, important long-run feedbacks are left out of some models. Another reason is that the long-run behaviour of most large econometric models can be studied only by simulation on a computer. Such simulations have shown that some of these models are incapable of generating plausible long-run behaviour except for special paths of the exogenous variables (see the articles contained in Hickman 1972). But, since they are not amenable to mathematical analysis, it is difficult to determine which structural features of the model are responsible for this.

The model that will be used in this paper (see Bergstrom and Wymer 1976 [ch. 10 of this volume]) was constructed primarily in order to demonstrate the feasibility of methods of estimating the structural parameters of continuous-time dynamic models (see Bergstrom 1976; and Gandolfo 1981). But, in addition to being a continuous-time model, it has other features which distinguish it from most earlier econometric models of complete economies. One is that it is a closely integrated system that makes intensive use of economic theory in order to obtain cross equation restrictions on the parameters. This has resulted in very low asymptotic standard errors for the parameter estimates, particularly for those parameters that have an important influence on the long-run behaviour of the system. Another feature is its amenability to mathematical analysis. In particular, the steady-state solution and the conditions for its asymptotic stability can be expressed explicitly in terms of the structural parameters. These features of the model are very advantageous for the purpose for which the model will be used in this paper; that is, for the study of the long-run effects of introducing various policy feedbacks.

The model does contain an estimated monetary feedback relation which show the response of the money supply (M_1) to variations in the balance of payments and is capable of ensuring external balance, in the steady state, with a fixed exchange rate. It was not possible to obtain reliable estimates of the independent influence on the money supply of other variables such as employment and output. But the effects of allowing employment or output (in addition to the balance of payments) to influence the money supply were studied in a subsequent article (Bergstrom 1978). In this paper we shall extend the results given in that article by comparing the effects of several alternative policy rules involving three distinct policy variables: (1) the money supply, (2) the exchange rate, and (3) a fiscal policy variable, which will be precisely defined later.

These policy variables are best regarded as target control variables rather than instruments over which the government and the Bank of England have direct control. An extended version of the Bergstrom–Wymer model was produced later by Knight and Wymer (1978) who introduced variables which would be more directly controllable. But, in this paper, we shall work with the original model, the fact that the policy variables are only indirectly controllable being recognized by allowing these variables to adjust only gradually, rather than instantaneously, in response to changes in output and the balance of payments. A detailed discussion of the extent to which the three policy variables can be independently controlled is beyond the scope of this paper. But it should be observed that the Public Sector Borrowing Requirement (PSBR) can be financed in three distinct ways: (1) by domestic credit creation, (2) by selling bonds on the domestic market, and (3) by foreign borrowing (which is partly a reflection of official intervention in the foreign exchange markets). The division of the financing of the PSBR between these three ways provides two degrees of freedom. A third degree of freedom is provided by variations in the size of the PSBR. There are, therefore, three degrees of freedom altogether, and these should permit some independent control over the three policy variables.

The following section will be devoted to a brief description of the Bergstrom–Wymer model and its steady-state and stability properties. A full discussion of the properties of the model, the method of estimating its parameters and its predictive performance is given in Bergstrom and Wymer (1976). The results obtained in the later article (Bergstrom 1978) for the effects of policy feedbacks from employment or output to the money supply will be reproduced in the third section of this paper. The new results which relate to the effects of feedbacks involving a fiscal policy variable and the exchange rate will be presented in the fourth and fifth sections and some conclusions drawn in the final section.

The Estimated Model

The model of Bergstrom and Wymer (1976) is a closed system of 13 first-order differential equations based on a prototype cyclical growth model discussed in Bergstrom (1967: ch. 5). It is highly aggregative being, formally, a model of a single sector economy whose product can be used for consumption (private and public), capital formation (private or public) or export. Estimates were obtained, from 1955 to 1966 quarterly data, by first linearizing the differential equations about the sample means and then estimating the equivalent exact discrete model taking full account of the complicated non-linear, cross equation restrictions on the latter model implied by the structure of the continuous-time model. The estimated equations (with t values in parentheses) are as follows:

$$\frac{DC}{C} = \underset{(5.71)}{0.2454} \log \left[\frac{\overset{(215.18)}{0.9206}\, Y}{C} \right] \tag{1}$$

$$\frac{DL}{L} = \underset{(4.08)}{0.0990} \log \left[\frac{\overset{(108.04)}{0.0037} \exp(-\overset{(17.29)}{0.0058}\, t)(Y^{-1} - \overset{(24.07)}{1.2180}\, K^{-1})^{-1}}{L} \right] \tag{2}$$

$$Dk = \underset{(3.36)}{0.0886} \left[\underset{(3.87)}{1.1492} \left\{ \underset{(24.07)}{1.12180} \left(\frac{Y}{K}\right)^{2} - r + \frac{Dp}{p} \right\} + \underset{(1.52)}{0.0042} - k \right] \tag{3}$$

$$\mathrm{D}Y = \underset{(5.25)}{0.6437}\,[(1 - \underset{(76.46)}{0.1632})(C + \mathrm{D}K + E) - Y] \tag{4}$$

$$\begin{aligned}\mathrm{D}I = {} & \underset{(3.51)}{0.3486}\,[\,\underset{(75.46)}{0.1632}\,(C + \mathrm{D}K + E) - I] \\ & + \underset{(7.72)}{0.0448}\,[\,\underset{(66.15)}{1.5263}\,(C + \mathrm{D}K + E) - S]\end{aligned} \tag{5}$$

$$\frac{\mathrm{D}E}{E} = \underset{(4.18)}{0.2926}\log\left[\frac{\overset{(62.11)}{1365}p^{-1}\exp(\overset{(13.29)}{0.0113}t)}{E}\right] \tag{6}$$

$$\frac{\mathrm{D}p}{p} = \underset{(3.40)}{0.1312}\log\left[\frac{\overset{(108.04)}{0.0037}\,w\exp(\overset{(17.29)}{-0.0058}\,t)[1 - \overset{(24.07)}{1.2180}\,Y/K)]^{-2}}{p}\right] \tag{7}$$

$$\frac{\mathrm{D}w}{w} = \underset{(2.85)}{0.1047}\log\left[\frac{\overset{(108.04)}{0.0037}\,w\exp(\overset{(17.29)}{-0.0058}\,t)[Y^{-1} - \overset{(24.07)}{1.2180}\,K^{-1})]^{-1}}{\overset{(34.57)}{22.1379}\exp(\overset{(0.85)}{0.0008}t)}\right] \tag{8}$$

$$\frac{\mathrm{D}r}{r} = \underset{(1.33)}{0.1990}\log\left[\frac{\overset{(0.62)}{0.0537}\,pYr^{\overset{(2.08)}{-0.8289}}}{M}\right] \tag{9}$$

$$\mathrm{D}m = \underset{(2.69)}{0.0422}\log\left(\frac{E}{I}\right) + \underset{(4.68)}{0.1562}\,D\log\left(\frac{E}{I}\right) \tag{10}$$

$$\mathrm{D}S = Y + I - C - \mathrm{D}K - E \tag{11}$$

$$\frac{\mathrm{D}K}{K} = k \tag{12}$$

$$\frac{\mathrm{D}M}{M} = m \tag{13}$$

where C = real consumption, Y = real net income or output, K = amount of fixed capital, E = real exports, I = real expenditure on imports, S = stocks, L = employment, M = volume of money, r = interest rate, p = price level, w = wage rate, k = proportional rate of increase in amount of fixed capital, m = proportional rate of increase in volume of money, t = time, and $\mathrm{D} = \mathrm{d}/\mathrm{d}t$.

Equation (1) is a form of consumption function in which consumption adjusts gradually, rather than instantaneously, to a change in income. Corresponding to any level of income Y there is a partial equilibrium level of consumption $0.9206Y$, and, at each point of time, the proportional rate of increase in consumption is an increasing function of the proportional excess of partial equilibrium consumption over actual consumption. The equation implies that the logarithm of consumption depends, with an exponentially distributed time lag, on the logarithm of income, the mean of the time lag distribution being the reciprocal of the speed of adjustment parameter 0.2454, or about 12 months. It should be noted that, since C includes both private and public consumption, the parameter 0.9206 reflects not only the private propensity to save but also the taxation rate and the government propensity to spend during the sample period. The equation implies, therefore, that fiscal policy was, in a certain sense, neutral over the sample period. This assumption will be relaxed in the fourth section.

It is necessary, before discussing the next few equations, to say something about the underlying assumptions about firm behaviour. It is assumed that the economy is made up of a large number of monopolistic competitors with identical production functions and uniformly differentiated products. In a position of static equilibrium each firm would be employing labour and capital in proportions which make the marginal technical rate of substitution equal to the ratio of the real wage rate to the real interest rate, setting the price of its product equal to marginal cost plus a margin depending on the own-price elasticity of the partial demand function and producing the amount demanded at that price. Since the prices charged by the different firms are equal the economy can be treated as a single product economy, and it tends to a perfectly competitive economy as the degree of differentiation between the products tends to zero, i.e. as the elasticity of the partial demand function for each firm's product tends to minus infinity. (When the model was estimated we were unable to reject this limiting case.)

In order to obtain the dynamic behaviour of the firms we should, ideally, introduce adjustment costs into the production function and derive factor demand and output adjustment equations simultaneously by solving a dynamic optimization problem. Instead we asssume that the firm follows simpler rules which implicitly recognize the fact that output can be adjusted more easily (in the short run) than the number of persons employed, because of the possibility of overtime working, and the number of persons employed can be adjusted more easily than the amount of capital. We assume, therefore, that output adjusts in response to sales, that the number of persons employed adjusts in response to output (with the stock of capital taken as given) and that the stock of capital adjusts in response to the difference between the marginal product of capital and the real interest rate. These rules together with the pricing rule are consistent with the static equilibrium condition described in the previous paragraph.

The employment equation (2) is of the same general form as (1). The partial equilibrium level of employment is related to output and the stock of fixed capital by a form of production function in which there is an assumed constant elasticity of substiution 0.5 between labour and capital and Harrod neutral technical progress. The capital intensity parameter 1.218 implies that the average elasticity of output with respect to capital over the sample period, was 0.13. The technical progress parameter 0.0058 implies that there is a decrease at the instantaneous rate 2.3 per cent per annum in the amount of labour required to produce a given output with a given amount of capital. The speed of adjustment parameter 0.0990 implies a mean time lag of about $2\frac{1}{2}$ years for the adjustment of employment to a change in output. The high degree of precision with which all of these parameters are estimated is a result of the fact that they occur in equations (2),(3), (7) and (8), and the method of estimation makes full use of all cross equation restrictions.

Equation (3) is a second-order differential equation in K and implies, therefore, that the stock of fixed capital depends with a rather complicated distributed time lag on output Y and the real interest rate $r - Dp/p$. Corresponding to any real interest rate and capital output ratio, there is a partial equilibrium proportional rate of increase in the stock of fixed capital, and, at each point of time, the actual rate of increase in capital is increasing at a rate proportional to the excess of the partial equilibrium rate over the actual rate. The partial equilibrium rate of increase in capital is a linear function of the excess of the marginal product of capital 1.2180 $(Y/K)^2$ over the real interest rate. The parameter 0.0042, which reflects entrepreneurs' growth expectations and willingness to take risks, implies that when the marginal product of capital equals the real interest rate the partial equilibrium rate of increase in capital is about 1.6 per cent per annum. The parameter 1.1492 implies that an increase in the real interest rate from, say, 4 per cent per annum to 5 per cent per annum will reduce the partial equilibrium rate of increase in capital from say 4.5 per cent per annum to 3.35 per cent per annum, or by 25 per cent. The parameter 0.0886 implies a mean time lag of about 3 years for the adjustment of investment to a change in the real interest rate or the marginal product of capital.

Equations (4) and (5) are the production and import adjustment equations and are closely related. The term $C + DK + E$ equals total sales for consumption, capital formation and export. The parameters 0.1632 and 1.5262 are the partial equilibrium ratios of imports and stocks respectively to sales. Equation (4) implies, therefore, that the increase in output depends on the excess demand for home-produced goods to meet current sales, while equation (5) implies that the rate of increase in imports depends on the excess demand for imports to meet current sales and excess stocks. The coefficient 0.6437 implies a mean time lag of $4\frac{1}{2}$ months for the adjustment of output to a change in sales, while the coefficient 0.3486 implies a mean time lag of 9 months for the adjustment of imports to a change in sales. It was not possible to obtain reliable estimates of the independent influence of excess stocks on the rate of increase in output (the estimated adjustment coefficient being insignificant with the wrong sign). But the parameter 0.0448 measuring the influence of excess stocks on imports is very significant. Since the variable I, like all other real variables in the model, is obtained by deflating the value at current prices by the implicit price deflator of the gross domestic product, equation (5) implies that the elasticity of substitution between imports and home-produced goods is -1.0 and not 0 as it would do if I represented the volume of imports.

Equation (6) is essentially a demand equation for exports. The partial equilibrium level of exports depends on the domestic price level with an assumed elasticity -1.0, and a trend term which allows for the gradual increase in income and prices in the rest of the world. The trend coefficient 0.0113 implies a growth rate of 4.5 per cent per annum. The coefficient 0.2926 implies a mean time lag of 10 months for the adjustment of exports to a change in the domestic price level (or a change in the exchange rate). The parameter 1365 reflects the exchange rate during the sample period and was adjusted to allow for the 1967 devaluation when testing the post-sample predictive performance of the model. It will be treated as a variable in the fifth section when we consider a flexible exchange rate policy.

In the price adjustment equation (7) the partial equilibrium price level equals marginal cost $w(\partial L/\partial Y)$ where the partial derivative $\partial L/\partial Y$ is obtained from the production function in equation (2). The mark-up for imperfect competition was restricted to zero since a preliminary estimate of this parameter was negative and insignificant. The coefficient 0.1312 implies a mean time lag of 2 years for the adjustment of the price level to a change in marginal cost.

In the wage rate adjustment equation (8) the proportional rate of increase in the money wage rate is an increasing function of the proportional excess demand for labour, where the demand for labour is identified with the partial equilibrium level of employment in equation (2) and the supply of labour is represented by a trend term. The trend coefficient implies a rate of growth of the labour supply of 0.3 per cent per annum, while the adjustment parameter 0.1047 implies that a 1 per cent increase in the demand for labour increases the proportional rate of increase in the wage rate by 0.4 per cent per annum (e.g. from 5 per cent per annum to 5.4 per cent per annum). Although the model provides a satisfactory explanation of variations in the inflation rate over the sample period 1955–66 and in the post-sample forecasts up to 1970, it could not, with this wage adjustment equation, explain the large fluctuations in the inflation rate that have occurred during the last decade. The wage adjustment equation should be modified to take account of expectations (possibly rational expectations of the sort that are associated with a unique constant inflation rate level of employment). For this purpose it should be a second-order differential equation. Such a modification would not affect the steady-state solution of the model, although it would affect its stability. But I do not think that it would affect the main qualitative conclusions of this paper.

The interest rate adjustment equation (9) is a dynamic representation of the Keynesian liquidity preference theory of interest rate determination. At each point of time the proportional rate of increase in the interest rate is an increasing function of the proportional excess demand for money, the real demand for money depending on income and the interest rate. The income elasticity of demand for money is assumed to be 1.0, and the estimated interest elasticity of demand for money is −0.8289. The mean time lag for the adjustment of the interest rate in response to a change in the money supply or income is about $1\frac{1}{2}$.

Equation (10) represents the influence of the balance of payments on the money supply, this influence being partly automatic and partly a reflection of policy. The rate of increase in the proportional rate of increase in the money supply depends on both the ratio of exports to imports and the proportional rate of change in this ratio. The equation implies that if, for example, the money supply were increasing at 1 per cent per quarter, exports being equal to imports, then a once for all 1 per cent increase in the ratio of exports to imports would cause the money supply to increase by approximately 1.04 per cent next quarter, 1.08 per cent the following quarter and so on. But if the ratio of exports to imports started to increase by 1 per cent per quarter, then the money supply would increase by 1.20 per cent next quarter, 1.44 per cent the following quarter and so on.

The remaining equations in the model are definitional. Although equations (3) and (10) are, essentially, second-order differential equations in K and M respectively, it is mathematically convenient to define k and m by (12) and (13) respectively in order to express the complete model as a system of 13 first-order differential equations.

The system has a steady-state solution in which each variable grows at a constant proportional rate (which is zero for some variables). Introducing the notation

$$\mathbf{x}' = [x_1, \ldots, x_{13}] = [C, L, k, Y, I, E, p, w, r, m, S, K, M]$$

this solution can be written

$$x_i(t) = x_i^* e^{\rho_i t} \qquad i = 1, \ldots, 13.$$

The steady-state levels $x_1^*, \ldots, x_{13}^*$ and growth rates $\rho_1, \ldots, \rho_{13}$ can all be expressed as explicit functions of the structural parameters of the model and these formulae are given in Bergstrom and Wymer (1976). The steady-state solution is, essentially, neo-classical; employment grows at the same rate as the labour supply, output and other real variables grow at a rate equal to the rate of technical progress plus the rate of growth of the labour supply, the price level increases at a rate equal to the rate of increase in the money supply minus the rate of growth of output, and the rate of interest is constant. The steady-state rate of increase in the money supply is so related to the rate of increase in prices and income in the rest of the world as to keep the balance of payments in equilibrium at a fixed exchange rate.

A full analysis of the influence of various structural parameters on the steady-state solution is given in Bergstrom and Wymer (1976). It is shown, for example, that the steady-state interest rate is higher the lower is propensity to save and the more capital intensive the production function and that it is unaffected by the liquidity preference parameters in equation (9). The explanation of the latter result is that, in the steady state, the money wage rate is so adjusted to other monetary phenomena in the system that their effect is neutralized and the interest rate is determined by real phenomena. Monetary phenomena do, nevertheless, play an important role in the mechanism generating the transient paths of the real variables. A Keynesian type mechanism involving the liquidity preference relation, the investment function and the wage and price adjustment relations ensures that, provided certain stability conditions are satisfied, output will oscillate about a path that approximately follows the neo-classical full employment path over a long period. The way in which this mechanism works is described in Bergstrom and Wymer (1976, pp. 218–20 this volume).

An important property of the model, which greatly facilitates the asymptotic stability analysis, is that the logarithms of the ratios of the variables to their steady-state values satisfy a system of non-linear differential equations which does not involve t, i.e.

$$Dy(t) = \mathbf{f}[y(t)] \qquad (14)$$

where

$$y(t) = [y_1(t), \ldots, y_{13}(t)]$$

$\mathbf{f}$ is a vector of functions whose exact form is given in Bergstrom and Wymer (1976: pp. 220–1 this volume) and

$$y_i(t) = \log\left[\frac{x_i(t)}{x_i^* e^{\rho_i t}}\right\} \qquad i = 1, \ldots, 13.$$

It then follows from Taylor's theorem and a theorem of Perron (see Coddington and Levinson 1955, p. 314) that if

$$\mathbf{A} = \left(\frac{\partial \mathbf{f}(y)}{\partial y'}\right)_{y=0} \tag{15}$$

then $y(t) \to 0$ as $t \to \infty$ provided that the eigenvalues of $\mathbf{A}$ have negative real parts and the elements of $y(0)$ are sufficiently small. This means that if the eigenvalues of $\mathbf{A}$ have negative real parts then the proportional deviations of the variables $C(t), \ldots, M(t)$ from their steady-state values will tend to zero provided that the initial proportional deviations from the steady state are sufficiently small. In this case we say that the system is asymptotically stable. It should be noted that the stability analysis would be much more complicated if the functions f_i in (14) included t as a separate argument. For then we should have to verify the second condition of Perron's theorem, i.e. that as $|y|$ (the norm of y) tends to zero $\{\mathbf{f}(y, t) - \mathbf{A}y\}/|y|$ tends to zero, not only for each value of t (as Taylor's theorem ensures), but also uniformly in t. This point is often overlooked in the economic literature.

The estimated eigenvalues of $\mathbf{A}$ and their t values are shown in table 1.

There are five real eigenvalues, all of which are negative, and four pairs of complex eigenvalues, including three pairs with negative real parts and one pair with a positive real part. The t value associated with the real part of the last pair of complex eigenvalues indicates that the hypothesis that the steady state is asymptotically stable must be rejected at the 1 per cent level of significance. But the model can, nevertheless, produce plausible

TABLE 1. Eigenvalues for linear approximation about the steady state

Real part	t value	Imaginary part	t value
−0.8605	6.65		
−0.2946	4.13		
−0.1707	2.51		
−0.0990	4.08		
−0.0061	7.79		
−0.1791	5.30	±0.1425	3.92
−0.1826	4.52	±0.0599	2.10
−0.0179	5.23	±0.0514	4.55
0.0782	6.16	±0.0742	3.59

long-run behaviour if it is modified to take account of the sort of non-linearity introduced by Hicks (1950) in his theory of the trade cycle.

The equation that requires modification is the investment equation (3). Here it has been assumed that the partial equilibrium proportional rate of increase in the amount of fixed capital is a linear function of the excess of the marginal product of capital over the real interest rate. But, since gross investment cannot be negative, it is realistic to assume that k is bounded from below by $-\delta$ where δ is the rate of depreciation. It is realistic to assume, therefore, that the partial equilibrium proportional rate of capital formation $\hat{k}$ is related to the excess of the marginal product of capital over the real interest rate by a non-linear relation

$$\hat{k} = g\left(\frac{\partial Y}{\partial K} - r + \frac{\mathrm{D}p}{p}\right)$$

where the function g is bounded from below by $-\delta$ and is approximated in the neighbourhood of the steady-state value of

$$\frac{\partial Y}{\partial K} - r + \frac{\mathrm{D}p}{p}$$

by

$$1.1492\left(\frac{\partial Y}{\partial K} - r + \frac{\mathrm{D}p}{p}\right) + 0.0042.$$

With this modification the model can be expected to produce a limit cycle if undisturbed. The period of the explosive cycle generated by the approximate linear model is about 21 years. But the constraint imposed by the non-linearity in the investment function is likely to reduce this period.

The fact that the real eigenvalues are negative is important, for it ensures that the system will produce oscillations about the steady state rather than convergence to a permanent state of depression or inflation. If there is a change in the rate of technical progress or the rate of increase in the labour supply there will be a change in the steady-state growth rate. But the internal mechanism of the model (see Bergstrom and Wymer 1976) will ensure that, over a long period, demand follows supply so that the path of output will converge to a limit cycle about the new steady state.

Alternative Monetary Policies

Equation (10) plays an essential role in the model, for it ensures that, over a long period, the change in the domestic price level is so related to changes in income and prices in the rest of the world as to preserve balance-of-payments equilibrium with a fixed exchange rate. In a more general version of the model discussed in Bergstrom and Wymer (1976) the money supply adjustment equation takes the form

$$\mathrm{D}m = \gamma_1 \log\left(\frac{E}{I}\right) + \gamma_2 \mathrm{D} \log\left(\frac{E}{I}\right) + \gamma_3 \log\left(\frac{L}{\beta e^{\lambda t}}\right) + \gamma_4 \mathrm{D} \log\left(\frac{L}{\beta e^{\lambda t}}\right) \tag{16}$$

where $\beta e^{\lambda t}$ can be interpreted as the level of employment aimed at by the policy makers and λ is assumed to be equal to the rate of increase in the labour supply. Although it seems that, over the sample period, the level of employment did have some influence on monetary policy, it was not possible to obtain reliable estimates of γ_3 and γ_4, and in the estimated model these parameters were restricted to zero.

It was shown in Bergstrom and Wymer 1976 that the introduction of the last two terms in (16) has no effect on the steady-state level of employment. Moreover, balance-of-payments equilibrium will not be attained in the steady state unless either $\gamma_3 = 0$ or the level of employment aimed at is that which necessarily occurs in the steady state. The inclusion of the last two terms in equation (16) can have a beneficial effect, therefore, only if these terms have a stabilizing influence.

Experience with the prototype model in Bergstrom (1967: ch. 6) suggests that a negative value of γ_3 is likely to be destablizing, whereas the effect on stability of a negative value of γ_4 is less certain, being fairly sensitive to the values of the other parameters of the model. The following analysis is confined, therefore, to the effect on stability of variations in γ_4 when $\gamma_3 = 0$ and all other parameters of the model have their estimated values. It is concerned, that is to say, with the stability of the system of equations obtained by replacing equation (10) in the estimated model by

$$\mathrm{D}m = 0.0422 \log\left(\frac{E}{I}\right) + 0.1562\, \mathrm{D} \log\left(\frac{E}{I}\right) + \gamma_4\, (\mathrm{D} \log L - 0.0008) \tag{17}$$

in which the last term is the excess of the rate of increase in employment over the rate of increase in the labour supply.

A negative value of γ_4 implies that the rate of increase in the proportional rate of increase in the money supply is greater the lower is the proportional rate of increase in employment. Suppose, for example, that $\gamma_4 = -1.0$ and that employment is initially constant while the money supply is growing at 1 per cent per quarter. Then, if employment starts to decrease by 1 per cent per quarter the money supply will be increased by approximately 2 per cent next quarter, 3 per cent the following quarter, and so on, provided that the ratio of exports to imports is constant.

Although a variation in γ_4 has no effect on the steady state of the system, it does affect the vector **f** of functions in equation (14) and hence the matrix **A** defined by (15). The effect of a variation in γ_4 on the stability of the system depends on its effect on the eigenvalues of **A**. The eigenvalues of **A** when $\gamma_4 = -1.0$ are shown in table 2. A comparison with table 1, which

TABLE 2. Eigenvalues for linear approximation about the steady state when $\gamma_4 = -1.0$

Real part	Imaginary part
−0.8608	0.0000
−0.2946	0.0000
−0.1914	0.0000
−0.1124	0.0000
−0.0061	0.0000
−0.1798	±0.1459
−0.1756	±0.0743
−0.0121	±0.0415
0.0833	±0.0858

shows the corresponding eigenvalues when $\gamma_4 = 0$, shows that the negative feedback from employment to the money supply makes the system slightly more unstable. The pair of complex eigenvalues that dominates the steady-state behaviour changes from $0.0782 \pm 0.0742i$ to $0.0833 \pm 0.0858i$ so that the system generates a more explosive cycle with a shorter period.

The eigenvalues were calculated for various other values of γ_4 in the range -0.1 to -10.0. But in each case there was one pair of complex eigenvalues with a positive real part and in no case was the positive real part less than when $\gamma_4 = 0$. The eigenvalues with positive real parts for various values of γ_4 are shown in table 3.

We conclude, therefore, that a negative feedback from employment to the money supply of the type considered has a destabilizing influence on the estimated model. This result is undoubtedly explained by the long time lags in the causal chain of relations through which this type of feedback influences the behaviour of the system. We recall that there are mean time lags of $1\frac{1}{2}$ years for the adjustment of the interest rate to the money supply,

TABLE 3. Eigenvalues with positive real parts for selected values of γ_4

γ_4	Real part	Imaginary part
−0.10	0.0786	±0.0755
−0.50	0.0806	±0.0804
−1.00	0.0833	±0.0858
−5.00	0.1025	±0.1129
−10.00	0.1195	±0.1317

3 years for the adjustment of investment to the interest rate, 12 months for the adjustment of consumption to income, $4\frac{1}{4}$ months for the adjustment of output to sales and $2\frac{1}{2}$ years for the adjustment of employment to output.

The last of these lags could be eliminated from the causal chain through which monetary policy works by using a feedback directly from the output to the money supply. We shall consider, therefore, the effect of replacing (17) by the relation

$$\mathrm{D}m = 0.0422 \log\left(\frac{E}{I}\right) + 0.1562\,\mathrm{D} \log\left(\frac{E}{I}\right) = \gamma_5(\mathrm{D} \log Y - 0.0066) \tag{18}$$

in which the last term is the excess of the proportional rate of growth of output over the steady-state growth rate. A negative value of γ_5 implies that the rate of increase in the proportional rate of increase in the money supply is greater the lower is the proportional rate of increase in output.

The eigenvalues of **A** when equation (18) replaces equation (10) in the system of equations (1) to (13) and $\gamma_5 = -1.0$ are shown in table 4. A comparison with table 1 shows that the negative feedback on output also makes the system slightly more unstable. The pair of complex eigenvalues that dominate the behaviour of the system changes from $0.0782 \pm 0.0742i$ to $0.0798 \pm 0.0981i$, so that it generates a slightly more explosive cycle with a shorter period as compared with the cycle generated when there is no direct feedback from either employment or output to the money supply.

The eigenvalues were calculated for various values of γ_5 in the range -0.1 to -10.0. Again there was, in each case, one pair of complex eigenvalues with a positive real part. The eigenvalues with positive real parts are shown in table 5. Only in the case where $\gamma_5 = -0.1$ is the positive

TABLE 4. Eigenvalues for linear approximation about the steady state when $\gamma_5 = -1.0$

Real part	Imaginary part
−0.8582	0.0000
−0.2946	0.0000
−0.1557	0.0000
−0.0990	0.0000
−0.0059	0.0000
−0.1825	±0.1395
−0.1980	±0.0611
−0.0094	±0.0446
0.0798	±0.0981

TABLE 5. Eigenvalues with positive real parts for selected values of γ_5

γ_5	Real part	Imaginary part
−0.10	0.0781	±0.0768
−0.50	0.0783	±0.0867
−1.00	0.0798	±0.0981
−5.00	0.1022	±0.1557
−10.00	0.1253	±0.1943

real part less than when $\gamma_5 = 0.0$. But, in this case, the effect of the feedback is negligible, the change in the positive real part being from 0.782 when $\gamma_5 = 0.0$ to 0.781 when $\gamma_5 = -0.1$. We conclude, therefore, that a negative feedback from output to the money supply of the type considered has a destabilizing influence when applied with sufficient force to have a significant effect on the behaviour of the system.

In the above analysis we have not considered the effect of policy feedbacks in which the second derivative of the money supply depends on the second- or higher-order derivatives of employment or output. Such policies could, perhaps, have a stabilizing influence. But, since they assume that the money supply can be adjusted instantaneously in response to variations in the level of employment or output, they are rather impracticable.

Combined Monetary and Fiscal Policy

In the previous section we implicitly assumed that fiscal policy was, in a certain sense, neutral and that variations in monetary policy were achieved through open market operations. In this section we shall introduce a fiscal policy feedback relation in which taxation rates and government expenditure vary in response to deviations of output from its steady-state path. Although there are no taxation and government expenditure variables in the estimated model they can be allowed for, in a mathematically convenient way, by introducing a fiscal policy variable T defined by

$$T = \frac{0.9206}{g + (0.9206)(1 - \tau)}$$

where g = partial equilibrium ratio of government current expenditure to national income (i.e. gY = partial equilibrium level of public consumption) and τ = taxation rate (i.e. $(1 - \tau)Y$ = real private disposable income).

We then replace equation (1) in the estimated model by

$$\frac{DC}{C} = 0.2454 \log \left[\frac{0.9206Y}{TC}\right] \tag{19}$$

which, using the definition of T, can be written in the form

$$\frac{DC}{C} = 0.2454 \log \left[\frac{gY + (0.9206)(1 - \tau)Y}{C}\right].$$

The expression $gY + (0.9206)(1 - \tau)Y$ can then be interpreted as the partial equilibrium level of aggregate consumption, which is the sum of gY, the partial equilibrium level of public consumption, and $(0.9206)(1 - \tau)Y$, the partial equilibrium level of private consumption. Neutral fiscal policy can be defined as a policy in which $T = 1$, in which case (19) reduces to the estimated equation (1).

We shall now consider the effect of introducing into the model the fiscal policy feedback relation

$$\frac{DT}{T} = \gamma\left[\beta \log\left(\frac{Y}{6.056 \exp(0.0066t)}\right) - \log T\right] \tag{20}$$

in which β and γ are policy parameters and the term $6.056 \exp(0.0066t)$ is the steady-state growth path of output implied by the estimated parameters of the model (the steady-state growth rate being 0.66 per cent per quarter or about 2.6 per cent per annum). Equation (20) implies that $\log T$ depends with an exponentially distributed time lag on the logarithm of the ratio of output to its steady-state level. It implies, for example, that if g is constant then the taxation rate will be converging to a level which is higher the greater is the proportional excess of output over its steady-state level, while if τ is constant then the government propensity to spend will be converging to a level which is lower the greater is the proportional excess of output over it steady-state level. The parameter β measures the strength of the feedback while γ is a speed of adjustment parameter whose reciprocal is the mean of the distributed time lag with which $\log T$ responds to a change in $\log[Y/6.056 \exp(0.0066t)]$. The distributed time lag allows for the time taken for the government to adjust taxation rates and its expenditure plans, and for the lag with which information on output becomes available.

We shall consider the behaviour of the solution of the system comprising the 14 equations (2), (3), . . ., (13), (19), (20) for various combinations of the fiscal policy parameters β and γ. In this system we have two policy feedbacks: (i) the feedback from the balance of payments to the money supply represented by equation (10), and (ii) the feedback from output to the fiscal variable represented by equation (20). We shall hold the parameters of the former relation at their estimated levels.

It is obvious from the forms of equations (19) and (20) that the steady-state solution of the system does not depend on the values of β and γ; in the steady state we have $T = 1$ and all other variables following the same paths as in the steady-state solution of the estimated system (1) to (13). The matrix of the linear approximation about the steady state is a 14×14 matrix whose leading 13×13 minor is identical with the matrix $\mathbf{A}$, defined by (15), for the estimated system. The eigenvalues of this 14×14 matrix were computed for the various combinations of values of β and γ shown in table 6. The values $\gamma = 1.0$ and $\gamma = 0.5$ imply mean time lags of 3 months and 6 months respectively for the adjustment of taxation rates and government expenditure in response to a change of output. In judging the plausibility (political feasibility) of the assumed values of β it should be noted that if $\beta > 1$ then an increase in output will cause the government to increase the taxation rate and (or) reduce its planned expenditure to such an extent that the partial equilibrium level of aggregate consumption will be less than it would have been if output has not increased. It seems unrealistic, therefore, to assume a value of β much greater than 1.

For each assumed combination of values of β and γ there is one, and only one, pair of complex eigenvalues with a positive real part, while all the real eigenvalues are negative. The values of the complex eigenvalues with positive real parts are shown in table 6. In each case the real part of this dominant pair of eigenvalues is less than 0.0782 (the real part of the dominant pair of eigenvalues in the estimated model), indicating that the policy does have a stabilizing effect. Moreover, the real part of the dominant pair of eigenvalues is smaller the greater is each of the parameters β and γ, indicating that, over the range of parameter values considered, the system will be more stable the greater is the strength of the policy feedback and the greater the speed of adjustment of the control variables. The complete set of eigenvalues for the pair of parameter values $\beta = 2.0$, $\gamma = 0.5$ is shown in table 7.

TABLE 6. Eigenvalues with positive real parts for selected values of β and γ

		Eigenvalues	
β	γ	Real part	Imaginary part
0.2	1.0	0.0723	± 0.0763
0.5	1.0	0.0637	± 0.0781
1.0	1.0	0.0506	± 0.0783
0.4	0.5	0.0674	± 0.0782
1.0	0.5	0.0518	± 0.0800
2.0	0.5	0.0313	± 0.0744

TABLE 7. Eigenvalues for $\beta = 2.0, \gamma = 0.5$

Real part	Imaginary part
−1.0660	0.0000
−0.2920	0.0000
−0.1741	0.0000
−0.1491	0.0000
−0.0990	0.0000
−0.0046	0.0000
−0.1907	±0.1179
−0.1905	±0.3098
−0.0291	±0.0565
0.0313	±0.0744

We may conclude from these results that, under the assumptions of the model, a fiscal policy feedback of the type represented by equation (20) has a mild stabilizing effect. But it should be noted that there is an important implicit assumption in equation (19) and the definition of T. This is that the response of consumption to a change in real disposable income resulting from a change in the taxation rate is the same as it is for a change in real disposable income resulting from a change in output. It should be noted also that the model takes no account of the effect of taxation on the labour supply (i.e. an incentive to work).

The fact that a feedback from output to the fiscal policy variable is stabilizing whereas a feedback from output to the money supply (when fiscal policy is neutral) is destabilizing is not very surprising in view of the greater speed with which the fiscal policy variable affects the real variables in the system. It is, perhaps, more surprising that the stabilizing effect of the fiscal policy is not stronger. But it should be remembered that a change in the fiscal policy variable T has an indirect effect through the interest rate. The initial increase in income induced by a fall in T will cause an increase in the interest rate (equation (9)) which will tend to decrease the rate of investment (equation (3)) and hence output (equation (4)). The fiscal policy is stabilizing only because this indirect effect, through the interest rate, is slower and more diffused over time than the direct income effect.

In this model, neither monetary nor fiscal policy can have any permanent effect on employment. The only permanent real effect of a one for all decrease in T is to divert resources from capital formation to consumption (private or public) and hence decrease the steady-state ratio of capital to output and the steady-state level of output. And the only permanent effect of a once for all increase in the money supply (offset by a corresponding change in the exchange rate) is to increase the nominal wage rate and the price level. The relative effects of monetary and fiscal policy in a prototype, closed economy, version of the model are fully discussed in Bergstrom (1967: ch. 6).

Exchange Rate Policy

We have assumed, so far, that the exchange rate is fixed (as it was over the sample period). Under this assumption the model generates plausible long-run behaviour only if the money supply is allowed to adjust in such a way as to preserve balance-of-payments equilibrium over a long period. This is the function of equation (10) in the estimated model. We shall now relax the assumption of a fixed exchange rate and introduce an exchange rate adjustment relation which will play the same role as equation (10) does in the estimated model. At the same time we shall assume that the money supply is increased at an arbitrary constant rate.

Our aim is to compare the effects of two alternative policies. Under the first policy the exchange rate is fixed and long-run balance-of-payments equilibrium is achieved through adjustments in the money supply, while under the second policy the rate of growth of the money supply is fixed and long-run balance-of-payments equilibrium is achieved through variations in the exchange rate. The real long-run effects of the two policies are identical. The only difference between their long-run effects is that under the former policy the long-run inflation rate is tied to that in the rest of the world while under the latter policy it is determined domestically. But the effects of the two policies on the stability of the system can be very different.

In order to introduce the exchange rate as a variable we replace equation (6) by

$$\frac{\mathrm{D}E}{E} = 0.2926 \log\left[\frac{1365(Qp)^{-1}\exp(0.0113t)}{E}\right] \tag{21}$$

where Q = index of price of sterling in terms of foreign currency with $Q = 1$ over the sample period. When $Q = 1$ equation (21) is identical to the estimated equation (6).

The assumption that the money supply grows at a constant rate is introduced by replacing equation (9) by

$$\frac{\mathrm{D}r}{r} = 0.1990 \log\left[\frac{0.0537p\,Y\,r^{-0.8289}}{M^{*}\mathrm{e}^{m^{*}t}}\right] \tag{22}$$

where M^* and m^* are constant, m^* being the assumed growth rate of the money supply. Equation (22) is obtained by substituting $M = M^{*}\mathrm{e}^{m^{*}t}$ into the estimated equation (9).

Finally we replace equations (10) and (13) by equations (23) and (24) respectively:

$$\mathrm{D}q = \gamma_1 \log\left(\frac{E}{I}\right) + \gamma_2 \,\mathrm{D} \log\left(\frac{E}{I}\right) \tag{23}$$

$$\frac{\mathrm{D}Q}{Q} = q. \tag{24}$$

Equation (23) is the exchange rate adjustment equation while equation (24) defines q. The policy parameters γ_1 and γ_2 can be interpreted similarly to the estimated parameters 0.0422 and 0.1562 in equation (10) except that the policy variable is now the exchange rate rather than the money supply.

We shall consider the behaviour of the system of equations obtained by replacing equations (6), (9), (10) and (13) in the estimated model by (21), (22), (23) and (24) respectively.[1] It is easy to see that this system has a steady state in which Q changes at a constant proportional rate depending on m^* and that the steady state does not depend on γ_1 or γ_2. It is of particular interest to consider the case in which M^* and m^* are chosen so that $M^* e^{m^* t}$ is identical with the steady-state path of the money supply for the estimated model. In this case we have $Q = 1$ in the steady state while the steady-state paths of all other variables are the same as they are for the estimated model. If m^* is increased by a constant c, then the steady-state rate of increase in the wage rate w, the price level p and the price of foreign currency $1/Q$, and the steady-state level of the interest rate r will all increase by c, while the steady-state paths of all real variables will be unaffected.

The eigenvalues of the matrix of linear approximations about the steady state were computed for the six combinations of the policy parameters γ_1 and γ_2 shown in table 8. In each case two of the real eigenvalues are positive and, except in one case, the real values of all the complex eigenvalues are negative. The positive eigenvalues and the complex eigenvalues with positive real parts are shown in table 8, and the complete set of eigenvalues for the case $\gamma_1 = 0.05$, $\gamma_2 = 1.0$ are shown in table 9. These results suggest that, from a stability point of view, the policy of increasing the money supply at a constant proportional rate and relying on adjustments of the exchange rate to preserve long-run balance-of-payments equilibrium is inferior to a policy of keeping the exchange rate fixed and relying on adjustments in the rate of growth of the money supply to preserve long-run balance-of-payments equilibrium. The fact that, in all cases, the dominant eigenvalue is real and positive implies that the economy can fall into a permanent state of depression under the former policy.

This difference between the effects of the two policies on the stability of

[1] Because of the implicit assumption that the price elasticity of demand for imports if -1.0, equation (5) is unchanged.

TABLE 8. Eigenvalues with positive real parts for selected values of γ_1 and γ_2

		Eigenvalues	
γ_1	γ_2	Real part	Imaginary part
0.05	0.10	0.1071	
		0.0113	
0.05	0.20	0.1088	
		0.0113	
0.05	0.50	0.1125	
		0.0113	
0.10	0.10	0.1128	
		0.0113	
		0.0064	±0.2621
0.10	0.20	0.1136	
		0.0113	
0.10	0.50	0.1155	
		0.0113	

TABLE 9. Eigenvalues for $\gamma_1 = 0.05$, $\gamma_2 = 0.10$

Real part	Imaginary part
−0.8608	0.0000
−0.3466	0.0000
−0.1786	0.0000
−0.0990	0.0000
−0.0061	0.0000
0.0113	0.0000
0.1071	0.0000
−0.1765	±0.0565
−0.1343	±0.1492
−0.0197	±0.1943

the system might, at first sight, seem surprising. But there is a simple intuitive explanation. It is known (see Bergstrom 1967: ch. 3) that the introduction of a foreign trade sector will normally have a stabilizing effect on simple models of the trade cycle. This is because of the damping effect provided by the appearance of a balance-of-payments deficit in the boom and a balance-of-payments surplus in the depression. Or, more technically, it is because of the damping effect on the multiplier-accelerator mechanism of replacing the domestic multiplier by the smaller foreign trade multiplier. Conversely, if these cyclical fluctuations in the balance of payments are reduced through exchange rate variations then the fluctuations in output become more explosive. A similar effect is evidently taking place in the more complicated model being discussed in this paper. The money supply adjustment relation (10) in the estimated model also has a mild destabilizing effect (as indicated by unpublished computations of the eigenvalues for various assumed values of the parameters). But, because the effects of a change in the money supply are much more diffused over time than the effects of a change in the exchange rate, the money supply adjustment relation (10) is less destabilizing than the exchange rate adjustment equation (23). It is sufficient to ensure that, over a long period, the domestic price level is so adjusted to conditions in the rest of the world as to preserve long-run balance-of-payments equilibrium. But it does not effect the economy with sufficient speed and concentration to reduce significantly the stabilizing effect of internally generated fluctuations in the balance of payments.

Conclusions

We have been concerned, in this paper, with the effects of introducing into an econometric model of the United Kingdom simple policy feedback rules which might be suggested intuitively, and which appear to represent approximately actual or declared government policies during certain periods. We have followed an approach inspired by the pioneer work of Phillips (1954) who was the first economist to respond seriously to Hicks's warning statement quoted at the beginning of the paper. A more modern and sophisticated approach would be to use optimal control theory. A linear feedback rule could be obtained by assuming a quadratic criterion function and solving the Ricatti equation (see Aoki 1976, p. 164), treating the linear approximation about the steady state as if it were the true model. But such a rule would comprise a set of feedback relations in which each policy variable would depend on all other variables in the model, and it would be difficult to interpret these relations intuitively or make any judgement about the likely effects of a particular misspecification in the model.

Subject to qualifications concerning the specification of the model, the results obtained suggest the use of a simple policy rule in which the rate of growth of the money supply is adjusted in response to changes in the balance of payments in such a way as to preserve long-run balance-of-payments equilibrium with a fixed exchange rate, and fiscal variables (taxation and government expenditure) are adjusted in response to changes in output in such a way as to dampen fluctuations in output about the steady-state path. But it is important that the adjustment of the fiscal variables should not be allowed to affect the application of the money supply adjustment rule, for the effects of changes in the money supply are distributed over such a long period, and in such a way, that a policy of accelerating the expansion of the money supply during depressions and decelerating during booms amplifies the fluctuations in output. If fiscal adjustments are not allowed to affect the money supply they do, of course, affect interest rates. But the model does take account of this effect, and it is, presumably, because of the negative feedback through the interest rate that the fiscal policy has only a mild stabilizing effect and, for the range of parameter values that seems feasible, is unable to prevent the occasional deep recession whose end is partly dependent on the non-linearities in the system.

The results suggest also that a policy in which the exchange rate is fixed and long-run external balance is achieved through adjustments in the rate of growth of the money supply is preferable, from a stability point of view, to one in which the rate of growth of the money supply is fixed and external balance is achieved through adjustments in the exchange rate. This is, evidently, because the effects of a change in the money supply are much more diffused over time than the effects of a change in the exchange rate and hence adjustments in the rate of growth of the money supply can preserve long-run external balance without greatly reducing the stabilizing effects on output of internally generated fluctuations in the balance of payments.

A final qualification is that we have, implicitly, treated income and prices in the rest of the world as exogenous trends. If all countries were to adopt the sort of monetary feedback rule incorporated in our model there would be nothing to stop the world inflation rate from drifting, gradually, upwards. This could be avoided by international agreement on bounds (say 0 and 10 per cent per annum) between which the rate of growth of the money supply of each country should be restricted. Each country would then adjust the rate of growth of its money supply in response to changes in its balance of payments subject to the condition that the rate of growth of its money supply is restricted to the agreed range. An investigation of the way in which the international economy would behave under such a policy would, of course, require a linking of models of the economies of the different countries.

References

Aoki, M. (1976), *Optimal Control and Systems Theory in Dynamic Economic Analysis* (Amsterdam, North-Holland).

BERGSTROM, A. R. (1967), *The Construction and Use of Economic Models* (English Universities Press, London).

—— (ed.) (1976), *Statistical Inference in Continuous Time Economic Models* (Amsterdam, North-Holland).

—— (1978), 'Monetary policy in a model of the United Kingdom', in A. R. Bergstrom, A. J. L. Catt, M. H. Peston and B. D. J. Silverstone (eds), *Stability and Inflation* (New York, Wiley), pp. 89–102.

—— and WYMER, C. R. (1976), 'A model of disequilibrium neoclassical growth and its application to the United Kingdom', in A. R. Bergstrom (ed.), *Statistical Inference in Continuous Time Economic Models* (Amsterdam, North-Holland), pp. 267–327.

CODDINGTON, E. A. and N. LEVINSON, (1955), *Ordinary Differential Equations* (McGraw Hill, New York).

GANDOLFO, G. (1981), *Qualitative Analysis and Econometric Estimation of Continuous Time Dynamic Models* (Amsterdam, North-Holland).

HICKMAN, B. G. (ed.) (1972), *Econometric Models of Cyclical Behaviour* (Columbia University Press).

HICKS, J. R. (1950), *The Trade Cycle* (Oxford, Oxford University Press).

KNIGHT, M. D. and C. R. WYMER, (1978), 'A macroeconomic model of the United Kingdom', *IMF Staff Papers*, **25**, 742–78.

PHILLIPS, A. W. (1954), 'Stabilization policy in a closed economy', *Economic Journal*, **64**, 290–323.

12

Gaussian Estimation of a Continuous Time Model of Demand for Consumer Durable Goods with Applications to Demand in the United Kingdom, 1973–84[1]

A. R. Bergstrom and M. J. Chambers

1. Introduction

THERE is now an extensive literature on the estimation, from aggregate time series data, of models of demand derived from the theory of consumer behaviour (see Deaton 1986). But, in spite of the increasing complexity of these models, the econometric results have been somewhat disappointing. Even the simplest implications of the theory, particularly the homogeneity condition, are frequently rejected.

The results obtained in several recent studies suggest that one of the main reasons for these rejections is the inadequate dynamic specification of the models, especially the neglect of the influence of lagged dependent variables and consumers' stocks. For example, Klevmarken (1979) found that a dynamic linear expenditure system (incorporating a habit persistance effect) was superior, with regard to both predictive performance and consistency with the theory, to more general models which neglected the influence of lagged dependent variables; while Deaton and Muellbauer (1980) found evidence of serial correlation in the residuals of equations for which the homogeneity condition was rejected. The latter authors suggested that the neglect of the influence of consumers' stocks is likely to

[1] This paper is, partly, based on research funded by the Economic and Social Research Council (ESRC) reference number: RB0023 2215.

be of particular importance.

A dynamic model of consumer demand, taking account of consumers' stocks, was developed in a series of pioneer contributions by Stone and Rowe (1957, 1958, 1960) (see also Chow 1957 and Nerlove 1960). Although the model of Stone and Rowe was formulated in discrete time, they were, clearly, thinking in terms of an underlying continuous time model, and a natural development of their study was to formulate a dynamic model of demand in continuous time and estimate its parameters from discrete data. This was accomplished a few years later by Houthakker and Taylor (1966) whose model is similar to that of Stone and Rowe, except that it is formulated in continuous time. In both studies consumers' stocks are eliminated from the system in order to obtain a dynamic model in terms of observable time series, with the rate of depreciation (physical and psychological) of consumers' stocks occurring as one of the parameters to be estimated.

In order to obtain estimates of the parameters of their continuous time model (which reduces to a stochastic differential equation in consumers' purchases), Houthakker and Taylor make use of an approximate discrete model, obtained by replacing integrals of the time paths of the variables, over the unit observation period, by trapezoidal approximations. In a contemporaneous but independent study, Bergstrom (1966) made the first investigation of the sampling properties of estimates obtained by using this type of approximation (assuming a complete system of stochastic differential equations rather than a single equation), and the method has, since, been extensively used in the estimation of continuous time macroeconometric models.

Houthakker and Taylor showed that the disturbance term in their approximate discrete model has a moving average representation and derived the relation between the moving average coefficient and the parameters of the continuous time model. But they did not attempt to take account of this in their estimation procedure. Instead, they used the 'three pass least squares' estimation procedure developed by Taylor and Wilson (1964) for the estimation of models with lagged dependent variables and autoregressive disturbances.

With modern computing facilities it would be quite easy to obtain exact maximum likelihood estimates of the approximate discrete model derived by Houthakker and Taylor, assuming that this is the true model and that the innovations are Gaussian. But it would be possible now to do even better than this. We could obtain estimates that would take account of the exact restrictions on the distribution of the discrete data implied by the continuous time model and would be exact maximum likelihood estimates of the parameters of the continuous time model if the innovations were Gaussian and the unobservable continuous time paths of the exogenous variables satisfied certain conditions. As was shown by Phillips (1972), the gain in efficiency from taking account of the exact restrictions on the distribution of the discrete data implied by a continuous time model, rather than using an approximate discrete model, can be very considerable (i.e. a more than 50 per cent reduction in the root-mean-square errors of the estimates).

The main purpose of this paper is to derive the exact discrete analogue (Theorems 1 and 2) of a continuous time model of demand for consumer durable goods and use this in the development and application of an algorithm for computing the Gaussian estimators of the parameters of the continuous time model. The algorithm yields exact maximum likelihood estimates when the innovations are Gaussian and the exogenous variables (which are assumed to be observable only as integrals) are polynomials in time of degree not exceeding two and can be expected to yield very good estimates under much more general conditions. The continuous time model that we shall consider is very similar to that of Houthakker and Taylor [1966], except that it has a more sophisticated and realistic dynamic specification. Indeed, the Houthakker–Taylor model can be regarded as a limiting case of our model as one of the speed of adjustment parameters tends to infinity.

In a series of recent articles Bergstrom (1983, 1985, 1986) (Chapters 4, 5, and 6 of this volume) has developed a method of obtaining exact Gaussian estimates of the parameters of a higher order continuous time dynamic model from discrete stock and flow data. But the results obtained in these articles are not directly applicable to the model considered in this paper. The reason for this is that consumers' stocks are not directly observable and must be eliminated from the system as in the studies of Stone and Rowe (1960) and Houthakker and Taylor (1966). But, after eliminating the stock variable, we obtain a continuous time model (a second order stochastic differential equation) in which the innovation is not white noise as was assumed in Bergstrom (1983, 1985, 1986). The general methods (as distinct from the precise formulae) of these articles will, nevertheless, be used in this paper.

In Section 2 we shall formulate the continuous time model and derive the discrete analogue (Theorems 1 and 2). The latter will be used in Section 3, where we shall derive the likelihood function and describe the estimation procedure, and in Section 4 where we shall discuss, very briefly, the asymptotic sampling properties of the estimates, derive formulae (Theorem 3) to be used in the computation of their estimated standard errors, and outline the procedure for obtaining post-sample forecasts of the discrete observations. In Section 5 we shall demonstrate the new methodology with applications to the demand for consumer durable goods in the United Kingdom during the period 1973–84.

2. The Continuous Time Model and its Discrete Analogue

The methods that will be developed in this paper could be applied to a complete system of dynamic demand equations of the stock adjustment type, taking account of across equation restrictions implied by the theory of consumer behaviour. But, in order to concentrate on the dynamic aspects of the model in the simplest context, we shall, following Stone and Rowe (1957, 1958, 1960) and Houthakker and Taylor (1966) deal specifically with a model of demand for a single good. We assume the model[2]

$$\mathrm{d}q(t) = \gamma\{\alpha z(t) + s(t) - q(t)\}\mathrm{d}t + \zeta(\mathrm{d}t) \quad (t \geq 0), \tag{1}$$

$$\mathrm{d}s(t) = \{q(t) - \delta s(t)\}\mathrm{d}t \quad (\mathrm{t} \geq 0), \tag{2}$$

where

t = the continuous time parameter,

$q(t)$ = rate of consumers' purchases,

$s(t)$ = consumers' stocks,

$z(t)$ = an $m \times 1$ vector of non random functions whose elements include real disposable income, the real price of the good, and, possibly, other exogenous variables,

[2] Equation (1) can be, formally, obtained from more fundamental assumptions by a two-stage optimization procedure. In the first stage we obtain the optimal steady state levels of stocks $s^* = az(t)$ and purchases $q^* = \delta az(t)$ corresponding to $z(t)$, the elements of the $1 \times m$ vector a being functions of the parameters of some utility function. It is implicitly assumed that consumers treat $z(t)$ as the conditional expectation of $z(t+r)$ for all $r > 0$, i.e. that $\{z(t), -\infty < t < \infty\}$ is treated as a martingale process. In the second stage we minimize the cost function.

$$C = \int_t^\infty \left[\{s(r) - s^*\}^2 + c_1\{q(r) - q^*\}^2 + c_2\left\{\frac{\mathrm{d}q(r)}{\mathrm{d}r}\right\}^2\right]\mathrm{d}r$$

subject to the constraint of equation (2). The optimal feedback, which minimizes C, is equation (1) with:

$$[\gamma\beta, -\gamma] = [0, -c_2^{-1}]P, \; \alpha = \left(\frac{\delta - \beta}{\delta}\right)a,$$

where P is the non-negative definite 2×2 matrix satisfying the algebraic Riccati equation:

$$\begin{bmatrix} 1 & 0 \\ 0 & c_1 \end{bmatrix} + P\begin{bmatrix} -\delta & 1 \\ 0 & 0 \end{bmatrix} + \begin{bmatrix} -\delta & 0 \\ 1 & 0 \end{bmatrix}P$$
$$-P\begin{bmatrix} 0 & 0 \\ 0 & c_2^{-1} \end{bmatrix}P = 0.$$

$\zeta(dt)$ = a white noise innovation,

$\alpha = 1 \times m$ vector of parameters

β, γ, δ = scalar parameters.

Although the parameter β will normally be negative, the model could be applied to a perishable good, and, in that case $s(t)$ could be interpreted as a psychological stock (habit) and β would be positive. When the model is applied to a durable good δ reflects both physical depreciation and the strength of consumers' preferences for new goods.

With regard to $\zeta(dt)$ we make the following assumption, which is much weaker than the assumption that the innovations are generated by Brownian motion.

ASSUMPTION 1: *ζ is a random measure defined on all subsets of the half line $0 < t < \infty$ with finite Lebesgue measure, such that $E[\zeta(t)] = 0$, $E[\zeta(dt)]^2 = \sigma^2 dt$ and $E[\zeta(\Delta_1)\zeta(\Delta_2)] = 0$ for any disjoint sets Δ_1 and Δ_2 on the half line $0 < t < \infty$.* (See Bergstrom 1984, p. 1157 for a discussion of random measures and their application to continuous time stochastic models.)

We shall interpret (1) and (2) as meaning that $q(t)$ and $s(t)$ satisfy the stochastic integral equations

$$q(t) - q(0) = \gamma\int_0^t [\alpha z(r) + \beta s(r) - q(r)]dr + \int_0^t \zeta(dr) \qquad (t > 0), \tag{3}$$

$$s(t) - s(0) = \int_0^t [q(r) - \delta s(r)]dr \qquad (t > 0) \tag{4}$$

where $\int_0^t \zeta(dr) = \zeta[0, t]$ and the other integrals on the right-hand sides of (3) and (4) are defined in the wide sense (see Bergstrom 1984, p. 1152).

We shall also make the following assumption.

ASSUMPTION 2: q(0) *and* s(0) *are non-random.*

For reasons discussed in Bergstrom 1985, the assumption of a fixed initial state vector is usually more realistic, in econometric work, than the assumption of stationarity.

Before deriving a discrete analogue of the above model in terms of observable time series, it is of interest to compare it with the model used by Houthakker and Taylor (1966). Instead of (1) they assumed

$$q(t) = \alpha z(t) + \beta s(t) + u(t)$$

where $u(t)$ is a random disturbance function. Although they did not define $u(t)$ precisely it would be consistent with their subsequent analysis if this equation were written in the more precise form

$$q(t)\mathrm{d}t = [\alpha z(t) + \beta s(t)]\mathrm{d}t + \zeta(\mathrm{d}t), \tag{1'}$$

where $\zeta(\mathrm{d}t)$ is defined by Assumption 1 and equation (1′) is interpreted as meaning that

$$\int_0^t q(r)\mathrm{d}r = \int_0^t [\alpha z(r) + \beta s(r)]\mathrm{d}r + \int_0^t \zeta(\mathrm{d}r) \qquad (t > 0).$$

It can be regarded, therefore, as a limiting case of (1) as $\gamma \to \infty$.

The complete model of Houthakker and Taylor comprises equations (1′) and (2). Although this is more realistic than a model which ignores the influence of stocks, the dynamic specification is likely to be too simple for econometric purposes. It implies, for example, that if the system is initially in the steady state and there is a sudden increase in real disposable income (one of the elements of $z(t)$), then purchases will instantaneously jump to a maximum and then (assuming $\zeta = 0$) decline monotonically to a new steady state level as consumers' stocks are built up.

Casual observation suggests that the process of adjustment is more complicated than this. Shopping for new durable goods is a time-consuming activity. Most people will spend some time shopping around, comparing different brands, arranging finance, and so on before making purchases. In many cases purchases will rise gradually to a maximum some time after the sudden increase in income and then decline gradually, converging (possibly with oscillations) to a new steady state level as stocks are built up. In other cases purchases will approach the new steady state monotonically from below with the rate of increase in $q(t)$ rising to a maximum some time after the increase in real income and then declining as the new steady state is approached. Both of these processes of adjustment are allowed for by our more sophisticated model comprising equations (1) and (2).

The characteristic equation of the latter system is

$$x^2 + (\gamma + \delta)x + \gamma(\delta - \beta) = 0. \tag{5}$$

In the normal case (i.e. where $\beta < 0$) the roots of (5) have negative real parts and the system is stable.

By eliminating $s(t)$ from equations (1) and (2) we, formally, obtain the second order differential equation

$$\mathrm{d}[Dq(t)] = [-(\gamma + \delta)Dq(t) + \gamma(\beta - \delta)q(t) \\ + \gamma\alpha(D + \delta)z(t)]\mathrm{d}t + (D + \delta)\zeta(\mathrm{d}t), \tag{6}$$

where $D = \mathrm{d}/\mathrm{dt}$. But, since the two-dimensional random process $[q(t), s(t)]$ generated by the system (1) and (2) is not differentiable, even in the mean square sense, the operator D in equation (6) cannot be interpreted as the mean square differential operator. Moreover, the innovation $(D + \delta)\zeta(\mathrm{d}t)$ is not white noise but a more complicated improper random process. For these reasons equation (6) cannot be treated as a special case of the second order system.

$$d[Dx(t)] = \{A_1(\theta)Dx(t) + A_2(\theta)x(t) + Bz(t)\}dt + \zeta(dt) \qquad (t > 0) \quad (7)$$

dealt with in Bergstrom 1986. The system (7), in which D is the mean square differential operator, can be rigorously interpreted as meaning that $x(t)$ satisfies the stochastic integral equation

$$Dx(t) - Dx(0) = \int_0^t [A_1 Dx(r) + A_2 x(r) + Bz(r)]dr + \int_0^t \zeta(dr) \qquad (t > 0).$$

But a rigorous interpretation of (6) could be given only in terms of double integrals.

We shall, in fact, make no further use of equation (6). Our next aim is to derive a second order stochastic difference equation in terms of observable variables. This can, most easily, be done by obtaining, directly from equations (1) and (2), a system of first order stochastic difference equations in purchases and stocks, and then eliminating the unobservable stock variable to obtain a second order stochastic difference equation in purchases. For this purpose we define the discrete time series $\{q_t, s_t, z_t; t = 1, 2, \ldots\}$ by

$$\begin{bmatrix} q_t \\ s_t \\ z_t \end{bmatrix} = \begin{bmatrix} \int_{t-1}^t q(r)dr \\ \int_{t-1}^t s(r)dr \\ \int_{t-1}^t z(r)dr \end{bmatrix} \qquad (t = 1, 2, \ldots) \qquad (8)$$

We assume that the series $\{q_t\}$ and $\{z_t\}$ are observable while $s(t)$ is unobservable.

It is, of course, impossible to obtain an exact stochastic model relating to the series $\{q_t\}$ and $\{z_t\}$ without making some assumption about the unobservable continuous time path of $z(t)$. In Theorem 1 we shall, following Phillips 1974, 1976 and Bergstrom 1986, obtain a discrete stochastic model which holds exactly over any period in which the elements of $z(t)$ are polynomials in t of degree not exceeding two. But the practical importance of the theorem depends on the fact that the model is a very good approximation under much more general assumptions, which allow $z(t)$ to follow very different patterns of behaviour over different parts of the sample period.

THEOREM 1: *Let* [q(t), s(t)] *be the solution of the system of equations* (1) *and* (2) *on the interval* [0, T], *and assume that, over a subinterval* [t′ − 3, t′]

of [0, T], *the elements of* z(t) *are polynomials in* t *of degree not exceeding two. Then, under Assumptions* 1 *and* 2, *the random variables* $q_{t'}$, $q_{t'-1}$ *and* $q_{t'-2}$, *and non-random vectors* $z_{t'}$, $z_{t'-1}$ *and* $z_{t'-2}$ *defined by* (8) *satisfy the equation*

$$q_{t'} = k_1 q_{t'-1} + k_2 q_{t'-2} + l_0 z_{t'} + l_1 z_{t'-1}, + l_2 z_{t'-2} + \eta_{t'} \tag{9}$$

where

$$\eta_{t'} = \int_{t'-1}^{t'} \phi_1(t'-r)\zeta(dr) + \int_{t'-2}^{t'-1} \phi_2(t'-1-r)\zeta(dr) + \int_{t'-3}^{t'-2} \phi_3(t'-2-r)\zeta(dr), \tag{10}$$

and the constants k_1, k_2, l_0, l_1, l_2 *and functions* ϕ_1, ϕ_2, *and* ϕ_3 *are obtained from the parameters of the continuous time model through the following sequence of formulae.*

$$A = \begin{bmatrix} -\gamma & \gamma\beta \\ 1 & -\delta \end{bmatrix}$$

$$\begin{bmatrix} f_{qq} & f_{qs} \\ f_{sq} & f_{ss} \end{bmatrix} = F = e^A = \sum_{r=0}^{\infty} \frac{1}{r!} A^r,$$

$$k_1 = f_{qq} + f_{ss}$$

$$k_2 = f_{qs} f_{sq} - f_{ss} f_{qq}$$

$$B = \begin{bmatrix} \gamma\alpha \\ 0 \end{bmatrix} = \begin{bmatrix} \gamma\alpha_1 & \gamma\alpha_2 & \cdots & \gamma\alpha_m \\ 0 & 0 & \cdots & 0 \end{bmatrix},$$

$$Q = A^{-1}B,$$

$$L_{00} = (I + \tfrac{3}{2} A^{-1} + A^{-2})Q,$$

$$L_{01} = -2(A^{-1} + A^{-2})Q,$$

$$L_{02} = (\tfrac{1}{2} A^{-1} + A^{-2})Q,$$

$$L_{10} = (\tfrac{1}{2} A^{-1} + A^{-2})Q,$$

$$L_{11} = (I - 2A^{-2})Q,$$

$$L_{12} = (-\tfrac{1}{2} A^{-1} + A^{-2})Q,$$

$$L_{20} = (-\tfrac{1}{2} A^{-1} + A^{-2})Q,$$

$$L_{21} = 2(A^{-1} - A^{-2})Q,$$

$$L_{22} = (I - \tfrac{3}{2} A^{-1} + A^{-2})Q,$$

$$\begin{bmatrix} E_{00}^{q} \\ E_{00}^{s} \end{bmatrix} = E_{00} = FL_{10} - L_{00},$$

$$\begin{bmatrix} E^q_{01} \\ E^s_{01} \end{bmatrix} = E_{01} = FL_{11} - L_{01},$$

$$\begin{bmatrix} E^q_{02} \\ E^s_{02} \end{bmatrix} = E_{02} = FL_{12} - L_{02},$$

$$\begin{bmatrix} E^q_{10} \\ E^s_{10} \end{bmatrix} = E_{10} = FL_{20} - L_{10},$$

$$\begin{bmatrix} E^q_{11} \\ E^s_{11} \end{bmatrix} = E_{11} = FL_{21} - L_{11},$$

$$\begin{bmatrix} E^q_{12} \\ E^s_{12} \end{bmatrix} = E_{12} = FL_{22} - L_{12},$$

$$l_0 = E^q_{00} - f_{ss}E^q_{10} + f_{qs}E^s_{10},$$

$$l_1 = E^q_{01} - f_{ss}E^q_{11} + f_{qs}E^s_{11},$$

$$l_2 = E^q_{02} - f_{ss}E^q_{12} + f_{qs}E^s_{12},$$

$$\begin{bmatrix} f_{qq}(r) & f_{qs}(r) \\ f_{sq}(r) & f_{ss}(r) \end{bmatrix} = F(r) = \varepsilon^{rA},$$

$$\begin{bmatrix} [A^{-1}]_{qq} & [A^{-1}]_{qs} \\ [A^{-1}]_{sq} & [A^{-1}]_{ss} \end{bmatrix} = A^{-1},$$

$$\phi_1(r) = [A^{-1}]_{qq}[f_{qq}(r) - 1] + [A^{-1}]_{qs}f_{sq}(r)$$

$$\begin{aligned} \phi_2(r) = {} & [A^{-1}_{qq}]\,[f_{qq} - f_{qq}(r)] + [A^{-1}]_{qs}[f_{sq} - f_{sq}(r)] \\ & + \{f_{qs}[A^{-1}]_{sq} - f_{ss}[A^{-1}]_{qq}\}[f_{qq}(r) - 1] \\ & + \{f_{qs}[A^{-1}]_{ss} - f_{ss}[A^{-1}]_{qs}\}f_{sq}(r), \end{aligned}$$

$$\begin{aligned} \phi_3(r) = {} & \{f_{qs}[A^{-1}]_{sq} - f_{ss}[A^{-1}]_{qq}\}[f_{qq} - f_{qq}(r)] \\ & + \{f_{qs}[A^{-1}]_{ss} - f_{ss}[A^{-1}]_{qs}\}[f_{sq} - f_{sq}(r)] \end{aligned}$$

PROOF: The system of equations (1) and (2) can be written as

$$dx(t) = [Ax(t) + Bz(t)]dt = \bar{\zeta}(dt), \tag{11}$$

where

$$x(t) = \begin{bmatrix} q(t) \\ s(t) \end{bmatrix}, \bar{\zeta}(dt) = \begin{bmatrix} \zeta(dt) \\ 0 \end{bmatrix}.$$

Let

$$z(t) = a + b(t' - t) + c(t' - t)^2 \qquad (t' - 3 \leqslant t \leqslant t'), \tag{12}$$

where a, b, and c are constants.
Then, from (8) and Bergstrom (1986), equation (33), we have

$$a = \tfrac{11}{6} z_{t'} - \tfrac{7}{6} z_{t'-1} + \tfrac{1}{3} z_{t'-2},$$
$$b = -2z_{t'} + 3z_{t'-1} - z_{t'-2},$$
$$c = \tfrac{1}{2} z_{t'} - z_{t'-1} + \tfrac{1}{2} z_{t'-2}.$$

Now let u(t) be defined by

$$u(t) = x(t) + Qa - A^{-1}Qb + 2A^{-2}Qc \qquad (14)$$
$$+ (Q\mathrm{b} - 2A^{-1}Q\mathrm{c})(t' - t) + Qc(t' - t)^2$$
$$(t' - 3 \leq t \leq t').$$

Then, using (11) and (12), we obtain

$$\mathrm{d}u(t) = Au(t)\mathrm{d}t + \zeta(\mathrm{d}t) \qquad (t' - 3 \leqslant t \leqslant t'), \qquad (15)$$

which is interpreted as meaning that

$$u(t) - u(t' - 3) = A\int_{t'-3}^{t} u(r)\mathrm{d}r + \int_{t'-3}^{t} \zeta(\mathrm{d}r) \qquad (t' - 3 \leqslant t \leqslant t').$$

It follows from (15) and Bergstrom 1984, Theorem 8 that, if u_t is defined by

$$u_t = \int_{t-1}^{t} u(r)\mathrm{d}r,$$

then

$$u_{t'} = Fu_{t'-1} + \xi_{t'}, \qquad (16)$$
$$u_{t'-1} = Fu_{t'-2} + \xi_{t'-1} \qquad (17)$$

where

$$\xi_t = \int_{t-1}^{t} A^{-1}\{F(t - r) - I\}\zeta(\mathrm{d}r) \qquad (18)$$
$$+ \int_{t-2}^{t-1} A^{-1}\{F - F(t - 1 - r)\}\zeta(\mathrm{d}r).$$

By integrating (14) over the intervals $(t' - 1,\ t')$, $(t' - 2,\ t' - 1)$, and $(t' - 3,\ t' - 2)$ and using (13) we obtain equations (19), (20), and (21) respectively.

$$u_{t'} = x_{t'} + L_{00}z_{t'} + L_{01}z_{t'-1} + L_{02}z_{t'-2}, \qquad (19)$$
$$u_{t'-1} = x_{t'-1} + L_{10}z_{t'} + L_{11}z_{t'-1} + L_{12}z_{t'-2}, \qquad (20)$$
$$u_{t'-2} = x_{t'-2} + L_{20}z_{t'} + L_{21}z_{t'-1} + L_{22}z_{t'-2}, \qquad (21)$$

Then, from (16), (19), and (20) we obtain

$$x_{t'} = Fx_{t'-1} + E_{00}z_{t'} + E_{01}z_{t'-1} + E_{02}z_{t'-2} + \xi_{t'}, \qquad (22)$$

and from (17), (20), and (21)

$$x_{t'-1} = Fx_{t'-2} + E_{10}z_{t'} + E_{11}z_{t'-1} + E_{12}z_{t'-2} + \xi_{t'-1}. \tag{23}$$

Partitioning (22) we obtain

$$q_{t'} = f_{qq}q_{t'-1} + f_{qs}s_{t'-1} + E^q_{00}z_{t'} + E^q_{01}z_{t'-1} + E^q_{02}z_{t'-2} + \xi^q_{t'}, \tag{24}$$

$$s_{t'} = f_{sq}q_{t'-1} + f_{ss}s_{t'-1} + E^s_{00}z_{t'} + E^s_{01}z_{t'-1} + E^s_{02}z_{t'-2} + \xi^s_{t'}, \tag{25}$$

and partitioning (23)

$$q_{t'-1} = f_{qq}q_{t'-2} + f_{qs}s_{t'-2} + E^q_{10}z_{t'} + E^q_{11}z_{t'-1} + E^q_{12}z_{t'-2} + \xi^q_{t'-1}, \tag{26}$$

$$s_{t'-1} = f_{sq}q_{t'-2} + f_{ss}s_{t'-2} + E^s_{10}z_{t'} + E^s_{11}z_{t'-1} + E^s_{12}z_{t'-2} + \xi^s_{t'-1}. \tag{27}$$

Equation (9) is obtained by solving (24) and (26) for $s_{t'-1}$ and $s_{t'-2}$ respectively and substituting into (27).

End of Proof

It is important to notice that the coefficients k_1, k_2, l_0, l_1, and l_2 in (9) and the functions ϕ_1, ϕ_2, and ϕ_3 in (10) are all independent of the coefficients a, b, and c in (12). Thus, although the discrete model represented by equations (9) and (10) holds exactly over any part of the sample period in which the exogenous variables are quadratic functions of time, its parameters depend only on those of the continuous time model.

For this reason the discrete model is a good approximation under very general circumstances in which the exogenous variables are not quadratic functions of time and their behaviour varies widely as between different parts of the sample period. The error under these more general circumstances depends not on the accuracy with which the time path of each exogenous variable can be approximated by a single quadratic function of time over the whole interval $[0, T]$, but only on the accuracy within which it can be approximated by a series of quadratic functions over the overlapping intervals $[0, 3]$, $[1, 4]$, $[2, 5]$, . . ., $[T-3, T]$].

Indeed, it can be shown by an extension of the argument of Phillips [1976, p. 139] that, if $z(t)$ is a vector of thrice continuously differentiable functions of t, then the errors in (9) will be of the order h^4 as the unit observation period h tends to zero.

In order to derive the likelihood function for the structural parameter vector $[\alpha, \beta, \gamma, \delta]$ we shall need to supplement (9) and (10) by equations relating q_1 and q_2 to the initial state vector $[q(0), s(0)]$, which is treated as fixed by Assumption 2. It might be thought that, since $[q(0), s(0)]$ is unobservable, there is nothing to be gained from this and that equations (9) and (10), assuming that they hold for $t = 3, \ldots, T$, provide enough information for the formation of the likelihood function conditional on the observed values of q_1 and q_2. But this is not so. Since ε_3 and ε_4 are not independent of q_1 and q_2 the conditional distribution of $q_3, q_4, \ldots, q_T$ for given values of q_1 and q_2 depends, in an essential way, on $[q(0), s(0)]$, and the latter must be treated as a supplementary parameter vector to be estimated jointly with the structural parameter vector $[\alpha, \beta, \gamma, \delta]$ (see the comments in Bergstrom 1986, p. 371). The supplementary equations are given by the following theorem.

THEOREM 2: *Let* $[q(t), s(t)]$ *be the solution of the system of equations* (1) *and* (2) *on the interval* $[0, T]$, *and assume that, over the subinterval* $[0, 3]$, *the elements of* $z(t)$ *are polynomials in* t *of degree not exceeding two. Then, under Assumptions* 1 *and* 2, *the random variables* q_1 *and* q_2, *and non-random vectors* z_1, z_2, *and* z_3 *defined by* (8) *satisfy the equations*

$$q_1 = h_{11}q(0) + h_{12}s(0) + l_{11}z_2 + l_{12}z_3 + \eta_1, \tag{28}$$

$$q_2 = f_{qq}q_1 + h_{21}q(0) + h_{22}s(0) + l_{21}z_1 + l_{22}z_2 + l_{23}z_3 + \eta_2, \tag{29}$$

$$\eta_1 = \int_0^1 \phi_1(1-r)\,\zeta(\mathrm{d}r), \tag{30}$$

$$\eta_2 = \int_0^1 \psi(1-r)\,\zeta(\mathrm{d}r) + \int_1^2 \phi(2-r)\,\zeta(\mathrm{d}r), \tag{31}$$

the constants f_{qq} *and* f_{qs}, *and function* ϕ_1 *are defined as in Theorem* 1 *and the constants* $h_{11}, h_{12}, h_{21}, h_{22}, l_{11}, l_{12}, l_{13}, l_{21}, l_{22}$, *and* l_{23}, *and function* ψ *are obtained through the following sequence of formulae in which* $A, Q, F, E_{10}, E_{11}, E_{12}, L_{20}, L_{21}$, and L_{22} *are defined in Theorem* 1.

$$\begin{bmatrix} g_{qq} & g_{qs} \\ g_{sq} & g_{ss} \end{bmatrix} = G = A^{-1}[F - I),$$

$$h_{11} = g_{qq},$$

$$h_{12} = g_{qs},$$

$$h_{21} = f_{qs}g_{sq},$$

$$h_{22} = f_{qs}g_{ss},$$

$$N_1 = (\tfrac{11}{6}I - 2A^{-1} + A^{-2})Q,$$

$$N_2 = (-\tfrac{7}{6}I + 3A^{-1} - 2A^{-2})Q,$$

$$N_3 = (\tfrac{1}{3}I - A^{-1} + A^{-2})Q,$$

$$\begin{bmatrix} E_1^q \\ E_1^s \end{bmatrix} = E_1 = GN_1 - L_{22},$$

$$\begin{bmatrix} E_2^q \\ E_2^s \end{bmatrix} = E_2 = GN_2 - L_{21},$$

$$\begin{bmatrix} E_3^q \\ E_3^s \end{bmatrix} = E_3 = GN_3 - L_{20},$$

$$l_{11} = E_1^q$$

$$l_{12} = E_2^q$$

$$l_{13} = E_3^q$$

$$l_{21} = E_{12}^q + f_{qs}E_1^s$$

$$l_{22} = E_{11}^q + f_{qs}E_2^s$$

$$l_{23} = E_{10}^q + f_{qs}E_3^s$$

$$\psi(\mathrm{r}) = f_{qs}\{[A^{-1}]_{sq}[f_{qq}(r) - 1] + [A^{-1}]_{ss}f_{sq}(r)\} + [A^{-1}]_{qq}[f_{qq} - f_{qq}(r)] + [A^{-1}]_{qs}[f_{sq} - f_{sq}(r)]$$

PROOF: Let

$$z(t) = a + b(3 - t) + c(3 - \mathrm{t})^2 \qquad (0 \leq t \leq 3). \tag{32}$$

Then, putting $t' = 3$ in (13), we have

$$\begin{aligned} a &= \tfrac{11}{6}z_3 - \tfrac{7}{6}z_2 + \tfrac{1}{3}z_1, \\ b &= -2z_3 + 3z_2 - z_1, \\ c &= \tfrac{1}{2}z_3 - z_2 + \tfrac{1}{2}z_1. \end{aligned} \tag{33}$$

Now let $u(t)$ be defined by

$$u(t) = x(t) + Qa - A^{-1}Qb + 2A^{-2}Qc + (Qb - 2A^{-1}Qc)(3 - t) + Qc(3 - t)^2 \qquad (0 \leq t \leq 3). \tag{34}$$

Then, using (11) and (32), we obtain

$$du(t) = Au(t)\mathrm{d}t + \zeta(\mathrm{d}t) \qquad (0 \leq t \leq 3), \tag{35}$$

which is interpreted as meaning that

$$u(t) - u(0) = A\int_0^t u(r)\mathrm{d}r + \int_0^t \zeta(\mathrm{d}r) \qquad (0 \leq t \leq 3). \tag{36}$$

Putting $t = 1$ in (36) we have

$$u(1) - u(0) = Au_1 + \int_0^1 \zeta(\mathrm{d}r),$$

and hence

$$u_1 = A^{-1}\{u(1) - u(0)\} - A^{-1}\int_0^1 \zeta(\mathrm{d}r). \tag{37}$$

But, from the solution of (35) we obtain (see Bergstrom 1984, Theorem 3)

$$u(1) = Fu(0) + \int_0^1 F(1-r)\zeta(dr), \tag{38}$$

and, from (37) and (38)

$$u_1 = A^{-1}(F-I)u(0) + A^{-1}\int_0^1 \{F(1-r)-I\}\zeta(dr). \tag{39}$$

Puting $t' = 3$ in (21) we have

$$u_1 = x_1 + L_{20}z_3 + L_{21}z_2 + L_{22}z_1, \tag{40}$$

and from (33) and (34)

$$u(0) = x(0) + N_1 z_1 + N_2 z_2 + N_3 z_3. \tag{41}$$

Then, from (39), (40), and (41) we obtain

$$x_1 = Gx(0) + E_1 z_1 + E_2 z_2 + E_3 z_3 + \xi_1 \tag{42}$$

where

$$\xi_1 = A^{-1}\int_0^1 \{F(1-\mathrm{r})-I\}\zeta(\mathrm{dr}).$$

Partitioning (42) we obtain

$$q_1 = g_{qq}q(0) + g_{qs}s(0) + E_1^q z_1 + E_2^q z_2 + E_3^q z_3 + \xi_1^q, \tag{43}$$

$$s_1 = g_{sq}q(0) + g_{ss}s(0) + E_1^s z_1 + E_2^s z_2 + E_3^s z_3 + \xi_1^s, \tag{44}$$

and, putting $t' = 3$ in (26),

$$q_2 = f_{qq}q_1 + f_{qs}s_1 + E_{10}^q z_3 + E_{11}^q z_2 + E_{12}^q z_1 + \xi_2^q. \tag{45}$$

Equation (28) is obtained from (43), while equation (29) is obtained by substituting from (44) into (45).

End of Proof

3. The Likelihood Function and the Estimation Procedure

The discrete model comprising equations (9), (10), (28), (29), (30), and (31) provides the basis for an estimation procedure following the general method developed in Bergstrom 1985, 1986 and assuming a sample $q_1, q_2, \ldots, q_T, z_1, z_2, \ldots, z_T$. The procedure yields exact maximum likelihood estimates of the parameters of the continuous time model when the innovations are Gaussian and the exogenous variables are polynomials in time of degree not exceeding two, and it can be expected to yield very good estimates under much more general conditions. It is also highly efficient computationally since it completely avoids the computation and inversion of the $T \times T$ covariance matrix of the vector $[q_1, \ldots, q_T]$.

We start by deriving the exact Gaussian likelihood function under the assumption that the elements of $z(t)$ are polynomials in t of degree not exceeding two over the whole interval $[0, T]$. Let the 2×1 vector y and $T \times 1$ vectors, q, η, and c be defined by

$$y' = [q(0), s(0)],$$

$$q' = [q_1, q_2, \ldots, q_T],$$

$$\eta' = [\eta_1, \eta_2, \ldots, \eta_T],$$

$$c = [c_1, c_2, \ldots, c_T].$$

where

$$c_1 = l_{11}z_1 + l_{12}z_2 + l_{13}z_3, \tag{46}$$

$$c_2 = l_{21}z_1 + l_{22}z_2 + l_{23}z_3, \tag{47}$$

$$c_t = l_0 z_t + l_1 z_{t-1} + l_2 z_{t-2} \qquad (t = 3, \ldots, T), \tag{48}$$

and let the $T \times T$ matrix K and the $T \times 2$ matrix H be defined by

$$K = \begin{bmatrix} 1 & 0 & 0 & 0 & \cdot & \cdot & \cdot & 0 \\ -f_{qq} & 1 & 0 & 0 & \cdot & \cdot & \cdot & 0 \\ -k_2 & -k_1 & 1 & 0 & \cdot & \cdot & \cdot & 0 \\ 0 & -k_2 & -k_1 & 1 & \cdot & \cdot & \cdot & 0 \\ \cdot & \cdot & \cdot & \cdot & \cdot & \cdot & \cdot & \cdot \\ \cdot & \cdot & \cdot & \cdot & \cdot & \cdot & \cdot & \cdot \\ \cdot & \cdot & \cdot & \cdot & \cdot & \cdot & \cdot & \cdot \\ 0 & 0 & 0 & 0 & \cdot & -k_2 & -k_1 & 1 \end{bmatrix} \tag{49}$$

$$H = \begin{bmatrix} h_{11} & h_{12} \\ h_{21} & h_{22} \\ 0 & 0 \\ \cdot & \cdot \\ \cdot & \cdot \\ 0 & 0 \end{bmatrix} \tag{50}$$

Then the system of equations (9), (28), and (29) can be written

$$Kq - Hy - c = \eta. \tag{51}$$

Moreover, we have

$$E(\eta\eta') = \Omega$$

where $\Omega =$

$$\begin{bmatrix} \omega_{11} & \omega_{12} & \omega_2 & 0 & 0 & 0 & . & . & . & . & 0 \\ \omega_{21} & \omega_{22} & \omega_{23} & \omega_2 & 0 & 0 & . & . & . & . & 0 \\ \omega_2 & \omega_{32} & \omega_0 & \omega_1 & \omega_2 & 0 & . & . & . & . & 0 \\ 0 & \omega_2 & \omega_1 & \omega_0 & \omega_1 & \omega_2 & . & . & . & . & 0 \\ . & . & . & . & . & . & . & . & . & . & . \\ . & . & . & . & . & . & . & . & . & . & . \\ . & . & . & . & . & . & . & \omega_0 & \omega_1 & \omega_2 & 0 \\ . & . & . & . & . & . & . & \omega_1 & \omega_0 & \omega_1 & \omega_2 \\ . & . & . & . & . & . & . & \omega_2 & \omega_1 & \omega_0 & \omega_1 \\ 0 & 0 & 0 & 0 & 0 & 0 & 0 & 0 & \omega_2 & \omega_1 & \omega_0 \end{bmatrix} \tag{52}$$

$$\omega_0 = \sigma^2 \int_0^1 \{\phi_1^2(r) + \phi_2^2(r) + \phi_3^2(r)\}\mathrm{d}r,$$

$$\omega_1 = \sigma^2 \int_0^1 \{\phi_1(r)\phi_2(r) + \phi_2(r)\phi_3(r)\}\mathrm{d}r,$$

$$\omega_2 = \sigma^2 \int_0^1 \{\phi_1(r)\phi_3(r)\}\mathrm{d}r,$$

$$\omega_{11} = \sigma^2 \int_0^1 \phi_1^2(r)\mathrm{d}r,$$

$$\omega_{22} = \sigma^2 \int_0^1 \{\phi_1^2(r) + \psi^2(r)\}\mathrm{d}r,$$

$$\omega_{12} = \omega_{21} = \sigma^2 \int_0^1 \{\phi_1(r)\psi(r)\mathrm{d}r,$$

$$\omega_{23} = \omega_{32} = \sigma^2 \int_0^1 \{\phi_1(r)\phi_2(r) + \psi(r)\phi_3(r)\}\mathrm{d}r.$$

Let θ denote the $1 \times (m+3)$ vector of structural parameters of the continuous time model, excluding the variance σ^2 of the innovations, i.e.

$$\theta = [\alpha, \beta, \gamma, \delta].$$

Then the complete vector of parameters to be estimated, including the initial state vector, is the $1 \times (m+6)$ vector $[\theta, \sigma^2, y']$. Letting $L(\theta, \sigma^2, y')$ denote minus twice the logarithm of the Gaussian likelihood function, and noting that $|K| = 1$, we have

$$\begin{aligned} L(\theta, \sigma^2, y') &= \log|\Omega(\theta, \sigma^2)| + \eta\Omega^{-1}(\theta, \sigma^2)\eta \\ &= \log|\Omega(\theta, \sigma^2)| + \{K(\theta)q - H(\theta)y - c(\theta)\}'\Omega^{-1} \\ &\quad (\theta, \sigma^2)\{K(\theta)q - H(\theta)y - c(\theta)\}. \end{aligned} \tag{53}$$

As in Bergstrom 1985, 1986 an even simpler expression of L, which is very convenient for computational purposes, can be obtained by using the Cholesky factorization

$$\Omega = MM' \tag{43}$$

where M is a real lower triangular matrix with positive elements along the diagonal. From (53) and (54) we obtain

$$L = \sum_{i=1}^{T} (\varepsilon_i^2 + 2 \log m_{ii}), \tag{55}$$

where m_{ii} is the ith diagonal element of M and $\varepsilon = [\varepsilon_1, \varepsilon_2, \ldots, \varepsilon_T]$ is a vector whose elements can be computed recursively from

$$M\varepsilon = \eta. \tag{56}$$

The steps in computing the value of the likelihood functions for a given parameter vector $[\theta, \sigma^2, y']$ are then as follows.

(i) Compute K, H, c, and Ω using the formulae given in Theorems 1 and 2, equations (46) to (50), and equation (52).
(ii) Compute η from (51)
(iii) Compute the elements of M recursively from (54).
(iv) Compute the elements of ε recursively from (56).
(v) Compute L from (55).

Since M has only $3(T-1)$ non-zero elements, the total number of multiplications in the above algorithm is only of order T.

Now let $[\hat{\theta}, \hat{\sigma}^2, \hat{y}']$ be the Gaussian estimator of $[\theta, \sigma^2, y']$, i.e. the value of this vector that maximizes L. Then, as in Bergstrom 1985, equation (37) we have

$$\hat{y} = [H'(\hat{\theta})\Omega^{-1}(\hat{\theta}, \hat{\sigma}^2)H(\hat{\theta})]^{-1} H'(\hat{\theta})\Omega^{-1}(\hat{\theta}, \hat{\sigma}^2)\{K(\hat{\theta})q - c(\hat{\theta})\}. \tag{57}$$

The formula (57) can be used in an interative procedure (see Bergstrom 1985, p. 382) for computing $[\hat{\theta}, \hat{\sigma}^2, \hat{y}']$. In this procedure we alternately maximize L with respect to $[\theta, \sigma^2]$, for given y, using some optimization algorithm involving sucessive evaluations of L, and then use formula (57) to obtain a new estimate of y.

4. Asymptotic Sampling Properties and the Forecasting Procedure

The asymptotic sampling properties of Gaussian estimators of the structural parameters of a higher order continuous time dynamic model are discussed, briefly, in Bergstrom 1983, Section 7. But that article is concerned with a closed model, and the theorems relating to the asymptotic sampling properties of the estimators (Bergstrom 1983, Theorems 4 and 5) rely on the assumption of strict stationarity and the general theorems proved by Dunsmuir and Hannan (1976) and Dunsmuir (1979). A full and rigorous treatment of the asymptotic sampling properties of estimates of the parameters of open higher order continuous time models obtained by the methods of Bergstrom (1986) and Section 3 of this paper would require another long paper along the lines of Phillips (1976) who dealt with a first order system in which all variables are measured at equi-spaced points of time.

The difficulty of deriving the asymptotic sampling properties of estimators of the type developed in Bergstrom 1986 and Section 3 of this paper is that the discrete model comprising equations (9), (28), and (29), with the residual covariance matrix satisfying (52), does not hold exactly; although we shall refer to it as the 'exact discrete model' since it would be exact if all the exogenous variables were quadratic functions of time. For this reason the estimators will not be consistent, even under strong regularity conditions, but will have an asymptotic bias depending on the smoothness of the unobservable continuous time paths of the exogenous variables (i.e. the asymptotic bias will be smaller the smoother, in some sense, are the paths of the exogenous variables). Under suitable regularity conditions (similar to those assumed by Phillips (1976)) $\sqrt{T}$ times the deviations of the estimators from their probability limits will have a limiting normal distribution. But the covariance matrix of this limiting distribution will depend on the continuous time paths of the exogenous variables.

For the practical purpose of deriving approximate formulae for the asymptotic variances of the estimators, in terms of the observations and the parameter values, we shall proceed as if the 'exact discrete model' were really exact. When the exogenous variables are as smooth as those that will be used in Section 5 of this paper the error resulting from this assumption is likely to be very small (probably smaller than that resulting from errors in the data). If the 'exact discrete model' holds exactly, the parameter values are such that the system is stable, the innovations are independently and identically distributed, and the discrete observations of the exogenous variables satisfy suitable regularity conditions (for example, the conditions assumed by Malinvaud 1980, p. 535), then the Gaussian estimator of the parameter vector $[\theta, \sigma^2]$ is $\sqrt{T}$-consistent and asymptotically normal. If, in addition, the innovations are Gaussian, then the asymptotic covariance matrix of the estimator is $2[EL_{ij}]^{-1}$ where $[EL_{ij}]$ is the expectation of the Hessian of L (the objective function defined by equation (53)), evaluated at the true values of the parameters. If the innovations are not Gaussian then, because the coefficients of the 'exact discrete model' and the covariance matrix of its innovations both depend on the parameter vector θ, the asymptotic covariance matrix of the Gaussian estimator of $[\theta, \sigma^2]$ will have a more complicated form than that given above and will depend on the higher moments of the innovations. (See Robinson 1988 for a general discussion of this problem.)

Formulae expressing the elements of the matrix $[EL_{ij}]$ in terms of the structural parameters of the model, the initial state vector, and the observations of the exogenous variables are given by the following theorem.

THEOREM 3: *If the vector* $q' = [q_1, q_2 \ldots, q_T]$ *is generated by the system of equations* (28), (29), *and* (9) *in which the vector* $\eta = [\eta_1, \eta_2, \ldots, \eta_T]$ *has the mean vector zero and covariance matrix* Ω *defined by equation* (52), *then the elements of the expected Hessian of the function* L *defined by equation* (53) *are given by equation* (58) *to* (60) *in which* m *is defined by equation* (61) V *is defined by equation* (62) *and* W *is the lower triangular matrix satisfying equation* (67).

$$E\left[\frac{\partial^2 L}{\partial\theta_i \partial\theta_j}\right] = trW\left[\frac{\partial}{\partial\theta_i}V\right]W'W\left[\frac{\partial}{\partial\theta_j}V\right]W' \tag{58}$$

$$+ 2\left[\frac{\partial}{\partial\theta_j}m\right]' W'W\left[\frac{\partial}{\partial\theta_i}m\right] \qquad (i, j = 1, \ldots, m+3)$$

$$E\left[\frac{\partial^2 L}{\partial\theta_i \partial\sigma^2}\right] = trW\left[\frac{\partial}{\partial\theta_i}V\right]W'W\left[\frac{\partial}{\partial\sigma^2}V\right]W' \qquad (i = 1, \ldots, m+3) \tag{59}$$

$$E\left[\frac{\partial^2 L}{(\partial\sigma^2)^2}\right] = trW\left[\frac{\partial}{\partial\sigma^2}V\right]W'W\left[\frac{\partial}{\partial\sigma^2}V\right]W'. \tag{60}$$

PROOF: Let $m(\theta)$ and $V(\theta, \sigma^2)$ denote the mean and covariance-matrix of q. Then from equation (51) we obtain:

$$m(\theta) = K^{-1}[Hy + c], \tag{61}$$

$$V(\theta, \sigma^2) = K^{-1}\Omega K'^{-1}. \tag{62}$$

Moreover,

$$L(\theta, \sigma^2) = \log|V(\theta, \sigma^2)| + tr[V^{-1}(\theta, \sigma^2)]\{q - m(\theta)\}\{q - m(\theta)\}', \tag{63}$$

the expressions on the right-hand sides of equations (53) and (63) being identical as can easily be seen using equations (51), (61), and (62), together with the equation $|K| = 1$. Differentiating equation (63) with respect of θ_i we obtain

$$\frac{\partial L}{\partial \theta_i} = tr\, V^{-1}\left[\frac{\partial}{\partial \theta_i} V\right][I - V^{-1}(q - m)(q - m)'] \qquad (64)$$
$$-2\, tr\, V^{-1}\left[\frac{\partial}{\partial \theta_i} m\right](q - m)' \qquad (i = 1, \ldots, m + 3).$$

Then differentiating equation (64) with respect of θ_j we obtain

$$\frac{\partial^2 L}{\partial \theta_i \partial \theta_j} = -\, tr\, V^{-1}\left[\frac{\partial}{\partial \theta_j} V\right] V^{-1}\left[\frac{\partial}{\partial \theta_i} V\right][I - V^{-1}(q - m)(q - m)'] \quad (65)$$
$$+\, tr\, V^{-1}\left[\frac{\partial^2}{\partial \theta_i \partial \theta_j} V\right][I - V^{-1}(q - m)(q - m)']$$
$$+\, tr\, V^{-1}\left[\frac{\partial}{\partial \theta_i} V\right] V^{-1}\left[\frac{\partial}{\partial \theta_j} V\right] V^{-1}(q - m)(q - m)'$$
$$+\, 2tr\, V^{-1}\left[\frac{\partial}{\partial \theta_i} V\right] V^{-1}\left[\frac{\partial}{\partial \theta_j} m\right](q - m)'$$
$$+\, 2tr\, V^{-1}\left[\frac{\partial}{\partial \theta_j} V\right] V^{-1}\left[\frac{\partial}{\partial \theta_i} m\right](q - m)'$$
$$-2tr\, V^{-1}\left[\frac{\partial^2}{\partial \theta_i \partial \theta_j} m\right](q - m)'$$
$$+\, 2tr\, V^{-1}\left[\frac{\partial}{\partial \theta_i} m\right]\left[\frac{\partial}{\partial \theta_j} m\right]' \qquad (i, j = 1, \ldots, m + 3).$$

Since $E(q - m) = 0$ and $E(q - m)(q - m)' = V$, we obtain from equation (65)

$$E\left[\frac{\partial^2 L}{\partial \theta_i \partial \theta_j}\right] = tr\, V^{-1}\left[\frac{\partial}{\partial \theta_i} V\right] V^{-1}\left[\frac{\partial}{\partial \theta_j} V\right] \qquad (66)$$
$$+\, tr\, V^{-1}\left[\frac{\partial}{\partial \theta_i} m\right]\left[\frac{\partial}{\partial \theta_j} m\right]' \qquad (i, j = 1, \ldots, m + 3).$$

Now let W be the lower triangular $T \times T$ matrix satisfying

$$MW = K. \qquad (67)$$

Then from equations (54) and (62) we obtain

$$V^{-1} = W'W. \qquad (68)$$

From equations (66) and (68) we obtain equation (58), and in a similar way we obtain equations (59) and (60).

End of Proof

REMARK: The formulae given by equations (58) to (60) are very convenient for computation since they avoid the inversion of the $T \times T$ matrix V. The lower triangular matrix W is comparitively easy to compute. The matrices K and M will, already, have been computed as part of the estimation procedure, and the elements of W can be computed recursively from equation (67), taking advantage of the fact that M is a lower triangular matrix with only $3(T - 1)$ non-zero elements.

We conclude this section with a brief description of the procedure for obtaining optimal forecasts of the post-sample discrete observations. The procedure is essentially the same as that described in Bergstrom 1989 for obtaining forecasts of discrete stock and flow data generated by a higher order continuous time system of the type dealt with in Bergstrom 1986. We first extend the estimate of the matrix Ω to a $(T + 2) \times (T + 2)$ matrix $\hat{\Omega}^*$ which is the estimated covariance matrix of the vector $[\eta^*]' = [\eta_1, \eta_2, \ldots, \eta_{T+2}]$. This requires no additional computation since the estimates $\hat{\omega}_0$, $\hat{\omega}_1$, and $\hat{\omega}_2$ will have been computed as part of the estimation procedure. Next we complete the Cholesky factorization

$$\hat{\Omega}^* = \hat{\mathrm{M}}^*[\hat{\mathrm{M}}^*]'$$

where $\hat{\mathrm{M}}^*$ is a $(T + 2) \times (T + 2)$ lower triangular matrix, all except the last two rows of which will have been computed as part of the estimation procedure. Forecasts $\hat{\eta}_{T+1}$ and $\hat{\eta}_{T+2}$ of the first two post-sample residuals of the 'exact discrete model' can then be computed from the estimated extension of the system (56) obtained by putting M equal to $\hat{M}^*$ and z equal to a $(T + 2) \times 1$ vector whose last two elements are zero and whose remaining elements have the values obtained in the estimation procedure. Forecasts $\hat{q}_{t+1}, \hat{q}_{T+2}, \ldots, \hat{q}_{T+p}$ conditional on any assumed values of the post-sample observations of the exogenous variables can then be computed recursively from equation (9) using the forecasts $\hat{\eta}_{T+1}$, and $\hat{\eta}_{T+2}$ obtained by the above procedure and putting $\eta_{T+3}, \eta_{T+4}, \ldots, \eta_{T+p}$ equal to zero.

It has been shown (Bergstrom 1989, Theorem 2) that forecasts obtained by the above procedure are optimal in the sense that they are exact maximum likelihood estimates of the conditional expectations of the post-sample observations conditional on all the information in the sample when the innovations are Gaussian and the exogenous variables are polynomials in time of degree not exceeding two. The procedure is also highly efficient computationally when used in conjunction with the estimation procedure described in Section 3.

5. Applications to Consumer Demand in the United Kingdom

In order to demonstrate the applications of the methods developed in the previous sections we have fitted the model to United Kingdom data for the three main classes of consumer durable goods: (1) furniture and floor coverings, (2) cars and motorcycles, (3) other durable goods. The parameters of the model were estimated from quarterly data for the ten-year period 1973–82 and its post-sample predictive performance tested against quarterly data for the period 1983–4. The data and their sources are presented in Appendix B. The main reason for not using data for the period after 1984 is that the statistics relating to the demand for consumer durable goods are subject to large revisions during the first two or three years after publication (some figures being revised by as much as 5 per cent). Another reason is that quarterly estimates of the United Kingdom population were obtained by quadratic interpolation from the annual estimates, and the latter are published with a delay of over a year.

Following Stone and Rowe 1957, 1958, 1960 and Houthakker and Taylor 1966 we have used only two exogenous variables, real disposable income and the relative price index for the class of durable goods under consideration. The influence of population changes was allowed for by transforming the income and demand data to a per capita basis. The vector $z(t)$ in equation (1) was, for the purpose of this empirical study, defined as

$$z'(t) = [1, y(t), p(t)],$$

where

$$y(t) = \text{real per capital personal disposal income at time } t,$$

$$p(t) = \text{relative price index at time } t.$$

The parameter sub-vector associated with $z(t)$ is $\alpha = [\alpha_1, \alpha_2, \alpha_3]$ where α_1 is the constant term in the equation, α_2 measures the influence of income, and α_3 measures the influence of the relative price.

Estimates were obtained by the procedure described in Section 3, using the algorithm of Fletcher and Powell (1963) for the minimization of L with respect to the structural parameter vector $[\alpha_1, \alpha_2, \alpha_3, \beta, \gamma, \delta, \sigma^2]$, the gradient vector being evaluated numerically using two-sided differences. The estimate of the initial state vector $[q(0), s(0)]$ was updated, using the formula given by equation (57), after any iteration in which the length of the estimated parameter vector changed by less than a preassigned number.

The estimates of the structural parameters and their estimated standard errors are presented in Table 1. Only one of the twenty-one parameter estimates has a sign contrary to our prior expectations and all except two or three have plausible values. Moreover, plausible values of all the parameters are contained within their respective 95 per cent asymptotic confidence intervals. The parameters β, γ and δ are, on the whole, estimated with greater precision that the parameters α_2 and α_3 measuring

TABLE 1. Estimates of structural parameters

Parameter	Furniture and floor coverings		Cars and motorcycles		Other durable goods	
	Estimate	Standard error	Estimate	Standard error	Estimate	Standard error
α_1	18.688	92.554	−164.937	145.713	−25.718	26.585
α_2	0.011	0.100	0.687	0.403	0.080	0.036
α_3	−0.983	47.079	−3.074	60.427	−5.825	6.530
β	−0.125	0.089	−1.082	0.588	0.006	0.009
γ	0.125	0.071	0.303	0.150	0.523	0.147
δ	0.204	0.068	0.131	0.013	0.156	0.021
σ^2	2.207	0.516	74.245	19.636	2.605	0.669

the influence of income and prices, although the latter parameters have the correct signs for all three classes of durable goods.

The parameter estimate with a sign contrary to our prior expectations is the estimate of β for the 'other durable goods' class, which is positive but close to zero. Although this estimate is not statistically significant the smallness of the estimated standard error implies that we must, formally, reject the hypothesis that the level of consumers' stocks has a strong negative influence on the demand for this class of durable good. As was pointed out by Houthakker and Taylor [1966] and mentioned in Section 2 of this paper, a positive value of β could, in the case of perishable goods (or durable goods with a high rate of physical depreciation) be associated with a psychological stock representing the habit persistance effect. But there is another effect which is likely to be more important for the 'other durable goods' class in this study. Electrical goods are the main part of this class, and they include, in addition to long-established electrical goods, many new goods such as video systems, compact disc players, and personal computers. The possession of these new goods by some consumers undoubtedly induces others to buy them. This 'demonstration effect' as we might call it, could offset the normal negative effect of consumers' stocks on purchases and result in a near zero, and perhaps positive, value of β.

The estimates of the parameter γ are of particular interest since this is the key parameter in the more sophisticated dynamic specification of our model compared with earlier models of demand for consumer durable goods. As we pointed out in Section 2, the model of Houthakker and Taylor (1966) can be regarded as a limiting case of our model as $\gamma \rightarrow \infty$ and will be a good approximation to our model when γ is large. The results in Table 1 imply that, for all three classes of durable goods, the hypothesis that $\gamma > 1$ must be strongly rejected, while for furniture and floor covering even the hypothesis $\gamma > 0.5$ must be strongly rejected. We must, tentatively, conclude, therefore, that the model of Houthakker and Taylor is a rather poor approximation to our model for the classes of durable goods considered in this study. Further illumination of this point will be provided later in this section when we discuss the dynamic response of consumers' purchases to a change in income.

The parameter δ is, for all three classes of goods, estimated with greater precision than any of the other parameters except σ^2. The estimates of δ are, perhaps, a little higher than one might expect on the basis of prices in the second-hand markets. But there is no theoretical reason why δ (which is a measure of the effect of the age of goods on consumers' propensity to replace them with new ones) should be exactly equal to the rate of depreciation as measured by second-hand prices.

The plausibility of the estimates of the parameters α_1, α_2, and α_3 is most easily judged by reference to the implied estimates of the long-run income and price elasticities. These have been estimated at average 1984 income and price levels, using the formulae given by equations (A.2) and (A.3) of Appendix A and, together with their estimated standard errors, are presented in Table 2. Although all six elasticities have the correct signs, those for furniture and floor coverings are very unreliable. The reason for the large standard errors in this case is that the price index for furniture and floor covering showed very little independent variation (relative to income) over the sample period; it declined fairly steadily as income rose. The estimated elasticities for the other two classes of durable goods were estimated with much greater precision. The estimated long-run income elasticities are both very significant and much greater than one, putting cars and motorcycles and 'other durable goods' clearly into the luxury good class. The estimated long-run price elasticity of demand for cars and motor cycles is implausibly low, although plausible values are contained within an interval of one standard error from the estimate. Houthakker and Taylor (1966) obtained an estimate of −0.1525 for the long-run price elasticity of demand for new cars in the United States, and this was in accordance with

TABLE 2. Estimates of long-run elasticities

Parameter	Furniture and floor coverings		Cars and motorcycles		Other durable goods	
	Estimate	Standard error	Estimate	Standard error	Estimate	Standard error
e_y	0.294	2.570	1.600	0.433	2.394	0.959
e_p	−0.034	1.632	−0.010	0.196	−0.209	0.241

their prior expectations. We should, of course, expect a much lower price elasticity of demand for all cars than for any particular make of car.

We turn now to a discussion of the dynamic implications of the estimates. The eigenvalues of the estimated matrix A in equation (11), for each class of durable good, are presented in Table 3. For classes (1) and (2) these have complex values, implying that the paths of convergence of purchases to their partial equilibrium levels (associated with given income and price levels) are oscillatory, the period of the cycle being fifty-three quarters for furniture and floor coverings and eleven quarters for cars and motorcycles. There is, of course, nothing irrational about this. If a consumer's cost-utility function includes the rate of change of purchases, as well as the levels of purchases and stocks, then the optimal path of convergence to partial equilibrium can be oscillatory.

TABLE 3. Eigenvalues of estimated system

Furniture and floor coverings		Cars and motorcycles		Other durable goods	
Real part	Imaginary part	Real part	Imaginary part	Real part	Imaginary part
−0.164	+0.118	−0.217	+0.566	−0.148	0.000
−0.164	−0.118	−0.217	−0.566	−0.532	0.000

A clearer picture of the dynamic implications of the parameter estimates can be obtained by computing the dynamic response of purchases to a change in income. Using the formula given by the first equation of the system (A.5) in Appendix A, we have computed the effect of an instantaneous 10 per cent increase in income, assuming that income and prices are initially at their average 1984 levels, purchases and stocks are at the corresponding partial equilibrium levels, and that, after the increase, income and prices remain constant while $\zeta = 0$. The results are presented in Table 4 and Figures 1 to 3. It is clear from Figures 1 to 3 that the dynamic response generated by our model, with the estimated structure, is very different from that generated by the model of Houthakker and Taylor. Their model (equations (1′) and (2) of this paper) implies that, under the above assumptions, purchases jump instantaneously to a level above the new partial equilibrium and then converge to this equilibrium level along a downward path as stocks are built up. The dynamic response functions generated by our model, together with the low standard errors of the parameter estimates determining the form of these functions, particularly the parameter γ, provide strong empirical justification for the richer

TABLE 4. Dynamic response of consumers' purchases to 10 per cent increase in personal disposable income

Time delay (quarters)	Percentage increase in purchases		
	Furniture and floor coverings	Cars and motorcycles	Other durable goods
1	0.55	36.82	9.39
2	1.04	54.65	14.99
3	1.45	55.00	18.33
4	1.81	43.94	20.34
5	2.10	28.69	21.57
6	2.34	15.13	22.32
7	2.53	6.60	22.79
8	2.68	3.79	23.10
9	2.80	5.51	23.31
10	2.89	9.65	23.44
11	2.96	14.18	23.53
12	3.01	17.64	23.61
13	3.04	19.40	23.66
14	3.06	19.55	23.71
15	3.07	18.63	23.74
16	3.07	17.27	23.77
17	3.07	16.02	23.79
18	3.06	15.20	23.82
19	3.05	14.91	23.84
20	3.04	15.03	23.85
.	.	.	.
.	.	.	.
.	.	.	.
∞	2.94	16.00	23.94

dynamic specification incorporated in our model compared with that of earlier models.

It should, perhaps, be emphasized that the figures in Table 4 are the unobservable rates of purchase at points of time. In order to test the predictive power of the model we need to generate post-sample forecasts of the observations (which are integrals over quarters) conditional on the information in the sample and the post-sample observations of the exogenous variables. This can be done by using the exact discrete model, following the procedure outlined at the end of Section 4.

The estimated parameters of the exact discrete model, which are all complicated functions of the estimated structural parameters (Theorems 1

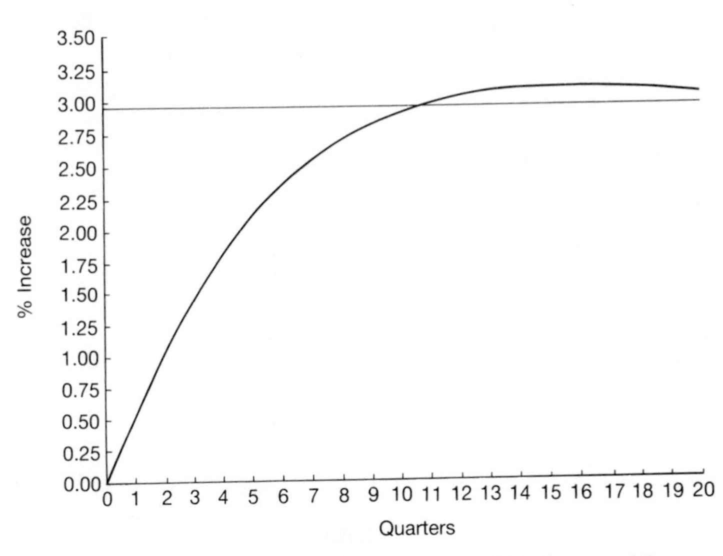

FIG. 1. Dynamic response of consumers' purchases of furniture and floor coverings to 10 per cent increase in personal disposable income.

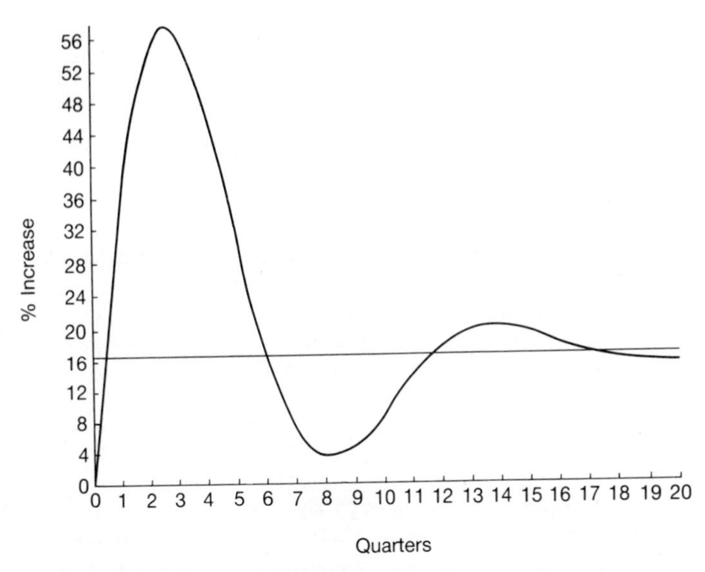

FIG. 2. Dynamic response of consumers' purchases of cars and motorcycles to 10 per cent increase in personal disposable income.

and 2), are set out in Table 5. For example, the estimated equation satisfied by the observed purchases of cars and motorcycles for $t = 3, 4, \ldots$ (equation 9) is

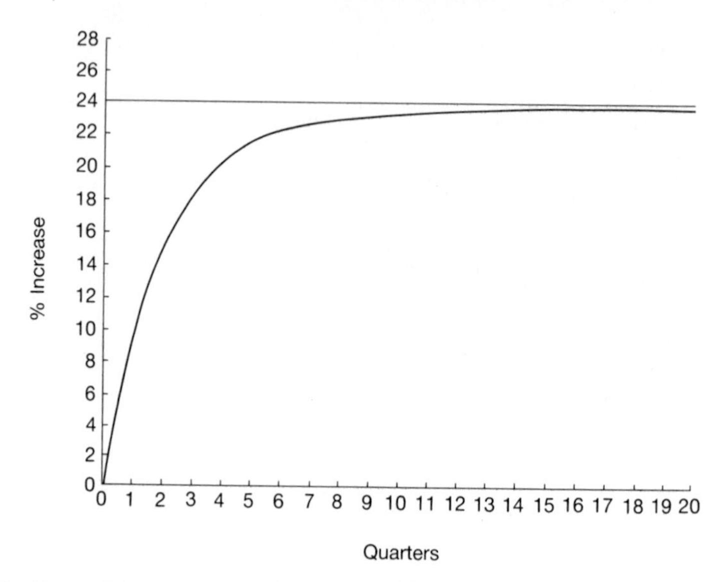

FIG. 3. Dynamic response of consumers' purchases of other durable goods to 10 per cent increase in personal disposable income.

TABLE 5. Estimates of parameters of exact discrete model

Parameter	Furniture and floor coverings	Cars and motorcycles	Other durable goods
	Estimates for Equation (9)		
k_1	1.6852	1.3583	1.4505
k_2	−0.7201	−0.6478	−0.5071
$\mathbf{l}'_0$	1.0849	−21.6684	−5.5673
	0.0007	0.0902	0.0173
	−0.0571	−0.4039	−1.2610
$\mathbf{l}'_1$	0.2275	−1.3774	−0.1642
	0.0001	0.0057	0.0005
	−0.0120	−0.0257	−0.0372
$\mathbf{l}'_2$	−0.9088	17.8983	4.2185
	−0.0006	−0.0745	−0.0131
	0.0478	0.3336	0.9555

TABLE 5. (*Cont.*):

Parameter	Furniture and floor coverings	Cars and motorcycles	Other durable goods
	Estimates for Equation (28)		
h_{11}	0.9377	0.8177	0.7791
h_{12}	−0.0070	−0.1386	0.0012
l'_{11}	1.3917	−27.1938	−6.9892
	0.0008	0.1132	0.0217
	−0.0732	−0.5069	−1.5831
l'_{12}	−0.3650	6.7853	1.7235
	−0.0002	−0.0282	−0.0054
	0.0192	0.1265	0.3904
l'_{13}	0.0908	−1.6827	−0.4271
	0.0001	0.0070	0.0013
	−0.0048	−0.0314	−0.0967
	Estimates for Equation (29)		
f_{qq}	0.8759	0.6133	0.5937
h_{21}	−0.0059	−0.1057	0.0009
h_{22}	−0.0120	−0.2230	0.0020
l'_{21}	0.8813	−12.8828	−3.9226
	0.0005	0.0536	0.0122
	−0.0464	−0.2401	−0.8885
l'_{22}	1.4857	−29.8561	−7.4293
	0.0009	0.1243	0.0231
	−0.0782	−0.5565	−1.6828
l'_{23}	−0.1830	3.6578	0.8688
	−0.0001	−0.0152	−0.0027
	0.0096	0.0682	0.1968
	Covariances of residuals		
ω_0	1.6303	45.5573	1.3864
ω_1	−0.5175	−14.5949	−0.4535
ω_2	−0.2647	−7.7912	−0.2232
ω_{11}	0.6686	18.6362	0.5977
ω_{22}	4.8713	37.5660	21.0937
ω_{12}	−1.5894	−11.8665	−3.4758
ω_{23}	0.0439	−12.6239	0.7820
	Durbin Watson statistics		
for $\hat{\eta}$	3.3442	3.4452	2.5832
for $\hat{\varepsilon}$	2.8710	3.2269	1.6868

$$q_t = -5.147 + 1.358\, q_{t-1} - 0.648\, q_{t-1} + 0.090 y_t$$
$$+ 0.006 y_{t-1} - 0.074\, y_{t-2} - 0.404\, p_t$$
$$- 0.026\, p_{t-1} + 0.334\, p_{t-2} + \eta_t$$
$$E(\eta_t)^2 = 45.557,\ E(\eta_t \eta_{t-1}) = -14.595$$
$$E(\eta_t \eta_{t-2}) = -7.791 \qquad t = 3, 4 \ldots$$

We could not, of course, hope to obtain reliable estimates of so many parameters for a single equation, from a sample of forty observations, without taking acount of the restrictions implied by the continuous time model.

The last section of Table 5 contains the Durbin Watson statistics for both the residual vector $\hat{\eta}$ of the exact discrete model and the vector $\hat{\varepsilon}$ obtained from $\hat{\eta}$ through the transformation defined by equation (56). Since $E(\varepsilon\varepsilon') = E(M^{-1}\eta\eta' M'^{-1}) = M^{-1}\Omega M = I$, the Durbin Watson statistic for $\hat{\varepsilon}$ will tend in probability to two as the sample size tends to infinity, if the model is correctly specified, whereas the Durbin Watson statistic for $\hat{\eta}$ will tend in probability to a number greater than two because of the negative value of ω_1. The average value of the statistic for the three different classes of consumer durable good is, in fact, 2.59 for $\hat{\varepsilon}$ and 3.12 for $\hat{\eta}$. The investigation of the sampling distribution of the Durbin Watson statistic and other test statistics when computed from the vector $\hat{\varepsilon}$ for the model of this paper or the corresponding vector for the mulivariate higher order continuous time model dealt with in Bergstrom 1986 is, of course, a major research problem for the future.

We turn, finally, to the post-sample forecasts. Forecasts for up to eight quarters beyond the end of the sample period, using the observed post-sample values of the exogenous variable, were obtained by the procedure outlined at the end of Section 4 and are presented in Table 6. As a check on the accuracy of these forecasts we have compared their root mean square errors with those of naïve single-period forecasts. The latter are obtained by assuming that purchases (seasonally adjusted) in each post-sample quarter will be the same as those in the previous quarter, and they would be optimal one step ahead forecasts if purchases followed a random walk. This is a fairly severe test, since the forecasts obtained from the estimated model make no use of observed purchases beyond the end of the sample period, whereas each naïve forecast uses the observed value of purchases in the preceding quarter. The forecasts from the model do, of course, make use of the post-sample observations of income and prices.

The results of the comparison are presented in Table 7. They show that the multi-period forecasts obtained from the estimated model were considerably better than the naïve single-period forecasts for furniture and floor coverings, slightly better for cars and motorcycles, and slightly worse

TABLE 6. Optimal post-sample multi-period forecasts

Quarter	Actual value	Forecast	Error
	Furniture and floor coverings		
1983 1	15.861	15.998	0.137
2	15.860	15.942	0.082
3	15.564	15.896	0.332
4	16.159	15.863	−0.296
1984 1	15.729	15.826	0.097
2	15.442	15.792	0.350
3	15.463	15.762	0.299
4	15.924	15.745	−0.179
	Cars and motorcycles		
1983 1	36.164	29.456	−6.708
2	31.980	28.601	−3.379
3	40.892	29.593	−11.299
4	32.158	31.557	−0.601
1984 1	34.698	31.964	−2.734
2	34.326	31.722	−2.604
3	34.537	31.409	−3.128
4	31.913	32.144	0.231
	Other durable goods		
1983 1	21.442	21.267	−0.175
2	20.669	20.923	0.254
3	21.434	21.015	−0.419
4	25.383	21.346	−4.037
1984 1	23.073	21.540	−1.533
2	22.139	21.735	−0.404
3	23.232	21.942	−1.290
4	29.459	22.340	−7.119

TABLE 7. Root mean square errors of forecasts

Class of consumer durable goods	RMSE of optimal multi-period forecasts	RMSE of naïve single-period forecasts
Furniture and floor coverings	0.244	0.341
Cars and motorcycles	5.106	5.110
Other durable goods	2.988	2.806

for other durable goods. On the whole, therefore, the forecasting performance of the model seems to be satisfactory, in spite of a few large errors, which are unavoidable when dealing with such volatile markets as those for cars and motorcycles, and other durable goods.

6. Conclusion

We have formulated a continuous time model of demand for consumer durable goods (generalizing, dynamically, the model of Houthakker and Taylor 1966), developed a procedure for the Gaussian estimation of its structural parameters, and applied the model and the estimation procedure in an econometric study of consumer demand in the United Kingdom. The estimation procedure relies on the general methods developed by Bergstrom (1983, 1985, 1986), but is complicated by the fact that consumers' stocks are unobservable and their rate of depreciation is unknown. The main methodological contribution of this paper is the derivation of an exact discrete model and, through this, an explicit formula for a pseudo-Gaussian likelihood function in terms of the discrete observations and the structural parameters of the continuous time model. Estimates obtained by maximizing this function are exact maximum likelihood estimates when the innovations are Gaussian and the exogenous variables are polynomials in time of degree not exceeding two, and they can be expected to have very good properties under much more general condition.

An alternative way of computing the pseudo-Gaussian likelihood for our model would be to use the Kalman filter algorithm of Harvey and Stock (1985), which could be extended, for this purpose to take account of exogenous variables. (See also Zadrozny 1988.) The unobservable stock variable would then be treated as part of the state vector, whose conditional mean and covariance matrix would be updated T times in each evaluation of the likelihood. The Kalman filter is a powerful algorithm which can take account of the simultaneous occurrence of such complications as missing observations and observations at varying frequencies. (See, for example, Stock 1988.) But, for the problem dealt with in this paper, our algorithm is computationally more efficient, since the elimination of the unobservable stock variable at an early stage reduces the dimension of the system and, thereby, reduces the number of calculations required in the evaluation of the likelihood. Our method also has advantages for the purpose of obtaining forecasts of the post-sample discrete observations, since these can be computed very efficiently by a direct recursive application of the exact discrete model.

The application of the methods developed in this paper, to consumer demand in the United Kingdom, has been very successful as judged by the plausibility of the estimates of the structural parameters, the consistency of their signs with prior expectations, and the accuracy of the forecasts of the post-sample discrete observations. It has also provided strong support for the richer dynamic specification of our model as compared with that of earlier models of demand for consumer durable goods of the stock adjustment type. More generally, it supports the view expressed in Section 1 concerning the need for more sophisticated and realistic dynamic specification in models used for the econometric analysis of consumer demand.

Appendix A

Derivation of Long-run Elasticities and Dynamic Response Functions

Let q and s be the steady state of levels of consumers' purchases and stocks corresponding to a given income level y and price p. Then from the system

$$Dx(t) = Ax(t) + Bz(t), \tag{A.1}$$

which is obtained from the stochastic system (11) by putting $\zeta = 0$, we have

$$\begin{bmatrix} q \\ s \end{bmatrix} = x = -A^{-1}Bz$$

$$= \begin{bmatrix} \delta/(\delta - \beta) \\ 1/(\delta - \beta) \end{bmatrix} [\alpha_1 + \alpha_2 y + \alpha_3 p].$$

It follows from the last equation that the long-run income elasticity of demand $\varepsilon_y(y, p)$ and price elasticity of demand $\varepsilon_p(y, p)$ are given by

$$\varepsilon_y(y, p) = \frac{\alpha_2 \delta y}{(\delta - \beta)q} = \frac{\alpha_2 y}{\alpha_1 + \alpha_2 y + \alpha_3 p} \tag{A.2}$$

$$\varepsilon_p(y, p) = \frac{\alpha_3 \delta y}{(\delta - \beta)q} = \frac{\alpha_3 p}{\alpha_1 + \alpha_2 y + \alpha_3 p}. \tag{A.3}$$

We shall assume now that the system is in stationary equilibrium at $t = 0$ and consider the dynamic response of $[q, s]$ to a 10 per cent increase in y when p is unchanged. We require the solution of the system (A.1) subject to the conditions

$$z(t) = \begin{bmatrix} 1 \\ y(0) \\ p(0) \end{bmatrix} = z(0) \qquad t = 0,$$

$$z(t) = \begin{bmatrix} 1 \\ (1.1)y(0) \\ p(0) \end{bmatrix} = z^* \qquad t > 0,$$

$$x(0) = -A^{-1}z(0).$$

The required solution is

$$x(t) = x^* + F(t)[x(0) - x^*] \tag{A.4}$$

where

$$x^* = -A^{-1}Bz^*.$$

From equation (A.4) we obtain

$$\begin{aligned} x(t) - x(0) &= [I - F(t)][x^* - x(0)] \\ &= [I - F(t)][-A^{-1}B][z^* - z(0)] \\ &= [I - F(t)](0.1)\alpha_2 y(0)\begin{bmatrix} \delta/(\delta - \beta) \\ 1/(\delta - \beta) \end{bmatrix} \end{aligned}$$

and have

$$\begin{bmatrix} \{q(t) - q(0)\}/q(0) \\ \{s(t) - s(0)\}/s(0) \end{bmatrix} = (0.1)\varepsilon_y(y(0), p(0))[I - F(t)]\begin{bmatrix} 1 \\ 1 \end{bmatrix}. \tag{A.5}$$

Appendix B

Data

The data are presented in Table B1, the variables being defined as follows.

y = seasonally adjusted personal disposable income, per capita, at 1980 prices;

q = seasonally adjusted expenditure, per capita, on the class of durable goods, at 1980 prices;

p = seasonally adjusted relative price index (relative to to the retail price index) of the class of durable goods (1980 = 1.0000).

Unadjusted data were obtained from the official sources below and adjusted for seasonable variation using the method of Durbin [1963]. Income and expenditure were reduced to a per capita basis using quarterly estimates of the United Kingdom population obtained by quadratic interpolation from the annual estimates. The price indices were obtained by dividing expenditure at current prices by expenditure at 1980 prices.

Sources: Economic Trends Annual Supplement (for income, expenditure, and prices) and Annual Abstract of Statistics (for population).

TABLE B1. Data

Quarter	Income	Furniture and floor coverings		Cars and motorcycles		Other durable goods	
	y	q	p	q	p	q	p
1973 1	567.834	16.694	1.019	31.772	0.852	12.038	1.425
2	570.325	14.213	1.034	27.134	0.838	12.592	1.347
3	569.253	14.720	1.068	24.464	0.830	12.922	1.347
4	565.704	14.403	1.058	27.661	0.846	12.370	1.302
1974 1	550.365	13.657	1.067	17.726	0.830	12.473	1.263
2	560.572	13.212	1.089	21.537	0.836	12.321	1.247
3	561.739	14.185	1.090	19.321	0.831	12.957	1.249
4	566.727	14.209	1.079	26.062	0.854	12.051	1.256
1975 1	557.775	14.440	1.057	18.492	0.857	12.847	1.229
2	564.065	14.405	1.033	23.586	0.873	13.426	1.280
3	554.048	14.418	1.028	22.100	0.858	10.842	1.297
4	551.474	14.088	1.023	24.037	0.856	9.831	1.297
1976 1	553.074	14.833	1.003	21.484	0.861	11.550	1.320
2	558.931	14.372	1.011	23.625	0.880	12.626	1.277
3	559.343	14.990	0.994	21.268	0.892	13.406	1.136
4	563.345	15.338	1.002	25.625	0.926	12.859	1.069
1977 1	552.251	14.377	1.019	21.566	0.925	12.516	1.163
2	551.174	13.506	1.024	20.522	0.940	12.471	1.150
3	554.237	14.107	1.026	21.582	0.951	12.878	1.168
4	567.187	14.170	1.038	23.086	1.011	12.724	1.181

TABLE B1. (*Cont.*):

Quarter	Income	Furniture and floor coverings		Cars and motorcycles		Other durable goods	
	y	q	p	q	p	q	p
1978 1	584.047	13.920	1.040	27.720	1.036	13.037	1.161
2	582.353	14.667	1.038	24.677	1.063	14.166	1.151
3	591.095	15.194	1.030	29.098	1.062	14.873	1.144
4	591.221	15.343	1.032	25.291	1.095	14.110	1.136
1979 1	602.118	15.352	1.023	28.801	1.116	15.110	1.103
2	628.082	18.745	1.016	38.509	1.093	17.997	1.082
3	602.713	15.462	1.030	24.473	1.166	15.515	1.072
4	612.257	16.155	1.033	29.021	1.096	15.349	1.065
1980 1	624.607	15.880	1.031	35.521	1.056	15.764	1.030
2	604.141	15.184	1.006	26.281	1.012	15.646	1.000
3	611.222	15.510	0.989	28.301	0.987	16.113	0.990
4	601.189	14.442	0.979	23.950	0.943	16.243	0.981
1981 1	612.612	15.804	0.974	28.987	0.929	16.646	0.955
2	607.242	15.059	0.948	28.374	0.902	16.533	0.924
3	609.228	14.963	0.926	30.292	0.891	16.843	0.910
4	607.638	14.254	0.917	28.022	0.900	17.349	0.899
1982 1	607.862	14.906	0.903	27.369	0.924	17.612	0.870
2	605.562	14.590	0.899	26.905	0.906	17.502	0.855
3	618.944	15.365	0.885	33.317	0.889	19.234	0.852
4	625.044	14.590	0.882	31.458	0.851	21.319	0.848

TABLE B1. (*Cont.*):

Quarter	Income	Furniture and floor coverings		Cars and motorcycles		Other durable goods	
	y	*q*	*p*	*q*	*p*	*q*	*p*
1983 1	630.008	15.861	0.877	36.164	0.861	21.442	0.827
2	629.690	15.860	0.884	31.980	0.871	20.669	0.824
3	657.377	15.564	0.885	40.892	0.880	21.434	0.835
4	646.585	16.159	0.872	32.158	0.867	25.384	0.814
1984 1	643.270	15.729	0.878	34.698	0.862	23.073	0.784
2	648.268	15.442	0.879	34.326	0.890	22.139	0.778
3	648.253	15.463	0.884	34.537	0.868	23.232	0.775
4	663.427	15.924	0.876	31.913	0.876	29.459	0.771

References

Bergstrom, A. R. (1966), 'Non-recursive models as discrete approximations to systems of stochastic differential equations', *Econometrica,* **34**, 173–82.

—— (1983), 'Gaussian estimation of structural parameters in higher order continuous time dynamic models', *Econometrica,* **51**, 117–52.

—— (1984), 'Continuous time stochastic models and issues of aggregation over time', in *Handbook of Econometrics,* vol. 2, ed. by Z. Griliches and M. D. Intriligator (Amsterdam: North-Holland).

—— (1985), 'The estimation of parameters in nonstationary higher-order continuous-time dynamic models', *Econometric Theory,* **1**, 369–85.

—— (1986), 'The estimation of open higher-order continuous-time dynamic models with mixed stock and flow data', *Econometric Theory,* **2**, 350–73.

—— (1989), 'Optimal forecasting of discrete stock and flow data generated by a higher-order continuous time system', *Computers and Mathematics with Applications,* **17**, 1203–14.

Chow, G. (1957), *Demand for Automobiles in the U.S.A. A Study in Consumer Durables* (Amsterdam: North-Holland).

Deaton, A. S. (1986), 'Demand analysis', in *Handbook of Econometrics,* vol. 3, ed. by Z. Griliches and M. D. Intriligator (Amsterdam: North-Holland).

—— and J. Muellbauer (1980), 'An almost ideal demand system', *American Economic Review,* **70**, 312–26.

Dunsmuir, W. (1979), 'A central limit theorem for parameter estimation in stationary vector time series and its application to models for a signal observed with noise', *Annals of Statistics,* **7**, 490–506.

—— and E. J. Hannan (1976), 'Vector linear time series models', *Advances in Applied Probability,* **8**, 339–64.

Durbin, J. (1963), 'Trend elimination for the purpose of estimating seasonal and periodic components in time series', in *Time Series Analysis,* ed. by M. Rosenblatt (New York: Wiley).

Fletcher, R. and M. J. D. Powell (1963), 'A rapidly convergent descent method for minimization', *The Computing Journal,* **6**, 163–8.

Harvey, A. C. and J. H. Stock (1985), 'The estimation of higher-order continuous time autoregressive models', *Economic Theory,* **1**, 97–118.

Houthakker, H. S. and L. D. Taylor (1966), *Consumer Demand in the United States, 1929–1970, Analysis and Projections* (Cambridge: Harvard University Press, 2nd edn. 1970).

Klevmarken, N. A. (1979), 'A comparative study of complete systems of demand functions', *Journal of Econometrics,* **10**, 165–92.

Malinvaud, E. (1980), *Statistical Methods of Econometrics* (Amsterdam: North-Holland).

Nerlove, M. (1960), 'The market demand for durable goods: A comment', *Econometrica,* **28** 132–42.

Phillips, P. C. B. (1972), 'The structural estimation of a stochastic differential equation system', *Econometrica,* **40** 1021–41.

—— (1974), 'The estimation of some continuous time models', *Econometrica,* **42**, 803–82.

—— (1976), 'The estimation of linear stochastic differential equations with exogenous variables', in *Statistical Inference in Continuous Time Economic Models,* ed. by A. R. Bergstrom (Amsterdam, North-Holland).

Robinson, P. M. (1988), 'Using Gaussian estimators robustly', *Oxford Bulletin of Economics and Statistics,* **50**, 97–106.

Stock, J. H. (1988), 'Estimating continuous-time processes subject to time deformation', *Journal of the American Statistical Association,* **83**, 77–85.

Stone, J. R. N. and D. A. Rowe (1957), 'The market demand for durable goods', *Econometrica,* **25**, 423–43.

—— —— (1958), 'Dynamic demand functions: Some econometric results', *Economic Journal,* **68**, 256–70.

—— —— (1960), 'The durability of consumers' goods', *Econometrica,* **28**, 407–16.

Taylor, L. D. and T. A. Wilson (1964), 'A method of estimating models with lagged dependent variables', *Review of Economics and Statistics,* **46**, 329–46.

Zadrozny, P. (1988), 'Gaussian likelihood of continuous-time ARMAX models when data are stocks and flows at different frequencies', *Econometric Theory,* **4**, 108–24.

Index